Progressive Thermochemical Biorefining Technologies

Progressive Thermochemical Biorefining Technologies

Edited by
Sonil Nanda and Dai-Viet N. Vo

CRC Press
Taylor & Francis Group
Boca Raton London New York

CRC Press is an imprint of the
Taylor & Francis Group, an **informa** business

First edition published 2022

by CRC Press
6000 Broken Sound Parkway NW, Suite 300, Boca Raton, FL 33487–2742

and by CRC Press
2 Park Square, Milton Park, Abingdon, Oxon, OX14 4RN

© 2022 selection and editorial matter, Sonil Nanda and Dai-Viet N. Vo; individual chapters the contributors.

CRC Press is an imprint of Taylor & Francis Group, LLC

ISBN: 978-0-367-56609-8 (hbk)
ISBN: 978-1-003-09859-1 (ebk)
ISBN: 978-0-367-56610-4 (pbk)

Typeset in Times
by Apex CoVantage, LLC

Contents

Preface

Considering the deleterious impacts of fossil fuels on the environmental and natural ecosystems, it has become imperative to make a paradigm shift towards renewable fuels, chemicals and materials. The exhaustive everyday usage of fossil fuels and processed petrochemical products is the leading cause for the increase in greenhouse gas emissions; global warming; climate changes; acid rain; ozone layer depletion; pollution of air, water and soil as well as accumulation of nonbiodegradable materials in the soil and oceans. On the contrary, biofuels, biochemicals and biomaterials derived from renewable wastes such as nonedible plant biomass (e.g., agricultural and forestry biomass), energy crops, microalgae, municipal solid waste, sewage sludge, and other biogenic residues seem to be carbon neutral. Therefore, the global interest in biorefining technologies, especially thermochemical and biological conversion processes, is gaining momentum in academic and industrial perspectives. This book covers the research and development of some cutting-edge thermochemical technologies related to the conversion of waste biomass, namely pyrolysis, liquefaction, torrefaction, carbonization, gasification, and reforming, to name a few.

Chapter 1 by Subathra et al. makes a state-of-the-art review of various biorefinery approaches for various biomasses. The survey identifies the vital role of biofuels to address the environmental issues caused by fossil fuels. Chapter 2 by Wu et al. reviews the production, characterization, and combustion of solid and liquid biofuels produced from renewable wastes. Chapter 3 by Patel et al. reviews various technologies for the conversion of municipal solid waste to biofuels through physical routes (e.g., refinement, briquetting, and pelletizing), thermochemical routes (e.g., torrefaction, gasification, liquefaction, pyrolysis, and incineration) and biological routes (e.g., anaerobic digestion and fermentation). Chapter 4 by Patel et al. discusses the pragmatic concerns of mixed plastic waste and its potential for conversion into high-energy fuels and chemicals by thermochemical processes such as pyrolysis, liquefaction, and gasification. Chapter 5 by Sarker et al. describes the torrefaction of lignocellulosic biomass, its process parameters and operating conditions as well as the physicochemical characteristics of torrefied materials. Chapter 6 by Anno-Kusi et al. gives an overview of the influence of binders on the densification of torrefied biomass. The application of pellets as well as techno-economic and lifecycle assessments of the torrefaction and pelletization process are discussed in this chapter. Chapter 7 by Monir et al. describes the conversion of lignocellulosic biomass into syngas using the co-gasification process with several feedstocks followed by a discussion on energy efficiency and environmental impacts. Chapter 8 by Phung et al. describes the clean technologies involved in the purification of glycerol along with the several reaction processes for glycerol conversion into value-added industrial products. Chapter 9 by Islam et al. reviews the substrates used in microbial fuel cells along with their performance, limitations, and prospects. Chapter 10 by Nguyen et al. attempts to provide insights on the relationship between oil price fluctuations, environmental quality, foreign direct investment, and renewable energy consumption, especially in East Asian countries.

We sincerely hope that this book will immensely benefit the students and researchers working on biofuels and biochemicals generated from the thermochemical conversion of waste biomass. We are grateful to all the authors for contributing their high-quality manuscripts to develop this book. We also express our sincere thanks to Ms. Renu Upadhyay and Ms. Jyotsna Jangra from CRC Press for their enthusiastic assistance and support in the preparation of this book.

Dr. Sonil Nanda
Research Associate
Department of Chemical and Biological Engineering
University of Saskatchewan
Saskatoon, Saskatchewan Canada

Dr. Dai-Viet N. Vo
Director
Center of Excellence for Green Energy and Environmental Nanomaterials
Nguyen Tat Thanh University
Ho Chi Minh City, Vietnam

Editors

Dr. Sonil Nanda is a Research Associate at the University of Saskatchewan in Saskatoon, Saskatchewan, Canada. He earned his PhD degree in Biology from York University, Canada; MSc degree in Applied Microbiology from Vellore Institute of Technology (VIT University), India; and BSc degree in Microbiology from Orissa University of Agriculture and Technology, India. Dr. Nanda's research areas are related to the production of advanced biofuels and biochemicals through thermochemical and biochemical conversion technologies such as gasification, pyrolysis, carbonization, torrefaction, and fermentation. He has gained expertise in hydrothermal gasification of various organic wastes and biomass including agricultural and forestry residues, industrial effluents, municipal solid wastes, cattle manure, sewage sludge, food waste, waste tires, and petroleum residues to produce hydrogen fuel. His parallel interests are also in the generation of hydrothermal flames for the treatment of hazardous wastes, agronomic applications of biochar, phytoremediation of heavy metal contaminated soils, as well as carbon capture and sequestration. Dr. Nanda has published 14 books, 60 book chapters, and more than 110 peer-reviewed journal articles. He is the editor of books titled *New Dimensions in Production and Utilization of Hydrogen* (Elsevier), *Recent Advancements in Biofuels and Bioenergy Utilization* (Springer Nature), *Biorefinery of Alternative Resources: Targeting Green Fuels and Platform Chemicals* (Springer Nature), *Fuel Processing and Energy Utilization* (CRC Press), *Bioprocessing of Biofuels* (CRC Press), and *Biotechnology for Sustainable Energy and Products* (I.K. International Publishing House Pvt. Ltd.). Dr. Nanda serves as a Fellow Member of the Society for Applied Biotechnology in India, as well as a Life Member of the Indian Institute of Chemical Engineers, Association of Microbiologists of India, Indian Science Congress Association, and the Biotech Research Society of India. He is also an active member of several chemical engineering societies across North America such as the American Institute of Chemical Engineers, the Chemical Institute of Canada, the Combustion Institute-Canadian Section, and Engineers Without Borders Canada. Dr. Nanda is Assistant Subject Editor of the *International Journal of Hydrogen Energy* (Elsevier) as well as Associate Editor of *Environmental Chemistry Letters* (Springer Nature) and *Applied Nanoscience* (Springer Nature). He has also edited several special issues in renowned journals such as the *International Journal of Hydrogen Energy* (Elsevier), *Chemical Engineering Science* (Elsevier), *Biomass Conversion and Biorefinery* (Springer Nature), *Waste and Biomass Valorization* (Springer Nature), *Topics in Catalysis* (Springer Nature), *SN Applied Sciences* (Springer Nature), and *Chemical Engineering & Technology* (Wiley).

Dr. Dai-Viet N. Vo is currently the Director of the Center of Excellence for Green Energy and Environmental Nanomaterials at Nguyen Tat Thanh University in Ho Chi Minh City, Vietnam. He earned his PhD degree in Chemical Engineering from the University of New South Wales in Sydney, Australia, in 2011. He has worked as a postdoctoral fellow at the University of New South Wales in Sydney and Texas A&M University in Qatar, Doha. Formerly, he was Senior Lecturer at the Faculty of Chemical & Natural Resources Engineering in the Universiti Malaysia Pahang in Kuantan, Malaysia (2013–2019). His research areas are in the production of green synthetic fuels via Fischer–Tropsch synthesis using bio-mass-derived syngas from various reforming processes. He is also an expert in advanced material synthesis and catalyst characterization. During his early career, he worked as the principal investigator and co-investigator for 21 different funded research projects related to sustainable and alternative energy. He has published five books, 20 book chapters, and more than 300 peer-reviewed journal articles and conference proceedings. He has served on the technical and publication committees of numerous international conferences in chemical engineering, catalysis, and renewable energy. Dr. Vo is the editor of books titled *New Dimensions in Production and Utilization of Hydrogen* (Elsevier), *Biorefinery of Alternative Resources: Targeting Green Fuels and Platform Chemicals* (Springer Nature), and *Fuel Processing and Energy Utilization* (CRC Press). Dr. Vo is Assistant Subject Editor for the *International Journal of Hydrogen Energy* (Elsevier) and Guest Editor for several special issues in high-impact factor journals such as the *International Journal of Hydrogen Energy* (Elsevier), *Comptes Rendus Chimie* (Elsevier), *Waste and Biomass Valorization* (Springer), *Topics in Catalysis* (Springer), *Journal of Chemical Technology & Biotechnology* (Wiley), *Chemical Engineering & Technology* (Wiley), and several others. He is also an Associate Editor of *Environmental Chemistry Letters* (Springer Nature) and *Applied Nanoscience* (Springer Nature). He is also Editorial Board Member of many international journals including *SN Applied Sciences* (Springer), *Scientific Reports* (Springer Nature), and *PLOS One*. Dr. Vo has been awarded as Top Peer Reviewer 2019 powered by Publons.

Contributors

Fuad Ameen
Department of Botany and Microbiology
King Saud University
Riyadh, Saudi Arabia

Jennifer Anno-Kusi
Department of Chemical and Biological Engineering
University of Saskatchewan
Saskatoon, Saskatchewan, Canada

Jayaseelan Arun
Centre for Waste Management
Sathyabama Institute of Science and Technology
Chennai, Tamil Nadu, India

Ramin Azargohar
Department of Process Engineering
Memorial University of Newfoundland
St. John's, Newfoundland and Labrador, Canada

Azrina Abd Aziz
Faculty of Civil Engineering Technology
Universiti Malaysia Pahang
Gambang, Malaysia

Guan-Bang Chen
Research Center for Energy Technology and
     Strategy
National Cheng Kung University
Tainan, Taiwan

Ajay K. Dalai
Department of Chemical and Biological Engineering
University of Saskatchewan
Saskatoon, Saskatchewan, Canada

Shanmuganantham Selvanantham Dawn
Centre for Waste Management
Sathyabama Institute of Science and Technology
Chennai, Tamil Nadu, India

Kannappan Panchamoorthy Gopinath
Department of Chemical Engineering
Sri Sivasubramaniya Nadar College of Engineering
Chennai, Tamil Nadu, India

Chao-Wei Huang
Department of Chemical and Materials
     Engineering
National Kaohsiung University of Science and
     Technology
Kaohsiung, Taiwan

M. Amirul Islam
Department of Electrical and Computer
     Engineering
Université de Sherbrooke
Sherbrooke, Québec, Canada

Ahasanul Karim
Department of Soil Sciences and Agri-Food
     Engineering
Université Laval
Québec City, Québec, Canada

Fatema Khatun
Faculty of Civil Engineering Technology
Universiti Malaysia Pahang
Gambang, Malaysia

Yueh-Heng Li
Department of Aeronautics and Astronautics
National Cheng Kung University
Tainan, Taiwan

Venkatesh Meda
Department of Chemical and Biological
     Engineering
University of Saskatchewan
Saskatoon, Saskatchewan, Canada

Nadzirah Mohd Mokhtar
Faculty of Civil Engineering Technology
Universiti Malaysia Pahang
Gambang, Malaysia

Minhaj Uddin Monir
Department of Petroleum and Mining
     Engineering
Jashore University of Science and Technology
Jashore, Bangladesh

Subhadeep Mukherjee
Centre for Management Studies
Dibrugarh University
Dibrugarh, Assam, India

Sonil Nanda
Department of Chemical and Biological
    Engineering
University of Saskatchewan
Saskatoon, Saskatchewan, Canada

Narasiman Nirmala
Centre for Waste Management
Sathyabama Institute of Science and
    Technology
Chennai, Tamil Nadu, India

Thu Thuy Nguyen
Thuongmai University
Hanoi, Vietnam

Quynh-Thy Song Nguyen
Industrial Development Center of Southern
    Vietnam
Ministry of Industry and Trade
Ho Chi Minh City, Vietnam

Van-Huy Nguyen
Faculty of Biotechnology
Binh Duong University
Thu Dau Mot, Vietnam

Van Chien Nguyen
Faculty of Economics
Thu Dau Mot University
Thu Dau Mot City, Vietnam

Ravi Patel
Department of Chemical and Biological
    Engineering
University of Saskatchewan
Saskatoon, Saskatchewan, Canada

Hong Duc Pham
Centre for Materials Science
Queensland University of Technology
Brisbane, Australia

Thanh Khoa Phung
Department of Chemical Engineering
Ho Chi Minh City International University
Ho Chi Minh City, Vietnam

Tumpa R. Sarker
Department of Chemical and Biological
    Engineering
University of Saskatchewan
Saskatoon, Saskatchewan, Canada

Sivaprasad Shyam
Department of Chemical Engineering
Sri Sivasubramaniya Nadar College of
    Engineering
Chennai, Tamil Nadu, India

Munusamy Subathra
Department of Bio-Technology
Aarupadai Veedu Institute of Technology
Paiyanoor, Tamil Nadu, India

Khoa Dang Tong
Department of AquaScience
Ho Chi Minh City International University
Ho Chi Minh City, Vietnam

Vy Anh Tran
Institute of Research and Development
Duy Tan University
Danang, Vietnam

Dai-Viet N. Vo
Center of Excellence for Green Energy and
    Environmental Nanomaterials
Nguyen Tat Thanh University
Ho Chi Minh City, Vietnam

Khanh B. Vu
Department of Chemical Engineering
Ho Chi Minh City International University
Ho Chi Minh City, Vietnam

Fang-Hsien Wu
Research Center for Energy Technology and
    Strategy
National Cheng Kung University
Tainan, Taiwan

1

Thermochemical and Biological Conversion of Biomass into Biofuels and Biochemicals

Munusamy Subathra, Narasiman Nirmala, Shanmuganantham Selvanantham Dawn, Sivaprasad Shyam, Kannappan Panchamoorthy Gopinath and Jayaseelan Arun

CONTENTS

1.1 Introduction

Global energy demand has drastically increased due to rapid industrialization and explosion in the human population straining the conventional petroleum-based energy sector which contributes about 80–85% of energy needs of the world (Ong et al., 2020; Saravanan et al., 2020). Moreover, increasing environmental awareness among people and focus on achieving the United Nation's Sustainable Development Goals (SDGs) are propelling the need to utilize renewable energy from biomass (Omer and Noguchi, 2020; Pang, 2019). Biomass has an edge over other renewable sources as it is abundantly available and can be utilized for generating heat, electricity, transportation-grade fuels, platform chemicals, and other value-added products. In addition, utilizing biomass will create room for the growth of new biomass, and waste biomass can be potentially valorized for beneficial purposes (Ibarra-Gonzalez and Rong, 2019). There are several strategies for valorizing biomass as depicted in Figure 1.1. Thermochemical conversion of biomass essentially requires liquefaction, pyrolysis and gasification to produce biochar, bio-oil, syngas and other value-added chemical intermediates such as 5-hydroxymethylfurfural (HMF) and furfural (Gao et al., 2020; Yang et al., 2019). Biological conversion of biomass requires fermentation to produce bioethanol and requires anaerobic digestion to produce biogas (El-Dalatony et al., 2019; Ge et al., 2014; Hawkins et al., 2013).

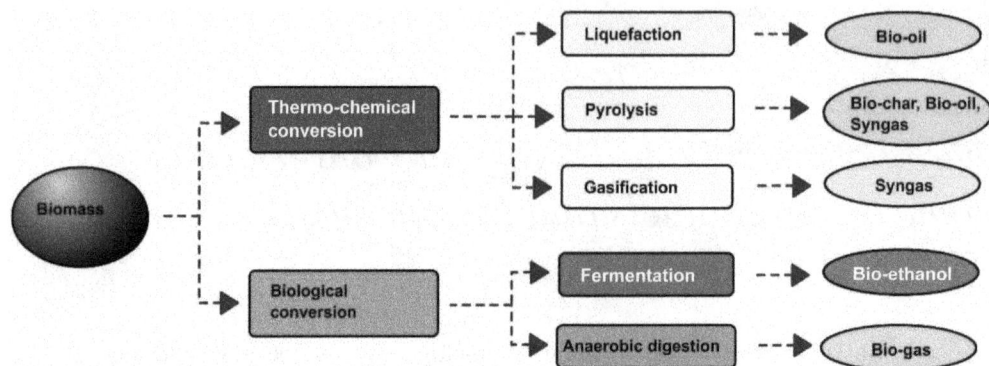

FIGURE 1.1 Biomass conversion pathways for the production of value-added compounds.

1.2 Waste Biomass

Biomass is widely used as a source for the production of renewable biofuels. Biomass is usually produced naturally, and it is considered an organic material, as it is obtained from the remains of plants and animals. Both plants and animals are the best sources of sugar residues. In general, the plant and algal materials that are considered as the primary sources of energy utilize the sunlight and produce carbohydrate and water by the process called photosynthesis. Carbohydrate and water are further converted into primary and secondary metabolites, which include cellulose, hemicelluloses, lignin, and starch that are widely used in various industrial sectors. The carbohydrate residues help in producing various biofuels which include bioethanol, biobutanol, heat energy, electrical energy, etc. (Huber et al., 2006; Naik et al., 2010). Renewable energy obtained from biomass thus not only fulfills the energy demand but also decreases the amount of waste disposed as all the organic waste can be utilized as the source for the production of bioenergy (Sims et al., 2008).

Wastes such as municipal solid waste, food waste, agricultural waste, and sewage waste come under the category of bio-renewable waste, which can be used as biomass due to the presence of high-energy-producing substances. Such waste can be utilized for the production of bioenergy by using various techniques such as fermentation process and thermochemical process. (Okolie et al., 2020; Arun et al., 2020). Food waste, fruit peel waste, agricultural waste, etc., are considered as the primary source of starch which can yield biogas, bioethanol, bio-crude, etc., by the anaerobic fermentation process. Sugar crops, oilseed crops, starch crops, etc., can be converted to biodiesel or bio-crude by the process of transesterification or hydrothermal liquefaction process. The residue can be used as the soil conditioner.

Forest products such as wood residues, tree barks, shrubs, sawdust can undergo thermal treatment, which results in the formation of heat energy that can be further converted into electrical energy. The ash that remains as a result of the thermal treatment is a rich source of minerals. On the application of the produced ash over soil, the mineral content increases, and it conditions the soil to increase its fertility and yields high crop growth. Most of the paper industry uses the process of the thermal treatment of wood residue to produce electricity for its use.

1.3 Composition and Properties of Biomass

1.3.1 Cellulose, Hemicellulose, and Lignin

The amount of bio-crude or bio-oil produced from the biomass depends majorly on the biodegradability of the cellulose or hemicelluloses or lignin content present. The cellulose has higher biodegradability than the lignin content and hemicelluloses, which acts as a major criterion for choosing the raw material for any thermal conversion or biological conversion process. The cellulose-rich biomass gives a higher

TABLE 1.1

Biomass and Its Category

Category of Biomass	Biomass Source
Bio-renewable wastes as biomass	Food waste
	Agricultural waste
	Urban organic waste
	Municipal solid waste
	Mill wood waste
	Sewage waste
Crops producing energy	Starch crops
	Oilseed crops
	Grasses
	Herbaceous wood plants
	Forage crops
Forest products as biomass	Trees, shrubs, and wood residues
	Logging residues
	Tree barks, sawdust, and roots
Aquatic crops as biomass	Algae
	Water hyacinth
	Waterweed
Other wastes as biomass	Industrial organic waste
	Lichens
	Mosses
	Landfills
	Reeds and rushes
	Cattle manure

References: Faaij (2006); Demirbas (2010)

yield, but the lignin-rich biomass provides higher energy on carrying out hydrolysis or enzymatic conversion process.

1.3.2 Moisture Content

The moisture content plays a major role in deciding the efficiency of the fuel produced. If the moisture content of the fuel is high, this will decrease the amount of heat produced per unit mass. On the other hand, decreasing the moisture content in the biofuel increases the amount of polluting gas expelled out into the atmosphere. This also creates dust, which thereby affects the equipment and its performance. The moisture content in the fuel can be calculated on a wet basis (mass of water divided by the total mass of fuel) and dry basis (mass of water divided by the dry mass of fuel). The moisture content is of two types in the biomass, that is, intrinsic and extrinsic moisture contents. Intrinsic moisture content will not be affected by the weather whereas extrinsic moisture content depends on the effect of weather (McKendry, 2002).

1.3.3 Ash and Alkali Metal Content

The ash content formed after the thermal process indicates the nature of biomass used. The ash formation occurs due to the presence of insoluble compounds present in the biomass. The soluble compounds present in the biomass are ionic and catalyze the thermal reaction, resulting in a higher yield of bio-crude substances. Sometimes, the ash formed starts to melt due to unexpected laboratory conditions, which may be deposited over the equipment and cause fouling, and get hardened to form chunks, a process called slagging. This can be avoided by maintaining a low temperature for the thermal reaction. The woody plants produce less ash content when compared to grassy crops. This is also affected by the harvest time and minerals present in the crops which may act as insoluble substances (Adler et al., 2006). The alkali metals present in the biomass react with the silica content and produce a liquid mobile phase, which stops the airflow in the thermal vessel and reduces the combustion process. A few examples of alkali metals are potassium, magnesium, calcium, and phosphorous.

1.3.4   Bulk Density

Bulk density acts as an important characteristic feature of biomass concerning volume, ease of handling, and transportation cost. The biomass bulk density affects the storage life of fuel. For example, pelletized or compressed forms of biomass are easier to handle than the raw form of biomass (harvested crop waste). By reducing the bulk density of the biomass, the transportation costs are reduced and handling is easier (McKendry, 2002).

1.4   Thermochemical Conversion Pathways for Biofuel Production

The thermochemical conversion routes to biofuels are performed with controlled heating of biomass as a part of several processes to harvest intermediate energy carriers alternatively. In developing countries, the thermochemical conversion routes signify a promising technology for the production of renewable electricity via biomass co-firing in coal power plants and decentralized electrifications and satisfy the standards endorsed in many countries (Bhaskar et al., 2011). Thermochemical conversion routes are classified according to their oxidation environment, heating rate, particle size, and the range of heating of biomass from endothermic to exothermic oxidation processes. The thermochemical conversion routes such as pyrolysis, gasification, and combustion technologies produce heat, electricity, as well as solid, gaseous, or liquid fuels and chemical precursors. Apart from the aforementioned technologies, the carbonization, hydrothermal process, torrefaction, and high-pressure technologies also come under thermochemical conversion routes.

The performance of these thermochemical conversion pathways depends on the use of high-quality biomass feedstock. The mass balance of a kilogram of biomass is, on many occasions, conceptualized by biochemical, proximate, or ultimate estimation. The biochemical evaluation refers to the relative abundance of a range of biopolymers such as cellulose and lignin present in the biomass, whereas ultimate evaluation refers to the relative abundance of particular elements like carbon, hydrogen, oxygen, nitrogen, and sulfur (Ibarra-Gonzalez and Rong, 2019). The proximate analysis involves the heating of biomass to quantify its thermal resistance through the relative proportions of constant carbon and volatile matter. Moisture and elemental ash complete the mass balance of a unit of freshly harvested biomass (Verma et al., 2012).

1.4.1   Liquefaction

Hydrothermal liquefaction is performed in the presence of alkali metal salt catalyst and hydrogen at a relatively low temperature of 300–400°C and a high pressure of 5–20 MPa to achieve the bio-oil production from the wet biomass. Sodium carbonate and potassium carbonate are the catalysts suggested by researchers for the mechanisms of biomass liquefaction. The proposed catalyst will help to hydrolyze the cellulose, hemicellulose, and lignin macromolecule materials present in the biomass into small micellar-like fragments (Bhaskar et al., 2011; Jayakishan et al., 2019). The hydrolyzed micellar-like fragments are further degraded into smaller components by performing any of the dehydration, dehydrogenation, deoxygenation, and decarboxylation mechanisms. During the process of hydrothermal liquefaction, other products such as gas and char formations also are produced. Approximately 10–73 wt.% of bio-oil, 8–20 wt.% of gases, and 0.2–0.5 wt.% of char formations are the product outcomes during the hydrothermal liquefaction process (Bhaskar et al., 2011).

Hydrothermal liquefaction involves two main processes such as depolymerization and decomposition (Arun et al., 2021). Initially, during the depolymerization mechanisms, the biomass polymers usually crack into small chains of hydrocarbons under significant temperature and pressure. Therein, the mechanism of dehydration takes place by the removal of the water molecules followed by decarboxylation by the removal of the carbon dioxide molecules, and then the deamination mechanism takes place by the removal of the amino acid molecules (Verma et al., 2012). Thus, the decomposition of polymer to oligomers and monomers takes place in the depolymerization mechanisms, which results in an increased yield of high-molecular weight compounds and carbonaceous molecules. An alkaline catalyst such as sodium carbonate, calcium hydroxide, and barium hydroxide is used to accelerate the reactions in hydrothermal liquefaction to enhance the yield of bio-oil production (Canabarro et al., 2013). Therefore, the end product of bio-oil contains monoaromatics (e.g., phenol, benzene, and cholesterol), fatty acids (e.g., meristic acid, palmitic acid, and stearic acid), alkanes (e.g., hexadecane, cycloalkane, and heptadecane), polyaromatics (e.g., naphthalene, fluorine, and indene), nitrogenous compounds (e.g., piperidines, pyrrols, and indoles), and oxygenates (e.g., esters, aldehydes, and ketones) (Wu et al., 2014).

In a comparison of the liquefaction process with the pyrolysis and gasification processes, hydrothermal liquefaction is the most expensive technique and produces less oxygen bio-oil. If the moisture content of biomass exceeds more than 90%, it could lead to an unfavorable energy balance of hydrothermal liquefaction. It also depends on the type of microalga and the quality of the microalgae, which decides the yield of bio-oil (Marulanda et al., 2019). The by-products of inorganic nutrients, which are produced during the hydrothermal liquefaction process, can also be used as fertilizers (Tanger et al., 2013).

In contrast to the carbonization process, the thermal liquefaction process can produce liquid fuels including petroleum products along with several high-value chemicals. The latest technique in biomass thermochemical conversion, the liquefaction process, could not be successful at professional scale due to the lower average yield of oil range between 20 wt.% and 55 wt.% compared to pyrolysis technique, inferior oil quality, firmer operational parameters, and requirements of catalyst and other reactants (Gollakota et al., 2018). Hydrothermal liquefaction can recover over 70% of the feedstock carbon content, which can be used for carbon sequestrations (Tanger et al., 2013).

1.4.2 Pyrolysis

Pyrolysis is a well-defined direct thermal decomposition technique of the organic components, which performs in the absence of oxygen to achieve solid, liquid, and gas as a product. Pyrolysis is a conventional technique, which performs as a slow, irreversible process of thermal decomposition of organic components present in the biomass (Bhaskar et al., 2011). Pyrolysis technique is performed for the preparation of fuels, solvents, chemicals, and biomass products. In this technique, the temperature of the pyrolysis process plays a vital role when it is in the range of 350–500°C for the conversion of organic components into the final pyrolyzed product. As the temperature increases, the decreased yield of charcoal is observed during the process. Suggestions to improve the yield of products can be a low temperature and low heating rate for the charcoal. Low temperatures, high heating rates, and short gas residence times are recommended for the liquid products. High temperature, low heating rate, and a long gas residence time are recommended for the fuel gas product of biomass pyrolysis (Gollakota et al., 2018). Depending on the process condition factors such as solid residence time, heating rate, particle size, and temperature, the pyrolysis process is divided into slow pyrolysis, fast pyrolysis, flash pyrolysis, and catalytic pyrolysis (Yaman, 2004).

Slow pyrolysis consumes quite a lot of hours to complete the process and results in biochar as a product. Slow pyrolysis is performed at slower heating rates. Fixed bed reactor and the tubular reactor are usually used for the slow pyrolysis process. This reaction process results in obtaining a less liquid and gaseous product and more biochar production with a lower temperature of 400–500°C. Fast pyrolysis of biomass is performed by heating to high temperatures in the absence of oxygen (Goswami et al., 2019). The fast pyrolysis process is categorized with a high heating transfer rate of about 600–1000 W/cm^2, short vapor residence times in seconds to prevent the biochar formation, controlled temperature of about 450–550°C, and fast cooling of the pyrolysis vapor. Pyrolysis oil, non-condensable gases, and biochar products are formed from the rapid quenching of the pyrolysis vapors and aerosols. The yield of each product depends on the operating conditions of the process, reactor design, and biomass characteristics like ash, cellulose, and lignin contents. The bio-oil obtained from the fast pyrolysis process after the vapor condensation is black or dark brown in color and contains less than 30 wt.% of water content and an enormous number of oxygenated compounds. The produced bio-oil mostly consists of hydrocarbon liquids, which can be hydro-treated to lower the oxygen content and to decrease the hydrophilicity, whereas the by-product biochar is full of carbon material, which can be separated by a cyclone from vapors and aerosols. During the vapor condensation, the non-condensable gases are collected, which can be recycled as fluidizing gas. Owing to its advantages such as storage and transported capacity, the liquid fuel obtained from fast pyrolyzed can replace the conventional fuel.

Flash pyrolysis is performed on the fine particles of biomass with the reaction time of several seconds and high heating rate in a specially configured fluidized bed reactor and flow reactor having biomass residence times of few seconds (Goswami et al., 2019). The two types of flash pyrolysis are flash hydro-pyrolysis and solar flash pyrolysis. Flash hydro-pyrolysis is performed under a hydrogen-rich atmosphere with a pressure of about 20 MPa, a short residence time in the range of 30 ms to 1.5 s, and the temperature range of 400–950°C. Depolymerization and thermal cracking of biomass can result in the desired end product as compared to conventional diesel fuel (Ibarra-Gonzalez and Rong, 2019). Concerning the solar flash pyrolysis process, solar radiations are used to perform the flash pyrolysis process (Yaman, 2004).

Catalytic pyrolysis is performed by using a catalyst for the efficient pyrolysis of biomass oil to bio-fuel via thermal exchange phenomena (Ibarra-Gonzalez and Rong, 2019). This catalytic pyrolysis was introduced due to the fact that the bio-oil produced during the slow pyrolysis, fast pyrolysis, and flash pyrolysis cannot be utilized as a direct fuel for the transportation purpose because of the presence of high oxygen and water content in it.

1.5 Biological Conversion Pathway for Biofuel Production

1.5.1 Syngas Fermentation

Syngas fermentation is a favorable conversion technique for the production of liquid biofuel from syngas (Acharya et al., 2014; Devarapalli et al., 2016; Phillips et al., 2017; Shen et al., 2017). In this process, microorganisms utilize syngas and produce liquid biofuels. The syngas is produced from thermochemical process such as gasification reaction by using biomass as its raw material (Asimakopoulos et al., 2018; Monir et al., 2018, 2020; Sikarwar et al., 2017). Syngas is a mixture of CO and H_2, which can be utilized by bacteria to produce organic substances including acetic acid and alcohols (Kennes et al., 2016; Müller, 2003). These products can be used to produce electricity or they can be used as a transportation fuel or as laboratory chemicals (Woolcock and Brown, 2013). The microorganisms used for syngas fermentation are listed in Table 1.2.

TABLE 1.2

Different Microorganisms and Their Mode of the Fermentation Process to Produce Bioproducts from Syngas

Microorganisms	Mode of Fermentation	Products Formed	Reference
Mesophilic bacteria			
Acetobacterium woodii	Batch	Acetate	Munasinghe and Khanal (2011)
Clostridium autoethanogenum	Batch and fed-batch	Acetic acid, ethanol, and bioethanol	Liew et al. (2017); Xu et al. (2017)
Clostridium carboxidivorans	Batch	Acids and alcohols	Lagoa-Costa et al. (2017)
Clostridium ljungdahlii	Batch	Acetic acid, ethanol	Kim et al. (2014)
Clostridium ragsdalei	Batch, fed-batch and semi-continuous	Acetate, ethanol	Sun et al. (2018a)
Eubacterium limosum	Batch	Acetate	Munasinghe and Khanal (2011)
Thermophillic bacteria			
Clostridium thermocellum	Batch	Ethanol	Tian et al. (2016)
Moorella thermoacetica	Batch	Acetone	Molitor et al. (2017)
Moorella thermoautotrophica	Batch	Acetone	Sakai et al. (2004)

The biomass used in this process cannot be directly converted into biofuel by using microorganisms since the breakdown mechanism of biomass cannot be achieved efficiently by the usage of microorganisms. Hence, a thermochemical conversion process to produce syngas carries out the preliminary breakdown of biomass (lignin content). The produced syngas contains impurities after the gasification process, which includes dust particles, tar ammonia, mercury, sulfur impurities, hydrochloric acid, and hydrogen cyanide (Adhikari et al., 2017; Beagle et al., 2018; Ramos et al., 2018; Susastriawan and Saptoadi, 2017; Zhang and Zheng, 2016). These impurities must be removed before the fermentation process to ensure maximum product yield. Some of the widely used syngas purification processes include catalytic tar removal, chemical absorption, wet scrubbing, and tar removal (Baidya et al., 2018; Hu et al., 2018; Pallozzi et al., 2018; Unyaphan et al., 2017).

The purified syngas undergoes fermentation process with optimal parameters, cell growth, and metabolic activity of the organism. There are various types of fermenters to carry out syngas fermentation, which include continuous stirred tank reactor, monolithic biofilm reactor, membrane-based system reactor, trickle bed reactor, and bubble column reactor (Heiskanen et al., 2007; Holland, 2015). The syngas fermentation process can be affected by factors such as pH level of the medium, gas flow rate, concentration of trace metals, concentration of reducing agent, mass transfer rate, the temperature of the medium, and organic source. Current research concentrates on the enhancement of yield produced from syngas fermentation by using nanoparticles as a catalyst, which increase the mass transfer rate from syngas to water and thus the organic substance is produced by fermentation of syngas.

1.5.2 Fermentation of Biomass

Fermentation of biomass is considered a promising method for the production of biofuels and biochemicals (Manish and Banerjee, 2008). Both the aerobic and anaerobic fermentations are possible with the use of microorganisms, which convert the biomass into the desired bioproduct by both the dark and light fermentation processes (Sarangi and Nanda, 2020). The factors which affect the fermentation process include substrate concentration, microorganism use, pH, temperature, retention time, and pressure. The bioproducts produced from the fermentation process are listed in Table 1.3.

Cellulose and hemicellulose content present in biomass are rich in carbohydrates, which are depolymerized by acids or enzymes to monosaccharides. These monosaccharides can be further converted into a wide range of value-added biochemicals (Arun et al., 2017; Jing et al., 2019). Table 1.4 shows the detailed biochemical potential of biomasses available easily.

TABLE 1.3

Different Biomasses and Organisms Utilized in a Different Mode of Reaction to Produce Bioproducts by Fermentation of Biomass

Organism Used	Biomass Utilized	Mode of Fermentation	Products Formed	References
Anaerobic and aerobic sludge	Food waste	Batch	Acetic acid, butyric acid	Alibardi and Cossu (2015)
Anaerobic bacteria	Raw cassava starch	Batch	Butyric acid, acetic acid, ethanol	Wang et al. (2017)
Anaerobic mixed culture	By-product of biodiesel from sunflower	Continuous stirred tank reactor	Butyric acid, ethanol	Dounavis et al. (2015)
Anaerobic sludge	Mixed food waste	Batch	–	Pan et al. (2008)
Caldicellulosiruptor saccharolyticus	Wheat straw	Batch	–	Ivanova et al. (2009)
Caldicellulosiruptor saccharolyticus	Sugarcane bagasse	Batch	–	Ivanova et al. (2009)
Caldicellulosiruptor saccharolyticus	Switchgrass	Batch	Acetic acid, succinic acid	Talluri et al. (2013)

(Continued)

TABLE 1.3 (Continued)

Organism Used	Biomass Utilized	Mode of Fermentation	Products Formed	References
Clostridium sp.	Dairy manure	Continuously stirred anaerobic bioreactor	Butyric acid. acetic acid. ethanol, propionic acid, butanol	Xing et al. (2010)
Clostridium thermocellum	Barley hulls	Batch	Acetic acid, formic acid, ethanol	Magnusson et al. (2008); Ren et al. (2016)
Immobilized mixed culture	Waste glycerol	Upflow aerobic sludge blanket reactor, continuous	Propionic acid, formic acid	Reungsang et al. (2013)
Mesophilic bacteria	Organic municipal waste	Erlenmeyer flask	Caproic acid, butyric acid	Gomez et al. (2006)
Mixed culture	Cheese whey and whey water	Batch	Butyric acid, acetic acid	Venetsaneas et al. (2009)
Mixed culture	Wastewater of citric acid	Upflow aerobic sludge blanket reactor, continuous	Acetic acid, propionic acid	Yang et al. (2006)
Mixed culture	Soybean straw	Batch	Butyric acid	Han et al. (2012)
Mixed culture from activated sludge	Brewery wastewater	Batch	Butyric acid, acetic acid	Chu et al. (2013)
Mixed culture from cow dung	Palm oil	Upflow aerobic sludge blanket reactor, Continuous	–	Vijayaraghavan and Ahmad (2006)
Mixed culture, anaerobic sludge	Kitchen waste	Continuous stir tank reactor	Butyric acid, acetic acid	Jayalakshmi et al. (2009)
Sewage sludge	Cattle wastewater	Batch	Butyric acid, acetic acid, ethanol, propionic acid	Tang et al. (2008)
Thermoanaerobacterium, Thermosaccharolyticum	Cornstalk	Batch	Acetic acid. butyric acid, ethanol	Sheng et al. (2015)
Thermotoga neapolitana (mixed culture)	Rice straw	Batch	Acetic acid, butyric acid	Chen et al. (2012)

TABLE 1.4

Biochemicals Produced from Biomasses

Biomass	Conversion Method	Biochemicals Produced	Reference
Ash tree	Thermal catalytic	Monophenols, diols	Li et al. (2012)
Beechwood	Thermal catalytic	Lignin monomers, glucose, furfural, xylose	Shuai et al. (2016)
Brewers spent grain	Hydrothermal	Biobutanol	López-Linares et al. (2019)
Birch tree	Thermal catalytic	Pentanes, hexane and alkyl cyclohexanes	Xia et al. (2016)
Birch tree	Thermal catalytic	Furfural, HMF, C_7-C_9 hydrocarbons	Guo et al. (2018)
Birch tree	Thermal catalytic	Aromatic monomers, xylose, glucose	Matson et al. (2011)
Birch tree	Thermal catalytic	Phenolic monomers, aromatic dimmers	Li et al. (2018a)
Bone dry wood, recycled cardboard	Thermal deoxygenation	Furfural, levulinic acid	Gunukula et al. (2018)
Bone dry wood, recycled cardboard	Thermal deoxygenation	Oil	Gunukula et al. (2018)

(Continued)

TABLE 1.4 (Continued)

Biomass	Conversion Method	Biochemicals Produced	Reference
Chlorella species	Hydrothermal	Bioethanol	Ngamsirisomsakul et al. (2019)
Corncob	Hydrothermal decomposition	Levulinic acid, furfural	Li et al. (2018b)
Corncob	Hydrothermal decomposition	Bio-jet fuel	Li et al. (2018b)
Cornstalk	Thermal catalytic	Alkylcyclohexanes, phenols, polyols	Van den Bosch et al. (2015)
Defatted microalgae	Hydrothermal liquefaction	Bio-crude	Ou et al. (2015)
Food waste	Fermentation	Lactic acid	Kwan et al. (2018)
Glucose or carbohydrate	Hydrothermal	Lactic acid	Cantero et al. (2015)
Grape seeds	Hydrothermal	Bio-oil	Yedro et al. (2014)
Hydrolyzed C5 sugar syrup	Fermentation	Xylitol	Mountraki et al. (2017)
Lignocellulosic biomass	Fermentation	Ethanol	Han et al. (2015)
Lignocellulosic fraction of municipal solid waste	Biological treatment	Levulinic acid	Sadhukhan et al. (2016)
Oil palm empty fruit bunches	Steam explosion	Lignin, xylitol, ethanol	Medina et al. (2018)
Orange peel	Hydrothermal	Hesperidin, narirutin	Lachos-Perez et al. (2018)
Pine tree	Thermal catalytic	Aromatic monomers	Sun et al. (2018b)
Pine tree	Thermal catalytic	Cyclohexanols	Liu et al. (2015)
Potato starch, linoleic acid, bovine serum albumin	Hydrothermal	Bio-oil	Posmanik et al. (2017)
Raw sugarcane bagasse and trash feed	Hydrothermal liquefaction	Xylitol, citric acid and glutamic acid	Ou et al. (2015)
Sugarcane bagasse	Hydrothermal	5-Hydroxymethyl furfural	Iryani et al. (2013)
Switchgrass, corn stover, Durian seed waste, wheat straw	Hydrothermal	Ethanol	Kumar et al. (2011); Purnomo et al. (2016)
Thymbra spicata	Hydrothermal	p-Cymene, E-3-caren-2-ol, carvacrol	Ozel et al. (2003)

1.6 Conclusions

Biomasses could be a key candidate in the future for renewable biofuel production. Apart from biofuels, biomasses also possess the ability to provide valuable chemical compounds. Resource recovery through the biorefinery approach is the hot topic of research around the globe in recent years and years to come. Various modes of operation with numerous parameters were studied in thermochemical processes for the production of green biofuel from biomasses. However, the main challenge lies in the storage, supply, and meeting the public demand in the modern world.

Acknowledgments

The authors wish to thank the Management of Sathyabama Institute of Science and Technology for providing support in the completion of this chapter.

REFERENCES

Acharya, B., Roy, P., and Dutta, A. 2014. Review of syngas fermentation processes for bioethanol. *Biofuels* 5(5): 551–564.

Adhikari, S., Abdoulmoumine, N., Nam, H., and Oyedeji, O. 2017. Biomass gasification producer gas cleanup. In: *Bioenergy Systems for the Future*. Woodhead Publishing, pp. 541–557. United Kingdom.

Adler, P. R., Sanderson, M. A., Boateng, A. A., Weimer, P. J., and Jung, H. J. G. 2006. Biomass yield and biofuel quality of switchgrass harvested in fall or spring. *Agronomy Journal* 98(6): 1518–1525.

Alibardi, L., and Cossu, R. 2015. Composition variability of the organic fraction of municipal solid waste and effects on hydrogen and methane production potentials. *Waste Management* 36: 147–155.

Arun, J., Avinash, U., Arun Krishna, B., Pandimadevi, M., and Gopinath, K. 2017. Ultrasound assisted enhanced extraction of lutein (β, ε-carotene-3, 3′-diol) from Mircroalga (*Chlorella pyrenoidosa*) grown in wastewater: Optimization through Response Surface Methodology. *Global Nest Journal* 19(4): 574–583.

Arun, J., Gopinath, K. P., Sivaramakrishnan, R., Madhav, N. V., Abhishek, K., Ramanan, V. G. K., and Pugazhendhi, A. 2021. Bioenergy perspectives of cattails biomass cultivated from municipal wastewater via hydrothermal liquefaction and hydro-deoxygenation. *Fuel* 284: 118963.

Arun, J., Gopinath, K. P., SundarRajan, P., Felix, V., JoselynMonica, M., and Malolan, R. 2020. A conceptual review on microalgae biorefinery through thermochemical and biological pathways: Bio-circular approach on carbon capture and wastewater treatment. *Bioresource Technology Reports* 100477.

Asimakopoulos, K., Gavala, H. N., and Skiadas, I. V. 2018. Reactor systems for syngas fermentation processes: A review. *Chemical Engineering Journal* 348: 732–744.

Baidya, T., Cattolica, R. J., and Seiser, R. 2018. High performance Ni-Fe-Mg catalyst for tar removal in producer gas. *Applied Catalysis A: General* 558: 131–139.

Beagle, E., Wang, Y., Bell, D., and Belmont, E. 2018. Co-gasification of pine and oak biochar with sub-bituminous coal in carbon dioxide. *Bioresource Technology* 251: 31–39.

Bhaskar, T., Bhavya, B., Singh, R., Naik, D. V., Kumar, A., and Goyal, H. B. 2011. Thermochemical conversion of biomass to biofuels. In: *Biofuels*. Elsevier, pp. 51–77.

Canabarro, N., Soares, J. F., Anchieta, C. G., Kelling, C. S., and Mazutti, M. A. 2013. Thermochemical processes for biofuels production from biomass. *Sustainable Chemical Processes* 1(1): 22.

Cantero, D. A., Álvarez, A., Bermejo, M. D., and Cocero, M. J. 2015. Transformation of glucose into added value compounds in a hydrothermal reaction media. *The Journal of Supercritical Fluids* 98: 204–210.

Chen, C.-C., Y.-Chuang, S., Lin, C.-Y., Lay, C.-H., and Sen, B. 2012. Thermophilic dark fermentation of untreated rice straw using mixed cultures for hydrogen production. *International Journal of Hydrogen Energy* 37(20): 15540–15546.

Chu, C.-Y., Tung, L., and Lin, C.-Y. 2013. Effect of substrate concentration and pH on biohydrogen production kinetics from food industry wastewater by mixed culture. *International Journal of Hydrogen Energy* 38(35): 15849–15855.

Demirbas, A. 2010. Biorefinery technologies for biomass upgrading. *Energy Sources, Part A: Recovery, Utilization, and Environmental Effects* 32(16): 1547–1558.

Devarapalli, M., Atiyeh, H. K., Phillips, Lewis, R. S., and Huhnke, R. L. 2016. Ethanol production during semi-continuous syngas fermentation in a trickle bed reactor using *Clostridium ragsdalei*. *Bioresource Technology* 209: 56–65.

Dounavis, A. S., Ntaikou, I., and Lyberatos, G. 2015. Production of biohydrogen from crude glycerol in an upflow column bioreactor. *Bioresource Technology* 198: 701–708.

El-Dalatony, M. M., Saha, S., Govindwar, S. P., Abou-Shanab, R. A., and Jeon, B.-H. 2019. Biological conversion of amino acids to higher alcohols. *Trends in Biotechnology* 37(8): 855–869.

Faaij, A. P. 2006. Bio-energy in Europe: Changing technology choices. *Energy Policy* 34(3): 322–342.

Gao, N., Kamran, K., Quan, C., Williams, P. T. 2020. Thermochemical conversion of sewage sludge: A critical review. *Progress in Energy and Combustion Science* 79: 100843.

Ge, X., Yang, L., Sheets, J. P., Yu, Z., and Li, Y. 2014. Biological conversion of methane to liquid fuels: Status and opportunities. *Biotechnology Advances* 32(8): 1460–1475.

Gollakota, A., Kishore, N., and Gu, S. 2018. A review on hydrothermal liquefaction of biomass. *Renewable and Sustainable Energy Reviews* 81: 1378–1392.

Gomez, X., Moran, A., Cuetos, M., and Sanchez, M. 2006. The production of hydrogen by dark fermentation of municipal solid wastes and slaughterhouse waste: A two-phase process. *Journal of Power Sources* 157(2): 727–732.

Goswami, G., Makut, B. B., and Das, D. 2019. Sustainable production of bio-crude oil via hydrothermal liquefaction of symbiotically grown biomass of microalgae-bacteria coupled with effective wastewater treatment. *Scientific Reports* 9(1): 1–12.

Gunukula, S., Klein, S. J., Pendse, H. P., DeSisto, W. J., and Wheeler, M. C. 2018. Techno-economic analysis of thermal deoxygenation based biorefineries for the coproduction of fuels and chemicals. *Applied Energy* 214: 16–23.

Guo, T., Li, X., Liu, X., Guo, Y., and Wang, Y. 2018. Catalytic Transformation of Lignocellulosic Biomass into Arenes, 5-Hydroxymethylfurfural, and Furfural. *ChemSusChem* 11(16): 2758–2765.

Han, H., Wei, L., Liu, B., Yang, H., and Shen, J. 2012. Optimization of biohydrogen production from soybean straw using anaerobic mixed bacteria. *International Journal of Hydrogen Energy* 37(17): 13200–13208.

Han, J., Luterbacher, J. S., Alonso, D. M., Dumesic, J. A., and Maravelias, C. T. 2015. A lignocellulosic ethanol strategy via nonenzymatic sugar production: Process synthesis and analysis. *Bioresource Technology* 182: 258–266.

Hawkins, A. S., McTernan, P. M., Lian, H., Kelly, R. M., and Adams, M. W. 2013. Biological conversion of carbon dioxide and hydrogen into liquid fuels and industrial chemicals. *Current Opinion in Biotechnology* 24(3): 376–384.

Heiskanen, H., Virkajärvi, I., and Viikari, L. 2007. The effect of syngas composition on the growth and product formation of *Butyribacterium methylotrophicum*. *Enzyme and Microbial Technology* 41(3): 362–367.

Holland, D. 2015. Applications of tomography in bubble column and trickle bed reactors. In: *Industrial Tomography*. Elsevier, pp. 477–507.

Hu, M., Laghari, M., Cui, B., Xiao, B., Zhang, B., and Guo, D. 2018. Catalytic cracking of biomass tar over char supported nickel catalyst. *Energy* 145: 228–237.

Huber, G. W., Iborra, S., and Corma, A. 2006. Synthesis of transportation fuels from biomass: Chemistry, catalysts, and engineering. *Chemical Reviews* 106(9): 4044–4098.

Ibarra-Gonzalez, P., and Rong, B.-G. 2019. A review of the current state of biofuels production from lignocellulosic biomass using thermochemical conversion routes. *Chinese Journal of Chemical Engineering* 27(7): 1523–1535.

Iryani, D. A., Kumagai, S., Nonaka, M., Sasaki, K., and Hirajima, T. 2013. Production of 5-hydroxymethyl furfural from sugarcane bagasse under hot compressed water. *Procedia Earth and Planetary Science* 6: 441–447.

Ivanova, G., Rákhely, G., and Kovács, K. L. 2009. Thermophilic biohydrogen production from energy plants by *Caldicellulosiruptor saccharolyticus* and comparison with related studies. *International Journal of Hydrogen Energy* 34(9): 3659–3670.

Jayakishan, B., Nagarajan, G., and Arun, J. 2019. Co-thermal liquefaction of Prosopis juliflora biomass with paint sludge for liquid hydrocarbons production. *Bioresource Technology* 283: 303–307.

Jayalakshmi, S., Joseph, K., and Sukumaran, V. 2009. Bio hydrogen generation from kitchen waste in an inclined plug flow reactor. *International Journal of Hydrogen Energy* 34(21): 8854–8858.

Jing, Y., Guo, Y., Xia, Q., Liu, X., and Wang, Y. 2019. Catalytic production of value-added chemicals and liquid fuels from lignocellulosic biomass. *Chem* 5(10): 2520–2546.

Kennes, D., Abubackar, H. N., Diaz, M., Veiga, M. C., and Kennes, C. 2016. Bioethanol production from biomass: Carbohydrate vs syngas fermentation. *Journal of Chemical Technology & Biotechnology* 91(2): 304–317.

Kim, Y.-K., Park, S. E., Lee, H., and Yun, J. Y. 2014. Enhancement of bioethanol production in syngas fermentation with *Clostridium ljungdahlii* using nanoparticles. *Bioresource Technology* 159: 446–450.

Kumar, S., Kothari, U., Kong, L., Lee, Y., and Gupta, R. B. 2011. Hydrothermal pretreatment of switchgrass and corn stover for production of ethanol and carbon microspheres. *Biomass and Bioenergy* 35(2): 956–968.

Kwan, T. H., Hu, Y., and Lin, C. S. K. 2018. Techno-economic analysis of a food waste valorisation process for lactic acid, lactide and poly (lactic acid) production. *Journal of Cleaner Production* 181: 72–87.

Lachos-Perez, D., Baseggio, A. M., Mayanga-Torres, P., Junior, M. R. M., Rostagno, M., Martínez, J., and Forster-Carneiro, T. 2018. Subcritical water extraction of flavanones from defatted orange peel. *The Journal of Supercritical Fluids* 138: 7–16.

Lagoa-Costa, B., Abubackar, H. N., Fernández-Romasanta, M., Kennes, C., and Veiga, M. C. 2017. Integrated bioconversion of syngas into bioethanol and biopolymers. *Bioresource Technology* 239: 244–249.

Li, C., Zheng, M., Wang, A., and Zhang, T. 2012. One-pot catalytic hydrocracking of raw woody biomass into chemicals over supported carbide catalysts: Simultaneous conversion of cellulose, hemicellulose and lignin. *Energy & Environmental Science* 5(4): 6383–6390.

Li, X., Guo, T., Xia, Q., Liu, X., and Wang, Y. 2018a. One-pot catalytic transformation of lignocellulosic bio-mass into alkylcyclohexanes and polyols. *ACS Sustainable Chemistry & Engineering* 6(3): 4390–4399.

Li, Y., Zhao, C., Chen, L., Zhang, X., Zhang, Q., Wang, T., Qiu, S., Tan, J., Li, K., and Wang, C. 2018b. Production of bio-jet fuel from corncob by hydrothermal decomposition and catalytic hydrogenation: Lab analysis of process and techno-economics of a pilot-scale facility. *Applied Energy* 227: 128–136.

Liew, F., Henstra, A. M., Köpke, M., Winzer, K., Simpson, S. D., and Minton, N. P. 2017. Metabolic engi-neering of *Clostridium autoethanogenum* for selective alcohol production. *Metabolic Engineering* 40: 104–114.

Liu, Y., Chen, L., Wang, T., Zhang, Q., Wang, C., Yan, J., and Ma, L. 2015. One-pot catalytic conversion of raw lignocellulosic biomass into gasoline alkanes and chemicals over LiTaMoO6 and Ru/C in aqueous phosphoric acid. *ACS Sustainable Chemistry & Engineering* 3(8): 1745–1755.

López-Linares, J. C., García-Cubero, M. T., Lucas, S., González-Benito, G., and Coca, M. 2019. Microwave assisted hydrothermal as greener pretreatment of brewer's spent grains for biobutanol production. *Chemical Engineering Journal* 368: 1045–1055.

Magnusson, L., Islam, R., Sparling, R., Levin, D., and Cicek, N. 2008. Direct hydrogen production from cel-lulosic waste materials with a single-step dark fermentation process. *International Journal of Hydrogen Energy* 33(20): 5398–5403.

Manish, S., and Banerjee, R. 2008. Comparison of biohydrogen production processes. *International Journal of Hydrogen Energy* 33(1): 279–286.

Marulanda, V. A., Gutierrez, C. D. B., and Alzate, C. A. C. 2019. Thermochemical, biological, biochemi-cal, and hybrid conversion methods of bio-derived molecules into renewable fuels. In: *Advanced Bioprocessing for Alternative Fuels, Biobased Chemicals, and Bioproducts*. Elsevier, pp. 59–81.

Matson, T. D., Barta, K., Iretskii, A. V., and Ford, P. C. 2011. One-pot catalytic conversion of cellulose and of woody biomass solids to liquid fuels. *Journal of the American Chemical Society* 133(35): 14090–14097.

McKendry, P. 2002. Energy production from biomass (part 1): Overview of biomass. *Bioresource Technology* 83(1): 37–46.

Medina, J. D. C., Woiciechowski, A. L., Zandona Filho, A., Brar, S. K., Júnior, A. I. M., and Soccol, C. R. 2018. Energetic and economic analysis of ethanol, xylitol and lignin production using oil palm empty fruit bunches from a Brazilian factory. *Journal of Cleaner Production* 195: 44–55.

Molitor, B., Marcellin, E., and Angenent, L. T. 2017. Overcoming the energetic limitations of syngas fermenta-tion. *Current Opinion in Chemical Biology* 41: 84–92.

Monir, M. U., Abd Aziz, A., Kristanti, R. A., and Yousuf, A. 2018. Gasification of lignocellulosic biomass to produce syngas in a 50 kW downdraft reactor. *Biomass and Bioenergy* 119: 335–345.

Monir, M. U., Abd Aziz, A., Kristanti, R. A., and Yousuf, A. 2020. Syngas production from co-gasification of forest residue and charcoal in a pilot scale downdraft reactor. *Waste and Biomass Valorization* 11(2): 635–651.

Mountraki, A., Koutsospyros, K., Mlayah, B. B., and Kokossis, A. 2017. Selection of biorefinery routes: The case of xylitol and its integration with an organosolv process. *Waste and Biomass Valorization* 8(7): 2283–2300.

Müller, V. 2003. Energy conservation in acetogenic bacteria. *Applied and Environmental Microbiology* 69(11): 6345–6353.

Munasinghe, P. C., and Khanal, S. K. 2011. Biomass-derived syngas fermentation into biofuels. In: *Biofuels*. Elsevier, pp. 79–98.

Naik, S. N., Goud, V. V., Rout, P. K., and Dalai, A. K. 2010. Production of first and second generation biofuels: A comprehensive review. *Renewable and Sustainable Energy Reviews* 14(2): 578–597.

Ngamsirisomsakul, M., Reungsang, A., Liao, Q., and Kongkeitkajorn, M. B. 2019. Enhanced bio-ethanol pro-duction from *Chlorella* sp. biomass by hydrothermal pretreatment and enzymatic hydrolysis. *Renewable Energy* 141: 482–492.

Okolie, J. A., Nanda, S., Dalai, A. K., Berruti, F., and Kozinski, J. A. 2020. A review on subcritical and super-critical water gasification of biogenic, polymeric and petroleum wastes to hydrogen-rich synthesis gas. *Renewable and Sustainable Energy Reviews* 119: 109546.

Omer, M. A., and Noguchi, T. 2020. A conceptual framework for understanding the contribution of building materials in the achievement of Sustainable Development Goals (SDGs). *Sustainable Cities and Society* 52: 101869.

Ong, H. C., Chen, W.-H., Singh, Y., Gan, Y. Y., Chen, C.-Y., and Show, P. L. 2020. A state-of-the-art review on thermochemical conversion of biomass for biofuel production: A TG-FTIR approach. *Energy Conversion and Management* 209: 112634.

Ou, L., Thilakaratne, R., Brown, R. C., and Wright, M. M. 2015. Techno-economic analysis of transportation fuels from defatted microalgae via hydrothermal liquefaction and hydroprocessing. *Biomass and Bioenergy* 72: 45–54.

Ozel, M. Z., Gogus, F., and Lewis, A. C. 2003. Subcritical water extraction of essential oils from Thymbra spicata. *Food Chemistry* 82(3): 381–386.

Pallozzi, V., Di Carlo, A., Bocci, E., and Carlini, M. 2018. Combined gas conditioning and cleaning for reduction of tars in biomass gasification. *Biomass and Bioenergy* 109: 85–90.

Pan, J., Zhang, R., El-Mashad, H. M., Sun, H., and Ying, Y. 2008. Effect of food to microorganism ratio on biohydrogen production from food waste via anaerobic fermentation. *International Journal of Hydrogen Energy* 33(23): 6968–6975.

Pang, S. 2019. Advances in thermochemical conversion of woody biomass to energy, fuels and chemicals. *Biotechnology Advances* 37(4): 589–597.

Phillips, J. R., Huhnke, R. L., and Atiyeh, H. K. 2017. Syngas fermentation: A microbial conversion process of gaseous substrates to various products. *Fermentation* 3(2): 28.

Posmanik, R., Cantero, D. A., Malkani, A., Sills, D. L., and Tester, J. 2017. Biomass conversion to bio-oil using sub-critical water: Study of model compounds for food processing waste. *The Journal of Supercritical Fluids* 119: 26–35.

Purnomo, A., Y. Yudiantoro, A. W., Putro, J. N., Nugraha, A. T., Irawaty, W., and Ismadji, S. 2016. Subcritical water hydrolysis of durian seeds waste for bioethanol production. *International Journal of Industrial Chemistry* 7(1): 29–37.

Ramos, A., Monteiro, E., Silva, V., and Rouboa, A. 2018. Co-gasification and recent developments on waste-to-energy conversion: A review. *Renewable and Sustainable Energy Reviews* 81: 380–398.

Ren, N.-Q., Zhao, L., Chen, C., Guo, W.-Q., and Cao, G.-L. 2016. A review on bioconversion of lignocellulosic biomass to H$_2$: Key challenges and new insights. *Bioresource Technology* 215: 92–99.

Reungsang, A., Sittijunda, S., and Sompong, O. 2013. Bio-hydrogen production from glycerol by immobilized Enterobacter aerogenes ATCC 13048 on heat-treated UASB granules as affected by organic loading rate. *International Journal of Hydrogen Energy* 38(17): 6970–6979.

Sadhukhan, J., Ng, K. S., and Martinez-Hernandez, E. 2016. Novel integrated mechanical biological chemical treatment (MBCT) systems for the production of levulinic acid from fraction of municipal solid waste: A comprehensive techno-economic analysis. *Bioresource Technology* 215: 131–143.

Sakai, S., Nakashimada, Y., Yoshimoto, H., Watanabe, S., Okada, H., and Nishio, N. 2004. Ethanol production from H$_2$ and CO$_2$ by a newly isolated thermophilic bacterium, *Moorella* sp. HUC22–1. *Biotechnology Letters* 26(20): 1607–1612.

Sarangi, P. K., and Nanda, S. 2020. Biohydrogen production through dark fermentation. *Chemical Engineering & Technology* 43: 601–612.

Saravanan, A. P., Pugazhendhi, A., and Mathimani, T. 2020. A comprehensive assessment of biofuel policies in the BRICS nations: Implementation, blending target and gaps. *Fuel* 272: 117635.

Shen, Y., Brown, R. C., and Wen, Z. 2017. Syngas fermentation by *Clostridium carboxidivorans* P7 in a horizontal rotating packed bed biofilm reactor with enhanced ethanol production. *Applied Energy* 187: 585–594.

Sheng, T., Gao, L., Zhao, L., Liu, W., and Wang, A. 2015. Direct hydrogen production from lignocellulose by the newly isolated *Thermoanaerobacterium thermosaccharolyticum* strain DD32. *RSC Advances* 5(121): 99781–99788.

Shuai, L., Amiri, M. T., Questell-Santiago, Y. M., Héroguel, F., Li, Y., Kim, H., Meilan, R., Chapple, C., Ralph, J., and Luterbacher, J. S. 2016. Formaldehyde stabilization facilitates lignin monomer production during biomass depolymerization. *Science* 354(6310): 329–333.

Sikarwar, V. S., Zhao, M., Fennell, P. S., Shah, N., and Anthony, E. J. 2017. Progress in biofuel production from gasification. *Progress in Energy and Combustion Science* 61: 189–248.

Sims, R., Taylor, M., Saddler, J., and Mabee, W. 2008. From 1st-to 2nd-generation biofuel technologies: An overview of current industry and RD&D activities. *International Energy Agency*: 16–20.

Sun, X., Atiyeh, H. K., Kumar, A., and Zhang, H. 2018a. Enhanced ethanol production by Clostridium ragsdalei from syngas by incorporating biochar in the fermentation medium. *Bioresource Technology* 247: 291–301.

Sun, Z., Bottari, G., Afanasenko, A., Stuart, M. C., Deuss, P. J., Fridrich, B., and Barta, K. 2018b. Complete lignocellulose conversion with integrated catalyst recycling yielding valuable aromatics and fuels. *Nature Catalysis* 1(1): 82–92.

Susastriawan, A., and Saptoadi, H. 2017. Small-scale downdraft gasifiers for biomass gasification: A review. *Renewable and Sustainable Energy Reviews* 76: 989–1003.

Talluri, S., Raj, S. M., and Christopher, L. P. 2013. Consolidated bioprocessing of untreated switchgrass to hydrogen by the extreme thermophile *Caldicellulosiruptor saccharolyticus* DSM 8903. *Bioresource Technology* 139: 272–279.

Tang, G.-L., Huang, J., Sun, Z.-J., Tang, Q.-q., Yan, C.-h., and Liu, G.-q. 2008. Biohydrogen production from cattle wastewater by enriched anaerobic mixed consortia: Influence of fermentation temperature and pH. *Journal of Bioscience and Bioengineering* 106(1): 80–87.

Tanger, P., Field, J. L., Jahn, C. E., DeFoort, M. W., and Leach, J. E. 2013. Biomass for thermochemical conversion: Targets and challenges. *Frontiers in Plant Science* 4: 218.

Tian, L., Papanek, B., Olson, D. G., Rydzak, T., Holwerda, E. K., Zheng, T., Zhou, J., Maloney, M., Jiang, N., and Giannone, R. J. 2016. Simultaneous achievement of high ethanol yield and titer in *Clostridium thermocellum*. *Biotechnology for Biofuels* 9(1): 1–11.

Unyaphan, S., Tarnpradab, T., Takahashi, F., and Yoshikawa, K. 2017. Improvement of tar removal performance of oil scrubber by producing syngas microbubbles. *Applied Energy* 205: 802–812.

Van den Bosch, S., Schutyser, W., Vanholme, R., Driessen, T., Koelewijn, S.-F., Renders, T., De Meester, B., Huijgen, W., Dehaen, W., and Courtin, C. 2015. Reductive lignocellulose fractionation into soluble lignin-derived phenolic monomers and dimers and processable carbohydrate pulps. *Energy & Environmental Science* 8(6): 1748–1763.

Venetsaneas, N., Antonopoulou, G., Stamatelatou, K., Kornaros, M., and Lyberatos, G. 2009. Using cheese whey for hydrogen and methane generation in a two-stage continuous process with alternative pH controlling approaches. *Bioresource Technology* 100(15): 3713–3717.

Verma, M., Godbout, S., Brar, S., Solomatnikova, O., Lemay, S., and Larouche, J. 2012. Biofuels production from biomass by thermochemical conversion technologies. *International Journal of Chemical Engineering* 2012: 542426.

Vijayaraghavan, K., and Ahmad, D. 2006. Biohydrogen generation from palm oil mill effluent using anaerobic contact filter. *International Journal of Hydrogen Energy* 31(10): 1284–1291.

Wang, S., Ma, Z., Zhang, T., Bao, M., and Su, H. 2017. Optimization and modeling of biohydrogen production by mixed bacterial cultures from raw cassava starch. *Frontiers of Chemical Science and Engineering* 11(1): 100–106.

Woolcock, P. J., and Brown, R. C. 2013. A review of cleaning technologies for biomass-derived syngas. *Biomass and Bioenergy* 52: 54–84.

Wu, L. M., Zhou, C. H., Tong, D. S., and Yu, W. H. 2014. Catalytic thermochemical processes for biomass conversion to biofuels and chemicals. In: *Bioenergy Research: Advances and Applications*. Elsevier, pp. 243–254.

Xia, Q., Chen, Z., Shao, Y., Gong, X., Wang, H., Liu, X., Parker, S. F., Han, X., Yang, S., and Wang, Y. 2016. Direct hydrodeoxygenation of raw woody biomass into liquid alkanes. *Nature Communications* 7: 11162.

Xing, Y., Li, Z., Fan, Y., and Hou, H. 2010. Biohydrogen production from dairy manures with acidification pretreatment by anaerobic fermentation. *Environmental Science and Pollution Research* 17(2): 392–399.

Xu, H., Liang, C., Yuan, Z., Xu, J., Hua, Q., and Guo, Y. 2017. A study of CO/syngas bioconversion by *Clostridium autoethanogenum* with a flexible gas-cultivation system. *Enzyme and Microbial Technology* 101: 24–29.

Yaman, S. 2004. Pyrolysis of biomass to produce fuels and chemical feedstocks. *Energy Conversion and Management* 45(5): 651–671.

Yang, H., Shao, P., Lu, T., Shen, J., Wang, D., Xu, Z., and Yuan, X. 2006. Continuous bio-hydrogen production from citric acid wastewater via facultative anaerobic bacteria. *International Journal of Hydrogen Energy* 31(10): 1306–1313.

Yang, Z., Wu, Y., Zhang, Z., Li, H., Li, X., Egorov, R. I., Strizhak, P. A., and Gao, X. 2019. Recent advances in co-thermochemical conversions of biomass with fossil fuels focusing on the synergistic effects. *Renewable and Sustainable Energy Reviews* 103: 384–398.

Yedro, F. M., García-Serna, J., Cantero, D. A., Sobrón, F., and Cocero, M. J. 2014. Hydrothermal hydrolysis of grape seeds to produce bio-oil. *RSC Advances* 4: 30332–30339.

Zhang, Y., and Zheng, Y. 2016. Co-gasification of coal and biomass in a fixed bed reactor with separate and mixed bed configurations. *Fuel* 183: 132–138.

2

Solid and Liquid Biofuels from Waste and Biomass: Production, Characterization and Combustion

Fang-Hsien Wu, Chao-Wei Huang, Yueh-Heng Li, Van-Huy Nguyen, Guan-Bang Chen

CONTENTS

Abbreviations and Nomenclature

Absolute temperature: T
Activation energy: Ea
Adaptive neuro-fuzzy inference system: ANFIS
American Society for Testing and Materials: ASTM
Artificial neural network: ANN
Average weight loss rate: DTG_{mean}
Biomass blending ratio: BBR
Burnout temperature: T_b
Carbon Capture and Storage: CCS
Coats and Redfern: CR
Combustion index: S
Conversion degree: α
Droplet's diameter at time t: $d_p(t)$
Environmental Protection Agency: EPA
Final mass: m_f
Fixed carbon: FC
Flammability index: C

Gas constant: R
Gasoline gallon equivalent: GGE
Greenhouse gas: GHG
Gross calorific value: GCV
Heavy fuel oil: HFO
Higher heating value: HHV
Hydrogen-to-carbon ratio: H/C
Hypothetical model of the reaction: $f(\alpha)$
Ignition temperature: T_i
Initial droplet diameter: d_p
Initial mass: m_0
International Energy Agency: IEA
Kissinger–Akahira–Sunose: KAS
Lower heating value: LHV
Maximum weight loss rate: DTG_{max}
Multilinear regression: MLR
Net calorific value: NCV
Oxygen-to-carbon ratio: O/C
Ozawa–Wall–Flynn: OWF
Particulate matter: PM
Pre-exponential factor: A
Pyrolysis castor oil: PCO
Reaction rate coefficient: k(T)
Sample mass at time t: m_t
Sludge pyrolysis oil: SPO
Sludge shiitake ratio: SSR
Thermogravimetric analysis: TGA
Thermogravimetric/Fourier transform infrared spectroscopy: TG-FTIR
Time: t
Vaporization/combustion rate: K
Volatile matter: VM
Wastewater treatment plants: WWTPs

2.1 Introduction

Energy supports the function and operation of the world. For example, the ecosystem and climate of the earth rely on the absorption of geothermal energy inside the earth and radiation energy from the sun. Besides, energy is also the engine of human civilization growth, influencing social and economic development. It exists in many forms, such as heat energy, kinetic energy, gravitational potential energy, electric potential energy, radiation energy, and chemical energy. Ideally, each energy form can be converted, stored, and transported, depending on its application. From the viewpoint of the supply and consumption of world energy sources for humans, the energy resources are changed with time.

Before the industrial revolution, wood and its waste were the main sources of energy for cooking, lighting, and civil heating. After the middle of the nineteenth century, fossil fuels such as coal, petroleum, and natural gas became the primary source of energy, as shown in Fig 2.1(a) (USEIA, 2020). The economics of fossil fuels, such as mining, distribution, and transportation were concentrated in a few countries only. Therefore, this situation caused wars and battles to control energy. With the breakthrough in the exploitation technology of shale gas, natural gas production increased significantly, reducing the dependence on oil from specific areas. Although the oil crisis or energy crisis has been resolved, the environmental issue caused by the use of fossil fuels continues to plague the human race. The atmospheric concentration of CO_2 is elevated from the initial 280 ppm level at the beginning of the industrial

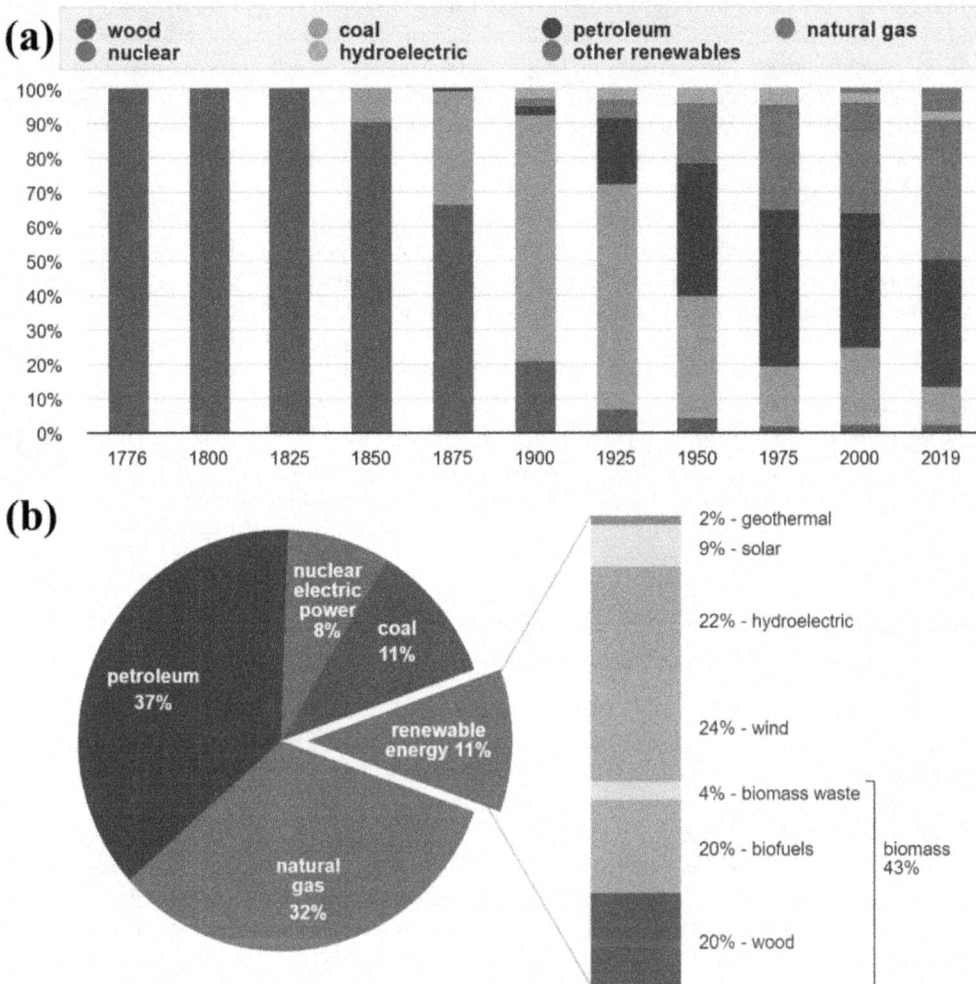

FIGURE 2.1 (a) Major sources of total US energy consumption in selected years (1776–2019) and (b) the distribution of energy source for US primary energy consumption in 2019. Source: U.S. Energy Information Administration, Monthly Energy Review (USEIA, 2020).

revolution to over 400 ppm at the present level (Buis, 2019). This caused disastrous problems such as global warming, climate changes, and the biological extinction of certain species.

Fossil fuels are still the largest energy source in 2019, including 11% coal, 32% natural gas, and 37% petroleum, which contribute the most amount of greenhouse gas emissions. Relatively, renewable energy only accounts for approximately 11% of overall US primary energy consumption. It indicates that the development of biomass energy technology for the application requires continuous progress. Fortunately, the variety and diversity of renewable energy sources, including 2% geothermal, 9% solar, 22% hydroelectric, 24% wind, and 43% biomass of renewable energy sources, exhibit the great potential to replace fossil fuel. Among these renewable energies, biomass energy, comprising 4% biomass waste, 20% biofuels, and 20% wood, accounts for the largest proportion of the renewable energy sources, as shown in Figure 2.1(b) (USEIA, 2020). Biomass energy promotes the renewability of the energy industry and helps local economic growth and circular economy development (Long et al., 2013). Thus, biomass energy can play an essential role in diminishing CO_2 emissions and facilitate the new economic transformation.

On the other hand, carbon neutrality is an essential characteristic of renewable energy. Mostly, biomass energy is regarded as a "net-zero-carbon-footprint" source to replace fossil fuels (Berndes et al.,

2016). Biomass is produced via photosynthesis, which can convert solar energy into chemical energy, followed by the light-independent Calvin cycle reactions to form the sugar or organisms (Dietz and Heber, 1984). For example, green plants can absorb CO_2 from the atmosphere, moisture, and minerals from the soil to produce organisms and necessary organic substances such as carbohydrates, lipids, proteins, and nucleic acids. On the other hand, some cyanobacteria and algae can conduct the reverse Krebs cycle to produce glucose or other carbohydrates from CO_2 (Buchanan and Arnon, 1990).

Through the photosynthesis of the plants, CO_2 can be converted into biomass or organic substances. Carbohydrate is one of the organic substances, which can be categorized as monosaccharide, disaccharide, polysaccharide, and starch. A monosaccharide is a basic unit and exhibits two isomers, including glucose and fructose, with the same molecular formula $C_6H_{12}O_6$. Glucose is the essential energy source as being an intermediate product of metabolism for organisms. It could also be stored in the plants as polysaccharide or cellulose with a series of sequential condensation reactions. For example, one glucose and one fructose atom can conduct a dehydration reaction to form a disaccharide, sucrose, which can be found in sugarcane or sugar beet. The synthesized glucose is adapted to produce complex carbohydrates and generate alternative organic compounds such as lipids, proteins, and nucleic acids. Among these organic substances, lipids enclose fat and fatty acid, exhibiting the potential of being the feedstock of biodiesel. These produced substances and even the matrix and components of the plants can be employed as the source of biomass. With appropriate treatments, such materials can be transformed into biofuels for combustion or other applications.

2.2 Overview of Solid and Liquid Biofuels

Most solid biofuels are derived from solid biomass, such as wood (Koponen and Hannula, 2017; Maryandyshev et al., 2015) and bark (Anupam et al., 2016; Nosek et al., 2016). The cell walls of the plants mainly construct solid biomass. These cell walls mainly comprise lignocellulose in the form of 30–50% cellulose, 10–40% hemicellulose, and 5–35% lignin (Pérez et al., 2002). The structural frameworks of the cell walls are mainly provided by the microfibers of cellulose, which are connected and tightly tethered by hemicellulose, while lignin interpenetrate and encase them (Gomez et al., 2008). The component of these lignocelluloses contains carbon and hydrogen elements, storing the chemical energy. After combustion or oxidation, such chemical energy would be released and converted into heat energy.

The common solid biofuels from biomass and its derivate include firewood (Karunanayake, 2018), wood chips (Perea-Moreno et al., 2016), wood pellets (Telmo and Lousada, 2011), and charcoal (Dobie and Sharma, 2015). Before fossil fuels, firewood was a major source of solid fuels for the use of individual households or private residences. It can release its inherent chemical energy by being burned at 220–300°C to generate fire, which provided human beings light and heat for daily use. Commonly, the unseasoned firewood has a high moisture content, up to 40–100% on its dry mass basis. After drying for a long time or by moisture movement, the moisture content can be reduced to 10–25%. Unseasoned firewood has lower energy content. The high moisture content reduces combustion efficiency since certain heat energy is absorbed to vaporize water during combustion. Therefore, firewood is typically dried by air drying and kiln drying before combustion. The dried firewood can provide an energy content of approximately 15 MJ/kg, which is 30–50% of that of fossil fuels (Guo et al., 2015). However, it is easy to cause incomplete combustion for firewood or other raw biomass in a conventional burner, leading to the formation of smoke and creosote, including carbon particulates, volatile organic compounds, and a dark brown oil containing phenols, cresols, and other organic compounds. The combustion efficiency of firewood in a typical furnace is 80–85% (Food and Nations, 1990). With the advancement of the industry, the consumption of firewood has reduced. More and more novel technologies are developed to enhance the combustion efficiency of firewood or raw biomass materials.

Firewood is usually packed in bundles, too bulky to be transported in an automated combustion system. Therefore, wood chips exhibit the advantages of a smaller piece, lower cost, and more availability. All biomass parts can be cut and broken into small chips for more comfortable transportation, storage, and usage. Besides, wood chips release lower concentrations of sulfur oxide and nitrogen oxide, as compared with coal (Guo et al., 2015). However, the heating value of the biomass, including chips, bark, and sawdust, lose 9%, 7%, and 3%, respectively, due to the deterioration by microorganisms (White et al., 1983). Moisture is an important factor to influence the survival of microorganisms. Accordingly, such loss can be mitigated by enlarging the particle size of biomass and conducting the pretreatment of conventional drying or even torrefaction. The thermal processes use heat to remove moisture from biomass, resulting in the death of the microorganisms due to temperature sterilization (Fagernäs et al., 2010).

After grinding wood chunks or chips into sawdust, a pelletizer is employed to compress the sawdust subsequently. During compressing, the pressure raises the temperature that plasticizes the lignin of biomass for gluing, followed by the formation of pellets. Wood pellets have a more uniform cylindrical shape with a similar diameter as compared to wood chips. Typically, if the feedstock is torrefied before pelletization and the moisture content be reduced below 10%, the energy content is improved by 9–18 MJ/m³ (Cocchi et al., 2011). In addition to wood, cultural and agricultural wastes, such as crop residues, nutshells, soybean pods, corn cobs, rice husks, and bagasse, can also be processed into pellets. Well-processed pellets should be transferred easily and stored compactly. Nevertheless, the production cost of wood pellets is higher than that of wood chips, so the use of pellets is limited, mainly in developed countries for residential heating (Guo et al., 2015).

As mentioned earlier, the energy contents of wood fuels are still much lower than that of fossil fuels. To improve this issue, people develop pyrolysis techniques that can convert wood into charcoal. Charcoal can be obtained by heating woods in a furnace in the absence of air at a high temperature of approximately 400°C called dry torrefaction. The energy content of charcoal is approximately 28–33 MJ/kg. To prepare the charcoal briquettes, sawdust, anthracite coal, mineral charcoal, and some additives are mixed. Additives as binders can improve ignition, facilitate combustion, and make manufacturing economical (Antal and Grønli, 2003). However, the addition of binders reduces the heating value, limits the carbon content, increases ash yield and higher pollutant emission; therefore, wet torrefaction is developed to reduce the amount of additives. During wet torrefaction, biomass reacts in compressed subcritical water at the temperature of 200–260°C to obtain biomass briquette. Such charcoal briquette exhibits higher mass density and higher strength due to its stronger bonding (Wu et al., 2018).

Nowadays, the primary liquid fuel is gasoline, which is derived from petroleum as a fossil fuel. Biofuel can be the fuel additive to replace gasoline partially. Some biofuels receive much attention, such as bioethanol, biodiesel, and pyrolysis bio-oil. Among these biofuels, bioethanol is employed as the fuel additive for gasoline. The gasoline gallon equivalent (GGE) of bioethanol is 1.5, which means the energy content of 1-gallon gasoline can be replaced by that of 1.5-gallon bioethanol (da Rosa et al., 2015). Bioethanol-additive gasoline is used in many countries such as the United States, Brazil, and Europe. Ethanol fuel mixtures with 10 vol.% anhydrous ethanol, noted as E10, are commonly employed in regular cars in the United States. The U.S. Environmental Protection Agency (EPA) agreed on E15 for use in conventional vehicles and flexible fuel vehicles. E25, E85, and E100 represent the ethanol fuel mixtures with 25 vol.%, 85 vol.% anhydrous ethanol, and 100 vol.% hydrous ethanol (containing average of 5.3 vol.% water), respectively. The use of ethanol fuel depends on its effect on the drivability of the vehicle, the emissions of the tailpipe, the design and the durability of catalyst, the operability, and the durability of the engine (West et al., 2012). Accordingly, not only the governmental policy regarding the content of ethanol additive but also the related technology of vehicles will affect the development of the bioethanol market.

Diesel is a vital fuel for heavy-duty machines and vehicles. Therefore, biodiesel is also developed to substitute for petroleum diesel. Through transesterification process, methanol, oil, or fat, such as vegetable oil, algal lipids, animal fats, and waste grease, can be converted into fatty acid methyl esters as the precursors for biodiesel (Colucci et al., 2005). Biodiesel's energy content is 10% lower than that of petroleum diesel since biodiesel has a higher oxygen content (Hoekman et al., 2012). Similarly, biodiesel can blend with diesel to form B20, B50, and B80 fuels. For example, B20 fuel comprises 20 vol.% biodiesel and 80 vol.% petroleum diesel (Yusuff et al., 2017).

Another liquid biofuel is pyrolysis bio-oil, which can be obtained from the plant biomass with pyrolysis. Thermal pyrolysis can convert up to 75% plant biomass into bio-oil intermediate (Carpenter et al., 2014). Wood is used comprehensively for producing bio-oil due to its low ash content. However, bio-oil contains high moisture, colloidal char particles, and various liquid compounds, such as alcohols, acids, esters, aldehydes, and aromatics. Such complex bio-oil contents may lead to some drawbacks, including corrosiveness, viscosity, unstableness, and low energy density (Junming et al., 2008). Atomized crude bio-oil can be burned directly in an industrial scale combustor. It needs to be upgraded via hydrogenation, hydrodeoxygenation, esterification, etc., to enhance the chemical stability and heat value and reduce the moisture and the acidity of bio-oil (Ruddy et al., 2014). However, there is a major challenge from economics and technologies to produce and utilize pyrolysis bio-oil commercially.

To sum up, the first-generation biofuels, referring to the food-crop-related biomass, such as starch, vegetable oil, and sugar, are prohibited due to their unsustainability and the occurrence of food crisis (Paschalidou et al., 2016). The second-generation biofuels are derived from wood materials, agricultural residues, and lignocellulosic waste, non-human food. In other words, biomass and its derivate, and biomass waste and human-created waste can be the feedstock of fuels. Human-created waste from domestic or industrial purposes includes municipal waste (Brzychczyk et al., 2018), waste tires (Hossain et al., 2017), organic wastewater organic solvents (Seyler et al., 2005), waste cooking oil (Yan et al., 2011), oil sludge (Huang et al., 2017), black liquor (Kang et al., 2012), waste paper, rice husk (Nizamuddin et al., 2018), wheat straw (Ríos-Badrán et al., 2020), etc. The second-generation biofuels from lignocellulosic biomass are ready for commercial utilization (Goel and Sharma, 2019). The third-generation biofuels are made of algal biomass, which is regarded as a promising renewable resource due to its high production rate (Chowdhury and Loganathan, 2019). It is expected that biofuel-related renewable energy technology can continuously improve to meet the growing demand of energy.

2.3   Production of Solid and Liquid Biofuels

The biomass treatment can be divided into two methods, one is thermochemical conversion and the other is biochemical conversion. According to the methods given before, the biomass can be converted into biofuel. For fermentation and some biological conversion, it belongs to biochemical conversion to produce biofuel. Besides, thermochemical conversion includes liquefaction, pyrolysis, torrefaction, gasification, and combustion (Cantrell et al., 2008). There are several advantages of thermochemical conversions, such as: (i) it does not require space for processing and reaction faster; (ii) it is applicable to a wide range of materials; and (iii) it can effectively remove active compounds or pathogens and can reduce harmful emissions (Cantrell et al., 2007). However, torrefaction is recommended to use for producing solid products. The operating heating temperature range of torrefaction is between 200°C and 300°C (Bergman and Kiel, 2005). If the temperature exceeds the maximum temperature of 300°C, it may induce the loss of lignin in biomass higher. The loss of lignin will make the biomass difficult to form pellets. The main purpose of torrefaction is increasing the composition ratio of carbon and reducing the oxygen-to-carbon ratio (O/C) and hydrogen-to-carbon ratio (H/C). It can enhance the energy density of biochar.

Besides, carbonization is a kind of thermochemical conversion and similar to torrefaction. The operating heating temperature range of carbonization is between 300°C and 600°C. The removal of volatile matter from biomass is important during carbonization and torrefaction (Shankar Tumuluru et al., 2011; Tumuluru et al., 2011). Carbonization has high energy density due to the high carbon containment than torrefaction. Otherwise, the pyrolysis is recommended to use for producing liquid products. The operating heating temperature range of pyrolysis is from 290°C to 750°C. The processing time is shorter, with a broader range of materials than that could be processed. Hence, it can produce high-yield pyrolysis liquid even in an environment with little or without oxygen. The different heating temperatures and residence time can produce a different kind of biofuel. For instance, longer residence time and lower temperatures can produce biochar. And longer residence time and higher temperature can produce biogas. Moreover, short residence times and suitable temperature will produce bio-liquids (Bridgwater, 2012; Klass, 1998).

2.3.1   Production of Solid Biofuels

In general, solid biofuels include charcoal, biochar, activated carbon, etc. Furthermore, charcoal has a high porosity of the pore structure. Thus, it is an excellent reducing adsorbent. The property of biomass and biofuels is shown in Table 2.1 (Wilk and Magdziarz, 2017). And the primary composition of treated

TABLE 2.1

The Comparison of Raw *Miscanthus* and Its Chars Derived at Different Hydrothermal Temperatures

Properties	Raw *Miscanthus*	Char	Char	Char
Temperature (C°)	–	180	200	220
Moisture (%)	7.38	1.48	2.37	1.36
Volatile (%)	84.10	81.70	75.01	73.23
Fixed carbon (%)	6.77	15.52	21.74	24.34
Ash (%)	1.75	1.30	0.88	1.05

Note: Reproduced with permission from Wilk and Magdziarz (2017).

Miscanthus is moisture, volatile, fixed carbon, and appropriate ash. Humans have started using charcoal as a fuel long ago. Raw *Miscanthus* could be treated into charcoal by several methods such as hydrothermal, torrefaction, and slow pyrolysis (Wilk and Magdziarz, 2017). Among these methods, torrefaction is much traditionally and comprehensively employed. It uses less oxygen to maintain *Miscanthus* combustion and provides heat for carbonization. Biochar is usually obtained by carbonizing agricultural waste or forest waste (e.g., wood, *Miscanthus*, shiitake substrate, and rice straw). Li et al. (2018) have investigated the combustion behavior of coal pellets blended with *Miscanthus* biochar. They filled nitrogen into the tubular furnace to produce the *Miscanthus* biochar.

Except for agricultural waste or forest waste, sewage sludge is also a common waste of industrial or municipal wastewater treatment plants. With technological progress and rapid urban development, the output in the form of sewage sludge has also increased. Whereas the sewage sludge of industrial processes is considered harmful owing to the fact that it contains heavy metals and alkaline and acidic materials, and other impurities. Consequently, sewage sludge should be appropriately treated to preserve the environment (Lundin et al., 2004; Muchuweti et al., 2006). Sewage sludge can be incinerated or subjected to thermochemical conversion to reduce the volume and recover its energy. Chen et al. (Chen et al., 2018a) reported that the pure sewage sludge and shiitake substrate offer good combustion characteristics and flammability index.

Some pieces of literature have reported the pyrolysis of sewage sludge and the subsequent utilization of solid products (biochar) as adsorbents (Abrego et al., 2009; Jeyaseelan and Qing, 1996; Jindarom et al., 2007; Lu and Lau, 1996). Nevertheless, based on previous studies, the adsorbent obtained from the pyrolysis of sewage sludge may eliminate the H_2S and NO_x pollutants during the thermochemical conversion process and even eliminate phenolic compounds, dyes, and metals (Pietrzak and Bandosz, 2007; Seredych and Bandosz, 2007; Smith et al., 2009; Yuan and Bandosz, 2007).

2.3.2 Production of Liquid Biofuels

Fast pyrolysis is the main approach to produce liquid biofuels (bio-oil). Fluidized beds are suitable to be used for fast pyrolysis, as shown in Figure 2.2 (Basu, 2018; Kaminsky and Kummer, 1989; Shen and Zhang, 2003; Stammbach et al., 1989). And some literature has studied the pyrolysis of sewage sludge using fewer conventional reactors. A semicontinuous vertical reactor is used by Pokorna et al. (2009). In the reactor, the sand that is continually moving through the Archimedes screw is used as a heat transfer medium.

There are some advantages of increasing the liquid biofuels yield by fast pyrolysis: (i) the heat temperature is 425°C to 600°C, and (ii) high heating rate and short residence time. The composition of bio-oil is dependent of which biomass it is made and which process is used. Typical bio-oil composition mainly contains aldehydes, ketones, carboxylic acids, furans, phenols, dehydration carbohydrates, etc. (Dobele et al., 2007; Heo et al., 2010). Table 2.2 shows the main components of bio-oil from pine pyrolysis (Basu, 2018; Bridgewater et al., 2001; Lyu et al., 2015).

Castor oil is the one bio-oil from the cold and hot squeezing castor beans. Castor oil obtained by hot squeezing is obtained by heating steam. Slow heating squeezing can reduce the viscosity of the

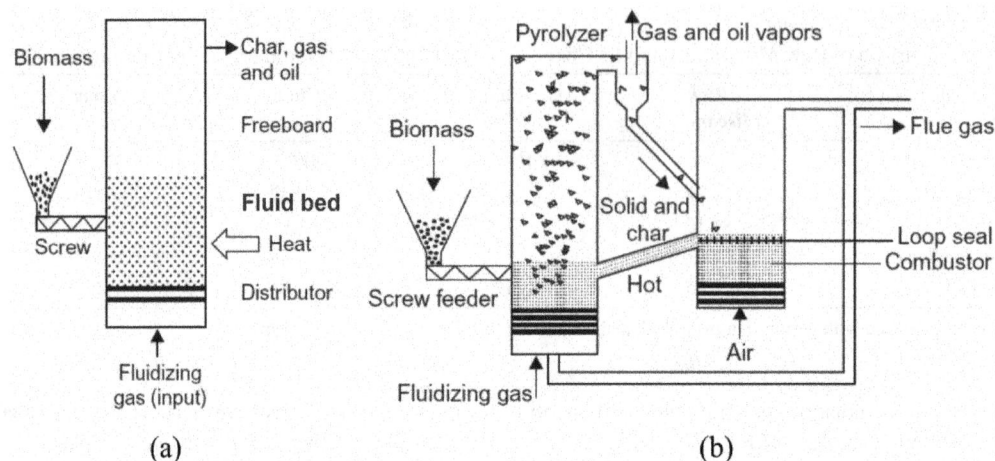

FIGURE 2.2 Different kinds of fluidizing bed pyrolysis: (a) bubbling and (b) circulating. Reproduced with permission from Basu (2018).

TABLE 2.2

The Compositions of Bio-Oil

Major Group	Formula	Area (%)
Levoglucosan	$C_6H_{10}O_5$	8.9
(E)-isoeugenol	$C_{10}H_{12}O_2$	5.9
Acetic acid	$C_2H_4O_2$	5.7
1-Hydroxy-2-propanone	$C_3H_6O_2$	4.9
2-Methoxy-4-methyl-phenol	$C_8H_{10}O_2$	4.6
2-Methoxy-4-vinylphenol	$C_9H_{10}O_2$	3.8
(Z)-Isoeugenol	$C_{10}H_{12}O_2$	3.1
1,2-Cyclopentanedione	$C_5H_6O_2$	3.3
2-Methoxy-phenol	$C_7H_8O_2$	3.3

Note: Reproduced with permission from Lyu et al. (2015); Open Access Resources.

remaining castor oil so that the oil that cannot be obtained by cold squeezing alone can be extracted (Koufopanos et al., 1991).

2.4   Characterization of Solid and Liquid Biofuels

2.4.1   Characterization of Solid Biofuels

Solid-fuel combustion attracts considerable attention due to its being used comprehensively in large-scale heat, steam generation, and electricity generation. Some properties of various solid biofuels from biomass and waste influences combustion performance, such as pollutant emissions, thermal efficiency, flame stability, combustion rates, and burnout times (Hurt, 1998). Therefore, it is important to characterize the basic properties of solid fuels, such as elemental analysis, proximate analysis, and heating value.

Elemental analysis, also called ultimate analysis, is an essential process to understand the property of fuels. The chemical elemental compositions of fuels, such as carbon (C), hydrogen (H), nitrogen (N), sulfur (S), and oxygen (O), can be determined qualitatively as well as quantitatively using an elemental analyzer. Commonly, elemental analysis is conducted through the process of combustion. It can be accomplished to analyze CHNSO elements in the same run. The elemental analysis can be based on the gravimetric determination (Parthasarathy et al., 2013), thermal conductivity (Islam et al., 2001), infrared spectroscopy detection (Zhang et al., 2017), mass spectrometric atomic spectroscopy (Ohno et al., 2010), or neutron activation analysis (Wasserman et al., 2001) of the combustion gases. Carbon and hydrogen can release chemical energy into heat energy after combustion. Compared with coal, biofuels have higher H/C and O/C ratios. Therefore, the raw feedstock of biomass must accomplish the pretreatment to lower the oxygen content. On the other hand, nitrogen and sulfur are the source of the emitting pollutants such as nitrogen oxides (NO_x) and sulfur oxides (SO_x), respectively. The gaseous pollution damages the ozone layer in the atmosphere, leading to global warming and the greenhouse effect.

Proximate analysis can be used for analyzing biomass and biofuels as the determination of moisture, volatile matter (VM), ash, and fixed carbon (FC), which are based on the standard test methods of ASTM (American Society for Testing and Materials), including ASTM E871, ASTM E872, ASTM E1755, and calculated by difference, respectively (Braga et al., 2014). The fuel sample can be particulate wood, sanderdust, sawdust, pellets, or chips. The volume of the fuel sample must be smaller than 16.39 cm³. According to ASTM E871, the sample should be placed in a drying oven with natural air circulation at the temperature of 103 ± 1°C for 16 h. By calculating the weight loss, the moisture content can be obtained (E871-82, 2006). Exclusive of moisture, ASTM E872 can determine the content of volatile matters. The fuel sample should be located in a crucible with a closely fitting cover, which is placed in a furnace. The temperature of the furnace is kept at 950 ± 20°C for 7 min. After that, the percentage of volatile matters is determined from the result of weight loss (E872–82, 2013). The ash content can esti- mate the percentage of mineral or other inorganic materials in the biomass. The fuel sample is put into a crucible and then heated at 575 ± 25°C in the muffle furnaces until the carbon is eliminated (at least for 3 h). The ash content can be calculated by the final mass of the ash, tare mass of container, and the initial mass of dried samples (ASTM, 1995). The content of fixed carbon can be obtained by calculating the difference of the residues. Among different contents from proximate analysis, volatile matters and fixed carbon are primarily attributed to the heat released from biofuels.

Heating value is a critical property, which presents the heat released from fuels. It primarily depends on the contents of carbon, hydrogen, moisture, and ash. Heating values can be obtained by measuring the heat of complete combustion using a bomb calorimeter. First, fuels and oxidizers are mixed in a stoi- chiometric ratio in a steel vessel. After ignition, the combustion starts, and it must allow the reaction to complete. Then, the container and the products are cooled down to ambient temperature. Accordingly, a higher heating value (HHV) can be evaluated as the heat released during the combustion. Two kinds of heating values, including HHV and lower heating value (LHV), are called gross calorific value (GCV) and net calorific value (NCV), respectively. The difference between HHV and LHV is the heat of vapor- ization of the water. HHV considers the latent heat of vaporization of water, leading to a higher value. It is based on the assumption that the systematic temperature before and after combustion is the same. In other words, the products are in the liquid state after combustion. On the other hand, LHV treats water as a vapor, indicating that the energy of the vaporization of water will not be released as heat (Sivaramakrishnan and Ravikumar, 2011).

Interestingly, the HHV of fuels also can be evaluated from the results of elemental analysis. Compared with the HHV determination by calorimetric methods, elemental analysis is time-saving due to automation availability. Therefore, some empirical equations were proposed based on the elemental contents for predicting the HHV of coal (Boie, 1953), fossil fuel liquids (Lloyd and Davenport, 1980), solid waste (Wilson, 1972), and plant biomass (Friedl et al., 2005), as shown in Eq. (2.1) to Eq. (2.4), respectively. Recently, Huang and Lo proposed a semi-empirical equation to evaluate the HHV of lignocellulosic biomass, as shown in Eq. (2.5) (Huang and Lo, 2020). They announced that the coef- ficient of determination is highest ($R^2 = 0.9939$), while the relative error is lowest (− 0.10 and 1.80 for mean percentage error and mean absolute percentage error, respectively), compared with others cor- relation in previous literature (Friedl et al., 2005; Sheng and Azevedo, 2005; Yin, 2011). Besides, they also represent that the HHV of sewage sludge, municipal solid waste, and industrial waste also can be predicted by modifying the coefficient before oxygen content. In future, the prediction of HHV from elemental analysis can be further improved when the characterization of elemental content becomes more precise and accurate.

$$HHV\left(coal, \frac{kcal}{kg}\right) = 83.2C + 22.4H + 25S + 15N - 25.80 \tag{2.1}$$

$$HHV\left(fossil\ fuel\ liquid, \frac{kcal}{kg}\right) = 357.77C + 917.58H - 84.51O - 59.38N - 111.87S \tag{2.2}$$

$$HHV\left(solid\ waste, \frac{kcal}{kg}\right) = 140.96C + 602.14\left(H - \frac{O}{8}\right) + 39.82S + 89.29\left(H - \frac{O}{8}\right)/2$$
$$+ 42.74\left(\frac{O}{2}\right) - 10.40N \tag{2.3}$$

$$HHV\left(plant\ biomass, \frac{kcal}{kg}\right) = 3.55C^2 - 232C - 2230H + 51.2C \times H + 131N + 20600 \tag{2.4}$$

$$HHV\left(lignocellulosic\ biomass, \frac{kcal}{kg}\right) = 0.3443C + 1.192H - 0.113O - 0.024N + 0.093S \tag{2.5}$$

Nevertheless, the elemental analysis equipment is still expensive; therefore, the elemental composition of biomass or fuels can be evaluated in advance from the proximate analysis. Parikh et al. derived the correlation for solid lignocellulosic biomass, as shown in Eq. (2.6) to Eq. (2.8). The correlation provides a simple and useful tool to obtain the proximate analysis elemental components, where $57.2\% \leq VM \leq 90.6\%$ and $4.7\% \leq FC \leq 38.4\%$. The correlations have an average absolute error of 3.21%, 4.79%, and 3.4% for the values of C, H, and O, respectively (Parikh et al., 2007). Recently, some correlations related to elemental composition from proximate analysis using different prediction models, such as ANFIS (adaptive neuro-fuzzy inference system), ANN (artificial neural network), MLR (multilinear regression) (Lawal et al., 2020), and other machine learning techniques (Ceylan and Sungur, 2020). Accordingly, the developed models for computing the elemental composition will still proceed to be cost-effective and efficient for practical use.

$$C(wt.\%) = 0.637FC + 0.455VM \tag{2.6}$$

$$H(wt.\%) = 0.052FC + 0.062VM \tag{2.7}$$

$$O(wt.\%) = 0.304FC + 0.476VM \tag{2.8}$$

2.4.2 Characterization of Liquid Biofuels

In addition to elemental analysis and calorimetry, it is important to inspect the basic properties of liquid fuels, such as density, viscosity, pH value, vapor pressure, flash point, ignition temperature, and pour point. The density and viscosity of liquid fuels can be obtained according to ASTM D941 and D445 test methods, respectively (Alptekin and Canakci, 2008). Besides, the temperature affects the measurement of fuel properties; therefore, the specific temperature is adopted for standard characterization. For instance, to inspect the density of the biodiesel or its blend with diesel fuel, a density meter is employed to measure the density of fuels at the temperature of 15°C (Alptekin and Canakci, 2008). Meanwhile, the viscosity of the fuels is evaluated at 40°C by using a capillary viscometer tube (Alptekin and Canakci, 2008).

On the other hand, the pH value can be determined on the basis of the ASTM D6423 method through standard pH meters (Gonçalves et al., 2011; Tzanetakis et al., 2008). Gonçalves et al. compared different glass electrodes of pH meters to compare with the specific electrode (Orion Ross Sure-flow® N° 8172 BN electrode) guided by the ASTM D6423. They found that the filling solution (aqueous KCl or ethanolic LiCl) of the combined glass electrodes could affect the inspection of pH measurement, whereas the reference electrode (Ag/AgCl or Pt/redox pair) and liquid junction (single or double) will not affect the results significantly. To strengthen the trade of biofuels in the global market, developing a general methodology to measure pH for liquid biofuel is necessary (Gonçalves et al., 2011).

2.5 Combustion of Solid and Liquid Biofuels

Combustion is one of the most promising fuel conversion techniques that provide energy in transportation, industrial application, and power generation. Traditionally, the energy supply was mainly from fossil fuels, which were obtained from ancient plants and animals under high-pressure and high-temperature conditions underground. However, fossil fuels are nonrenewable and are rapidly depleted. Burning of fossil fuel (especially coal and crude/heavy oil) leads to substantial greenhouse gas (GHG) emissions and other pollutants, including heavy metals, particulate matter (PM), NO_x, and SO_x, which have become a direct threat to human health and the environment. Owing to this emergent global economy and environmental concern, more and more stringent emission regulations and limits applied have become the impetus driving force for the innovative development of combustion technologies, such as alternative fuel, co-firing (Yang et al., 2019), and zero-carbon emission strategy (e.g., Carbon Capture and Storage, CCS).

Carbon-neutral renewable energy could be considered as alternative energy to partially supplant fossil fuel or direct utilization to sustain human development and preserve the environment. Among various alternative energy sources, biomass has attracted particular attention because of its abundance, inexhaustible nature, cost affordability, lower technical threshold, and versatility. Besides, the combustion of biomass can provide energy for usage and alleviate solid waste (from agricultural, forestry, and municipal) disposal and facilitate the circular economy. According to International Energy Agency (IEA) statistics, approximately 9.5% of the total primary energy supply in 2017 was provided by biofuels and waste, standing closely behind petroleum, coal, and natural gas in the world energy list.

The combustion process of biofuels is complicated because its physicochemical properties are quite different compared to traditional fossil fuels. The characterization of biomass combustion can be affected by fuel properties, structure, environment temperature, oxidizer concentration, flow regime, and specific experimental apparatus, resulting in various combustion phenomena. Herein, we choose sewage sludge and shiitake substrate as a representative solid biofuel and pyrolytic oil (from sewage sludge and castor) as liquid biofuel for combustion discussion to compare with fossil fuels by the following subsections.

2.5.1 Combustion of Solid Biofuels

In general, solid fuel properties can be measured using proximate, elemental, fiber, ash, and caloric analysis, following the ASTM test standard. Biomass typically possesses a higher atomic O/C ratio and H/C ratio compared to coal, according to the Van Krevelen chart. Hence, it usually contains low fixed carbon but high volatile matter content. The high volatile characteristics may enhance the gas-phase ignition or reduce the ignition delayed time during its combustion. Table 3 shows the fuel properties of biomass and coal (Saidur et al., 2011). It reveals that biomass has a lower fuel density, carbon content, heating value, and ignition temperature than that of coal. Also, the characterization of low ash and low sulfur content in biomass can effectively reduce the cost of ash disposal and SO_x emission prevention devices during the co-firing. It is noteworthy that biomass typically contains alkali and alkaline earth metal since it grows. After combustion, these would be in the metal oxide form in ash residue and occasionally reveal a lower melting point, resulting in ash fusion, agglomeration, and slagging problems near the furnace wall.

The combustion of solid fuel generally requires both homogeneous and heterogeneous reactions. The homogeneous reaction is dominated by the devolatilization of fuel and the diffusion of oxidizer. In contrast, the heterogeneous reaction process is more complicated which is governed by film diffusion of oxidizer, diffusion through the ash layer, adsorption on the surface of solid, chemical reaction at the surface, desorption of product gas from the surface, diffusion of product gas through the ash layer, and

TABLE 2.3

Fuel Properties of Biomass and Coal

Property	Unit	Biomass	Coal
Fuel density	kg/m³	500	1,300
Particle size	mm	3	100
C content	wt.%, dry fuel	43–54	65–85
O content	wt.%, dry fuel	35–45	2–15
S content	wt.%, dry fuel	Max 0.5	0.5–7.5
SiO_2 content	wt.%, dry fuel	23–49	40–60
K_2O content	wt.%, dry fuel	4–48	2–6
Al_2O_3	wt.%, dry fuel	2.4–9.5	15–25
Fe_2O_3	wt.%, dry fuel	1.5–8.5	8–18
Ignition temperature	°C	418–426	490–595
Heating value	MJ/kg	14–21	23–28

Note: Reproduced with permission from Saidur et al. (2011).

film diffusion back into the surrounding gas (Di Blasi, 2009). Therefore, several combustion models have been proposed to describe these transport (heat and mass) processes and the relevant reaction mechanism, such as the shrinking particle model, shrinking unreacted core model, and one-film and two-film combustion model (Ogle, 2016). However, solid waste models have seldom been studies because of their complex composition (inorganics and hazardous substances), which require to be studied further. The following techniques are often utilized to implement biomass combustion in the literature: (i) thermogravimetric/Fourier transform infrared spectroscopy (TG-FTIR) analysis, (ii) drop-tube furnace and hot-flow reactor, and (iii) medium-size furnaces (<100 kWth) such as fixed bed, fluidized bed, and entrained bed reactors. The apparatus of (i) and (ii) are often used to conduct a batch fuel combustion (transient process) (Huang et al., 2019; Li et al., 2018), and (iii) is suitable for determining the combustion performances (such as emissions, temperature profiles, and combustion performance) under a steady-state operation situation with continuously feeding the fuels (Sher et al., 2018).

Thermogravimetric analysis is extensively used to investigate the pyrolysis and oxidation process of biofuels. The weight loss process and the heat flow behaviors of the sample under a specific heating rate and carrier gas can be measured. Combustion parameters, including maximum weight loss rate (DTG_{max}), average weight loss rate (DTG_{mean}), ignition temperature (T_i), and burnout temperature (T_b), can be obtained from the TG and DTG curves. T_i and T_b are estimated from the intersection and conversion method, respectively (Lu and Chen, 2015). Combustion indices, such as flammability index (C) and combustion index (S), are calculated as follows (Chen et al., 2018a):

$$C = \frac{DTC_{max}}{T_i^2} \tag{2.9}$$

$$S = \frac{DTG_{max} DTG_{mean}}{T_i^2 \times T_b} \tag{2.10}$$

In general, a fuel with a larger value of C index and S index possesses a higher reactivity and higher combustion performance.

The kinetic parameters of the solid fuel reaction can be calculated as follows.

$$\frac{d\alpha}{dt} = k(T) \times f(\alpha) = A \exp\left(-\frac{E_a}{RT}\right) \times f(\alpha) \tag{2.11}$$

$$\alpha = \frac{m_o - m_t}{m_o - m_f} \tag{2.12}$$

where t is time, α is the conversion degree, k(T) is the reaction rate coefficient, f(α) represents the hypothetical model of the reaction, A is the pre-exponential factor, E_a the activation energy, R the gas constant, and T the absolute temperature; m_0, m_t, and m_f are initial mass, sample mass at time t, and the final mass, respectively.

For a non-isothermal reaction with a constant heating rate β = dT/dt under the thermogravimetric analysis (TGA), the reaction rate can be expressed as:

$$\frac{d\alpha}{dT} = \frac{A}{\beta}\exp(-\frac{E_a}{RT})\times f(\alpha) \tag{2.13}$$

Several models have been used to calculate the E_a and A from Eq. (2.13), such as Coats and Redfern (CR) integral method (Chen et al., 2018a), Ozawa–Wall–Flynn (OWF), Kissinger–Akahira–Sunose (KAS), and Kissinger methods as following (Zhang et al., 2019):

CR method:
$$\ln\left[\frac{-\ln(1-\alpha)}{T^2(1-n)}\right] = \ln\left(\frac{AR}{\beta E_a}\right) - \frac{E_a}{RT}; (n=1) \tag{2.14}$$

$$\ln\left[\frac{1-(1-\alpha)^{1-n}}{T^2(1-n)}\right] = \ln\left(\frac{AR}{\beta E_a}\right) - \frac{E_a}{RT}; (n\neq 1) \tag{2.15}$$

OWF: $\lg\beta = \lg\left(\frac{AE_a}{Rg(\alpha)}\right) - 2.315 - 0.4567\frac{E_a}{RT}$ \tag{2.16}

KAS: $\ln\left(\frac{\beta}{T^a}\right) = \ln\left(\frac{AE_a}{Rg(\alpha)}\right) - \frac{E_a}{RT}$ \tag{2.17}

Kissinger: $\ln\left(\frac{\beta}{T_p^2}\right) = \ln\left(\frac{AR}{E_a}\right) - \frac{E_a}{RT_p}$ \tag{2.18}

Take the CR method with reaction order n = 1, for example, the activation energy (E_a) and the pre-exponential factor (A) could be observed from the slope, $-E_a / R$, and the intercept of the regression line, respectively, by plotting $\ln\left[\frac{-\ln(1-\alpha)}{T^2(1-n)}\right]$ versus 1/T.

Shiitake mushrooms are widely grown in cultivation bags, filled with sawdust and rice straw chips. It is a potential agricultural waste that can be reused in a sustainable energy supply framework. On the other hand, sewage sludge is a by-product of urban wastewater treatment plants (WWTPs). This municipal solid waste also belongs to biomass. Chen et al. (2018a) investigated the co-combustion of sewage sludge with coal and shiitake substrate. Table 2.4 shows the fuel properties. It can be found that shiitake has the highest volatiles, and sewage sludge reveals high ash content and low carbon content compared to coal. Figure 2.3 shows the TGA results of individual fuels in air oxidation operating under various biomass blending ratio (BBR) and sludge shiitake ratio (SSR). The thermal decomposition temperature of coal revealed one stage (270–720°C). However, both sludge and shiitake showed two stages (190–370°C and 370–640°C for sludge; 190–410°C and 410–680°C for shiitake). The shiitake demonstrates a higher weight loss rate (especially in the first stage oxidation) compared to coal and sludge, which can be attributed to its high volatile matter characteristics of lignocellulosic biomass. The calculated combustion parameters of different coal-sludge ratios and shiitake-sludge ratios are shown in Table 2.5, respectively (Chen et al., 2018a). The results revealed that both C and S indexes rose with the addition of sludge to coal in the first oxidation stage. However, these data dropped in the next oxidation stage. The pure shiitake possesses the highest C and S index value, indicating high reactivity during combustion.

The sludge pellet's combustion process in a high-temperature surrounding typically undergoes dehydration, devolatilization, volatile combustion, char combustion, ash melting, and ash agglomeration, as shown in Figure 2.4 (Syed-Hassan et al., 2017) and Figure 2.5 (Chen et al., 2018a). In the devolatilization process, gaseous fuels (e.g., CH_4, CO, CO_2, and C_xH_y) are released from the pellet due to thermal degradation. Afterward, the gas-phase oxidation occurs and forms a diffusion flame to envelop the pellet. Once the devolatilization process is finished, the volatile flame gets extinct, and oxygen diffuses to the pellet surface for the char combustion stage. The exothermic reactions mainly originate from the volatile burning and char combustion stage. The combustion characteristics time such as ignition delay time t_{id}, volatiles burning duration t_f, char oxidation duration t_{char}, and total combustion time t_{tot} strongly depend on fuel properties. Generally, biomass contains high volatile matter, and high fixed carbon content may result in a longer t_f and longer t_{char}, respectively. Therefore, it is crucial to measure the combustion time for the optimum biomass combustor design and operation to achieve high-efficiency energy conversion and low emission production.

TABLE 2.4

Fuel Properties of Sewage Sludge, Shiitake, and Australian Coal

Item	Ultimate Analysis (wt.%)				
	C	H	O	N	S
DSSar	28.4	5.3	25.6	4.7	2.7
Shiitakear	40.5	6.0	29.0	3.2	0.4
Coal	73.1	4.3	5.3	1.7	0.5
Sample	**Proximate Analysis (wt.%)**				**HHV (MJ/kg)**
	Moisture	Volatile Matter	Ash	Fixed Carbon*	
DSSar	8.1	48.9	39.6	3.4	11.4
Shiitakear	13.3	69.1	13.5	4.2	15.5
Coal	2.0	30.8	17.2	50.1	26.1

Abbreviation: As-received basis (ar), calculated by difference (*). Note: Reproduced with permission from Chen et al. (2018a); Open Access Resources.

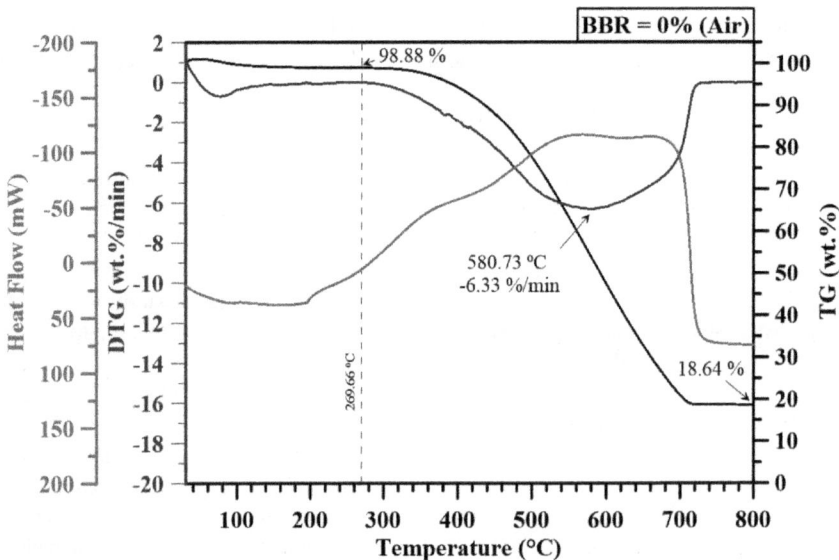

FIGURE 2.3A Thermogravimetric data of coal (BBR = 0%) under 20 °C/min, air environment. Reproduced with permission from Chen et al. (2018a); Open Access Resources.

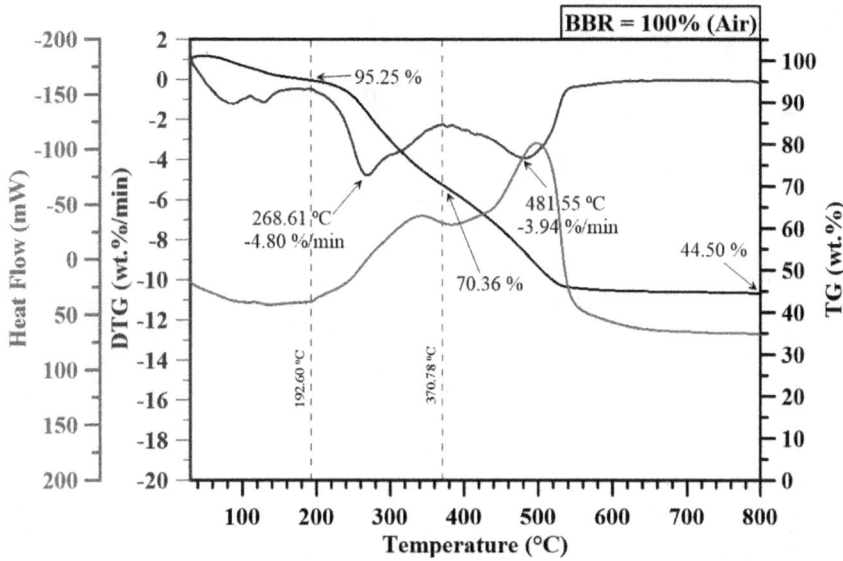

FIGURE 2.3B Thermogravimetric data of sewage sludge (BBR = 100%) under 20 °C/min, air environment. Reproduced with permission from Chen et al. (2018a); Open Access Resources.

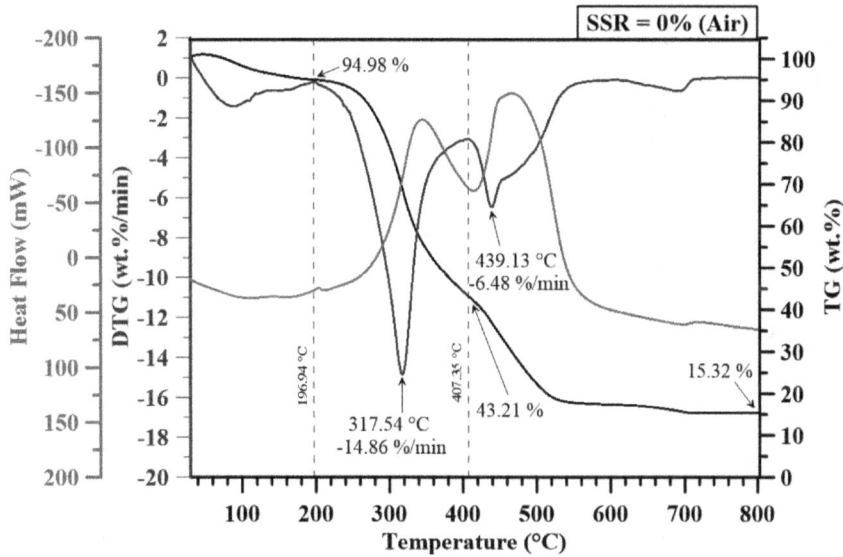

FIGURE 2.3C Thermogravimetric data shiitake substrate (SSR=0%) under 20 °C/min, air environment. Reproduced with permission from Chen et al. (2018a); Open Access Resources.

2.5.2 Combustion of Liquid Biofuels

Pyrolysis is an effective thermochemical conversion method to transform organic compounds. Herein, compounds are transformed from biomass/waste into bio-oil and biochar for further utilization, such as co-firing and direct combustion. In the combustor, liquid biofuel is usually atomized via a high-pressure nozzle into small droplets, then auto-ignited as spray combustion. However, the group combustion of droplet clouds is complex and present in different liquid–vapor distribution zones. Figure 2.6 shows the combustion of a typical fuel spray (Sánchez et al., 2015). Thereby, it is essential to understand a single droplet's fundamental burning characteristics for the estimation of spray combustion performance and combustor design. Because of the sewage sludge disposal problem and the resulting environmental issues,

TABLE 2.5

Combustion Characteristics for Various Coal–Sludge Ratios and Shiitake–Sludge ratios

BBR %	T_l (°C)		T_b (°C)		DTG_{max} (%/min)		DTG_{mean} (%/min)		$C \times 10^5$		$S \times 10^7$	
	T_{l1}	T_{l2}	T_{b1}	T_{b2}	Stage 1	Stage 2	Stage 1	Stage 2	C1	C2	S1	S2
0	–	438.29	–	709.15	–	6.33	–	5.15	–	3.30	–	2.39
25	247.30	458.42	306.90	693.62	1.13	5.68	0.96	4.68	1.85	2.70	0.58	1.82
50	240.97	462.41	357.28	665.57	2.11	4.74	1.77	3.93	3.64	2.22	1.81	1.31
75	238.31	446.67	367.44	629.45	3.25	4.29	2.51	2.95	5.72	2.15	3.91	1.01
100	239.40	439.32	368.57	633.29	4.80	3.94	3.48	1.65	8.37	2.04	7.90	0.53
SSR %	T_l (°C)		T_b (°C)		DTG_{max} (%/min)		DTG_{mean} (%/min)		$C \times 10^5$		$S \times 10^7$	
	T_{l1}	T_{l2}	T_{b1}	T_{b2}	Stage 1	Stage 2	Stage 1	Stage 2	C1	C2	S1	S2
0	277.43	426.59	404	680.82	14.86	6.48	7.02	1.83	19.30	3.56	33.56	0.96
25	272.38	457.96	417.81	666.85	10.84	4.80	5.56	1.68	14.61	2.29	19.42	0.58
50	266.82	469.67	413.85	688.17	6.95	3.52	4.49	1.67	9.76	1.59	10.59	0.39
75	239.40	436.20	379.99	620.62	5.88	4.14	4.33	1.89	9.12	2.18	10.39	0.66
100	239.40	439.32	368.57	633.29	4.80	3.94	3.48	1.65	8.37	2.04	7.90	0.53

Note: Reproduced with permission from Chen et al. (2018a); Open Access Resources.

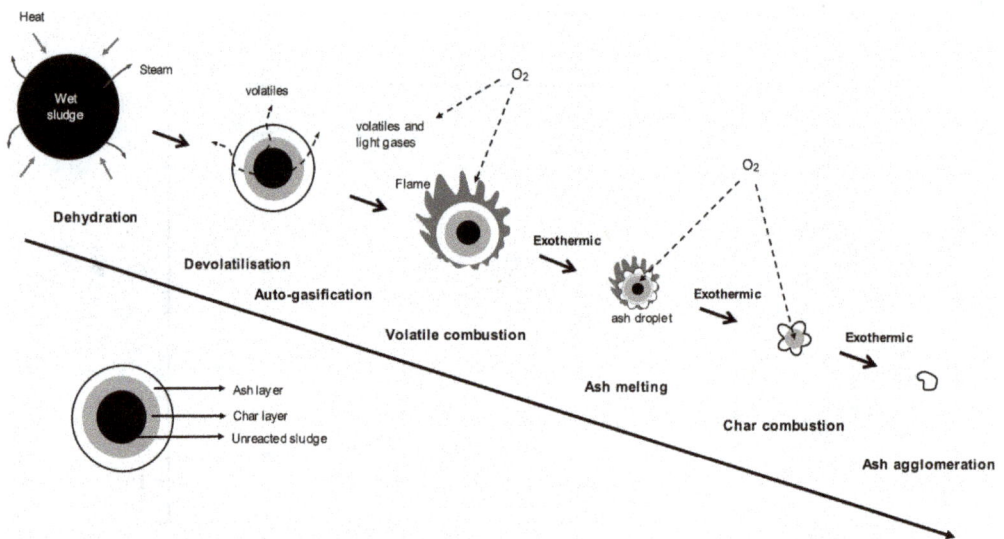

FIGURE 2.4 Combustion process of sewage sludge. Reproduced with permission from Syed-Hassan et al. (2017)

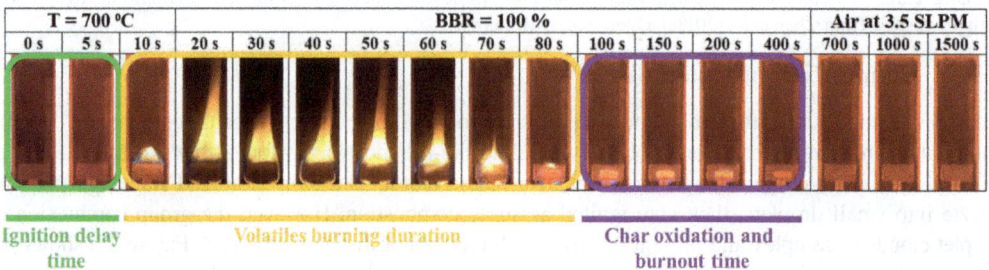

FIGURE 2.5 Combustion process of pure sewage sludge pellet at 700°C. Reproduced with permission from Chen et al. (2018a); Open Access Resources.

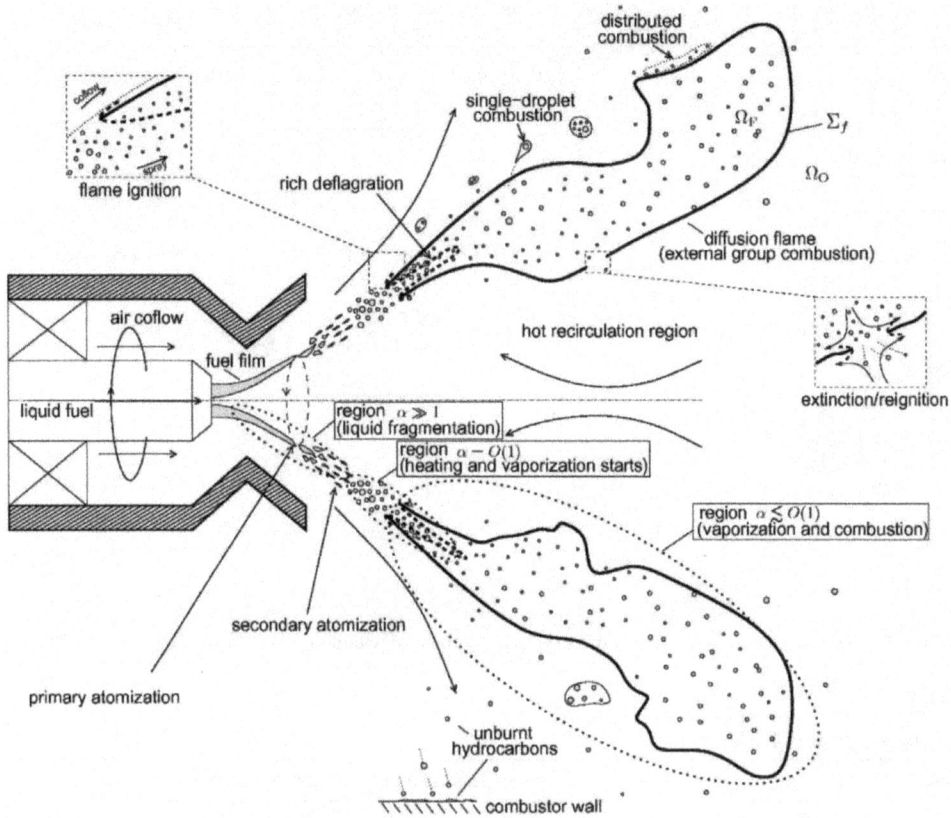

FIGURE 2.6 Schematic of liquid-fuel spray combustion. Reproduced with permission from Sánchez et al. (2015).

thermal pyrolysis is used to produce sludge pyrolysis oil (SPO), which is a good kind of renewable energy source. In this subsection, the combustion of SPO is discussed and compared to the pyrolysis castor oil (PCO) and conventional heavy fuel oil (HFO) (Chen et al., 2018b; Kuan et al., 2020).

Generally, the combustion process of homogeneous liquid fuel involves several stages: initial heating, vaporization, and combustion. The d^2-law, proposed by Spalding in 1953, is commonly used to estimate the vaporization/combustion rate of the droplet, which is expressed as: $d_p^2(t) = d_0^2 - Kt$, where $d_p(t)$ is the droplet's diameter at time t, d_p is the initial droplet diameter, and K is vaporization/combustion rate. According to the d^2-law, the droplet shrinkage varies linearly with the heating period. Therefore, K can be obtained from the slope of the reduction in the radius of the droplet. Unfortunately, bio-oil contains multi-components (e.g., lightweight or heavyweight components) with different boiling points, which might reveal sudden fragmentation events during the heating process, leading to the discrepancy between realistic vaporization/combustion rate and obtained K value. However, this micro-explosion phenomenon can promote a large droplet's fracture into small droplets, enhancing vaporization and combustion. Besides, after the gas-phase reaction, the solid impurity and residue in the bio-oil may burn in a heterogeneous-combustion form.

The suspended droplet system (Figure 2.7) is widely used to determine the burning behavior (e.g., evaporation, ignition, and combustion) of bio-oil under different surrounding temperatures (Chen et al., 2017; Hou et al., 2013; Shi et al., 2017). The droplet (size near 1 mm) is usually suspended on the thermocouple junction and fixed on a motorized stage, then entered into a high-temperature region in 1 second. Meanwhile, a high-speed camera and temperature data acquisition system are triggered and started to record the droplet's two-dimensional images and temperature profiles simultaneously, whereby the evaporation or combustion characteristics of the droplet during the heating process could be achieved.

Kuan et al. (2020) investigated the combustion behaviors of SPO and HFO. The SPO was produced from the dried sewage sludge with the operation condition based on their previous study (Chen et al.,

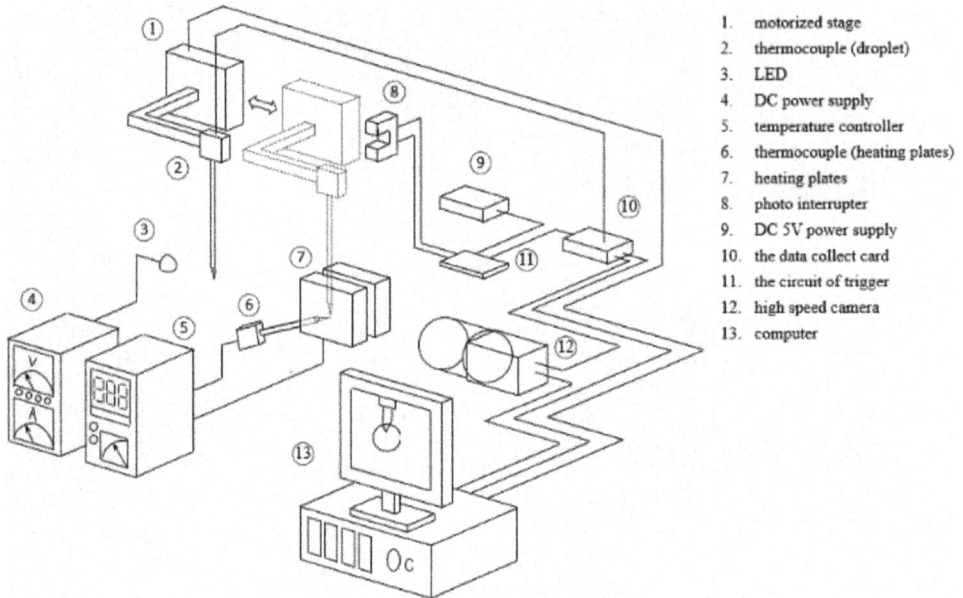

1. motorized stage
2. thermocouple (droplet)
3. LED
4. DC power supply
5. temperature controller
6. thermocouple (heating plates)
7. heating plates
8. photo interrupter
9. DC 5V power supply
10. the data collect card
11. the circuit of trigger
12. high speed camera
13. computer

FIGURE 2.7   Schematic of the suspended droplet system. Reproduced with permission from Chen et al. (2017).

TABLE 2.6

Elemental Analysis of Different Fuels

Sample	Elemental Analysis (wt.%)				
	N	**C**	**H**	**O**	**S**
SPO	5.88	58.85	9.29	12.31	1.52
HFO	0.46	86.7	11.56	0.49	0.46
Diesel	-	84.75	12.46	-	0.22
PCO	3.11	71.65	10.52	13.52	0

Note: Reproduced with permission from Kuan et al. (2020).

TABLE 2.7

The Properties of Different Fuels

Sample	Density (g/cc, @15 °C)	Viscosity (cSt, @40 °C)	Pour Point (°C)	Flash Point (°C)	Water Content (wt. %)	pH Value	HHV (MJ/kg)
SPO	1.04	64.86–66.81	6	>80	11.6	8.2	30.7
HFO	0.93	321.1	6	160	0.3	–	44.0
Diesel	0.833	3.02	–9	52–80	–	–	42.5
PCO	0.966	81.47	–6	37	2.5	–	34.8

Note: Reproduced with permission from Kuan et al. (2020).

2018b) using the Taguchi method. The blending fuels were prepared by mixing SPO and HFO in different ratios with surfactants. Table 2.6 and Table 2.7 show the properties of different fuels (Kuan et al., 2020). It can be found that both SPO and PCO had a lower carbon content but higher oxygen content when compared to HFO and diesel. In addition, SPO and PCO possess lower viscosity and low flash point as compared to HFO. Although the heating value of SPO and PCO is only two-thirds of that in HFO, it is higher than most lignocellulosic pyrolytic oil. It can be regarded as a potential alternative fuel for co-firing.

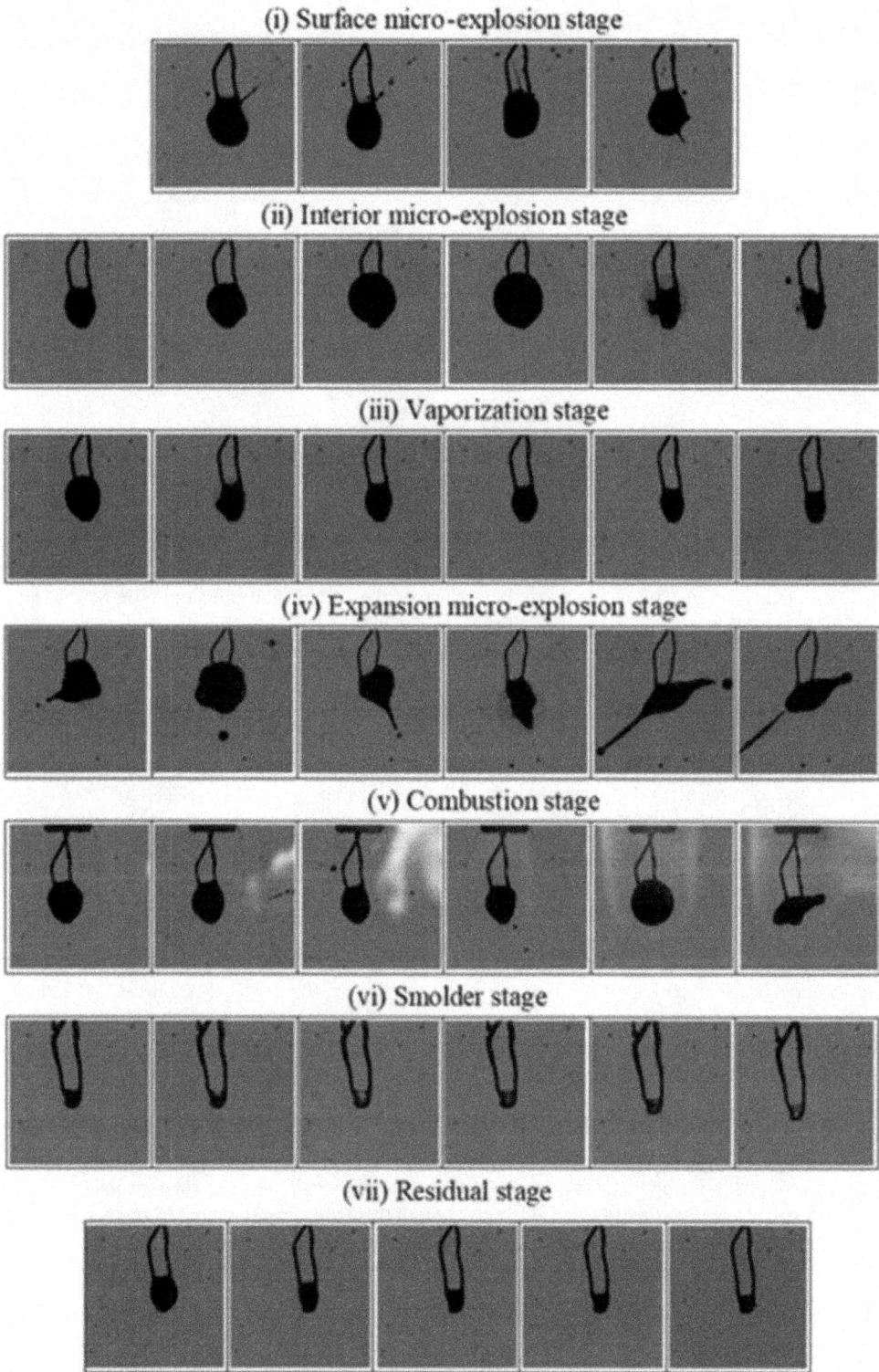

FIGURE 2.8 Different burning stages of SPO/HFO droplet. Reproduced with permission from Kuan et al. (2020).

TABLE 2.8

The Vaporization and Combustion Rates of Different Rof SPO/HFO Fuel Blends

Blending Ratio (% SPO)	Ambient Temperature (°C)		
	500	550	600
	Vaporization Rate (mm²/s)	Combustion Rate (mm²/s)	Combustion Rate (mm²/s)
20	–	–	2.102
40	–	–	2.212
60	–	2.226	2.831
80	0.582	2.556	2.839
100	0.607	3.047	3.078

Note: Reproduced with permission from Kuan et al. (2020).

On the whole, various distinct stages in the suspended droplet experiments, as seen in Figure 2.8 (Kuan et al., 2020), were identified for the SPO/HFO blends. They are surface micro-explosion stage, interior micro-explosion stage, vaporization stage, expansion micro-explosion stage, combustion stage, smoldering stage, and residual stage. Each stage's duration strongly depends on fuel properties. A higher SPO ratio in the blending fuel facilitates micro-explosion and fracture and reduces the final residue's size. Table 2.8 shows the estimated vaporization or combustion rates at different SPO/HFO blending fuels (Kuan et al., 2020). The increase of SPO enhances the combustion rate, and pure SPO exhibits a maximum combustion rate of 3.078 mm²/s, indicating its high reactivity in combustion.

In summary, biofuel combustion is currently one of the promising eco-friendly, energy conversion technologies for heat and power generation. However, biofuel's reaction mechanism is quite different as compared to conventional fossil fuels due to the diverse resources, especially for solid waste that includes hazardous substances and inorganics. The fuel database and systematic model should be further developed in the future to improve the reliability of biofuel utilization in the combustion system and operation to achieve high combustion efficiency and low pollutant emission.

2.6 Conclusions

This chapter has successfully reviewed the production, characterization, and combustion of renewable waste and biomass to form solid and liquid biofuels. Commercial conversion of renewable biomass and waste technologies is remarkably improving, leading to enhance the usage of resources and reduce the impact on the environment. Renewable biomass resources for conversion to biofuels usually require pretreatment, such as torrefaction and thermal pyrolysis, to improve feedstock quality. This chapter also thoroughly examines the conversion technologies to produce solid and liquid biofuels, the characterizations of solid and liquid biofuels, including heating value, approximate analysis, elemental analysis, pH value, etc.

Many technologies focusing on the combustion of solid and liquid biofuels have been proposed to reinforce the performance and stability of combustion and diminish emission by pollutants. Although significant progress has been made to enhance practical, economical, and efficient conversion processes, more effort still requires to be made in this field. Though there are several technical and economic challenges in promoting solid and liquid biofuels, there is still much attraction and potential for the available biofuels' market in the coming years.

REFERENCES

Abrego, J., Arauzo, J., Sánchez, J. L., Gonzalo, A., Cordero, T., and Rodríguez-Mirasol, J. 2009. Structural changes of sewage sludge char during fixed-bed pyrolysis. *Industrial & Engineering Chemistry Research* 48: 3211–3221.

Alptekin, E., and Canakci, M. 2008. Determination of the density and the viscosities of biodiesel – diesel fuel blends. *Renewable Energy* 33: 2623–2630.

Antal, M. J., and Grønli, M. 2003. The art, science, and technology of charcoal production. *Industrial & Engineering Chemistry Research* 42: 1619–1640.

Anupam, K., Sharma, A. K., Lal, P. S., Dutta, S., and Maity, S. 2016. Preparation, characterization and optimization for upgrading *Leucaena leucocephala* bark to biochar fuel with high energy yielding. *Energy* 106: 743–756.

ASTM E. 1995. *Standard Test Method for Ash in Biomass*. West Conshohocken, PA: ASTM International.

Basu, P. 2018. *Biomass Gasification, Pyrolysis and Torrefaction: Practical Design and Theory*. Academic Press.

Bergman, P. C., and Kiel, J. H. 2005. Torrefaction for biomass upgrading. *Proc. 14th European Biomass Conference*, Paris, France, pp. 17–21.

Berndes, G., Abt, B., Asikainen, A., Cowie, A., Dale, V., Egnell, G., Lindner, M., Marelli, L., Paré, D., and Pingoud, K. 2016. Forest biomass, carbon neutrality and climate change mitigation. *From Science to Policy* 3: 7.

Boie, W. 1953. Fuel technology calculations. *Energietechnik* 3: 309–316.

Braga, R. M., Melo, D. M., Aquino, F. M., Freitas, J. C., Melo, M. A., Barros, J. M., and Fontes, M. S. 2014. Characterization and comparative study of pyrolysis kinetics of the rice husk and the elephant grass. *Journal of Thermal Analysis and Calorimetry* 115: 1915–1920.

Bridgewater, A., Czernik, S., and Piskorz, J. 2001. An overview of fast pyrolysis. In: *Progress in Thermochemical Biomass Conversion*. Oxford: Blackwell Science, pp. 977–997.

Bridgwater, A. V. 2012. Review of fast pyrolysis of biomass and product upgrading. *Biomass and Bioenergy* 38: 68–94.

Brzychczyk, B., Hebda, T., and Giełżecki, J. 2018. Energy characteristics of compacted biofuel with stabilized fraction of municipal waste. In: *Renewable Energy Sources: Engineering, Technology, Innovation*. Springer, pp. 451–462.

Buchanan, B. B., and Arnon, D. I. 1990. A reverse KREBS cycle in photosynthesis: Consensus at last. *Photosynthesis Research* 24: 47.

Buis, A. 2019. *The Atmosphere: Getting a Handle on Carbon Dioxide. Global Climate Change*. NASA. https://climate.nasa.gov/news/2915/the-atmosphere-getting-a-handle-on-carbon-dioxide/ (accessed 7 December 2019)

Cantrell, K. B., Ducey, T., Ro, K. S., and Hunt, P. G. 2008. Livestock waste-to-bioenergy generation opportunities. *Bioresource Technology* 99: 7941–7953.

Cantrell, K. B., Ro, K., Mahajan, D., Anjom, M., and Hunt, P. G. 2007. Role of thermochemical conversion in livestock waste-to-energy treatments: Obstacles and opportunities. *Industrial & Engineering Chemistry Research* 46: 8918–8927.

Carpenter, D., Westover, T. L., Czernik, S., and Jablonski, W. 2014. Biomass feedstocks for renewable fuel production: A review of the impacts of feedstock and pretreatment on the yield and product distribution of fast pyrolysis bio-oils and vapors. *Green Chemistry* 16: 384–406.

Ceylan, Z., and Sungur, B. 2020. Estimation of coal elemental composition from proximate analysis using machine learning techniques. *Energy Sources, Part A: Recovery, Utilization, and Environmental Effects:* 1–17.

Chen, G.-B., Chatelier, S., Lin, H.-T., Wu, F.-H., and Lin, T.-H. 2018a. A study of sewage sludge co-combustion with Australian black coal and shiitake substrate. *Energies* 11: 3436.

Chen, G.-B., Li, J.-W., Lin, H.-T., Wu, F.-H., and Chao, Y.-C. 2018b. A study of the production and combustion characteristics of pyrolytic oil from sewage sludge using the Taguchi method. *Energies* 11: 2260.

Chen, G.-B., Li, Y.-H., Lan, C.-H., Lin, H.-T., and Chao, Y.-C. 2017. Micro-explosion and burning characteristics of a single droplet of pyrolytic oil from castor seeds. *Applied Thermal Engineering* 114: 1053–1063.

Chowdhury, H., and Loganathan, B. 2019. Third-generation biofuels from microalgae: A review. *Current Opinion in Green and Sustainable Chemistry* 20: 39–44.

Cocchi, M., Nikolaisen, L., Junginger, M., Goh, C. S., Heinimö, J., Bradley, D., Hess, R., Jacobson, J., Ovard, L. P., and Thrän, D. 2011. Global wood pellet industry market and trade study, IEA bioenergy task. *IEA Bioenergy*: 190.

Colucci, J. A., Borrero, E. E., and Alape, F. 2005. Biodiesel from an alkaline transesterification reaction of soybean oil using ultrasonic mixing. *Journal of the American Oil Chemists' Society* 82: 525–530.

da Rosa, G. M., Moraes, L., Cardias, B. B., and Costa, J. A. V. 2015. Chemical absorption and CO_2 biofixation via the cultivation of Spirulina in semicontinuous mode with nutrient recycle. *Bioresource Technology* 192: 321–327.

Di Blasi, C. 2009. Combustion and gasification rates of lignocellulosic chars. *Progress in Energy and Combustion Science* 35: 121–140.

Dietz, K.-J., and Heber, U. 1984. Rate-limiting factors in leaf photosynthesis. I. Carbon fluxes in the Calvin cycle. *Biochimica Et Biophysica Acta (BBA)-Bioenergetics* 767: 432–443.

Dobele, G., Urbanovich, I., Volpert, A., Kampars, V., and Samulis, E. 2007. Fast pyrolysis – effect of wood drying on the yield and properties of bio-oil. *BioResources* 2: 698–706.

Dobie, P., and Sharma, N. 2015. *Trees as a Global Source of Energy: From Fuelwood and Charcoal to Pyrolysis-driven Electricity Generation and Biofuels*. World Agroforestry Centre.

E871-82, A. 2006. *Standard Test Method for Moisture Analysis of Particulate Wood Fuels*. New York: ASTM.

E872–82, A. 2013. *Standard test method for volatile matter in the analysis of particulate wood fuels*. West Conshohocken, PA: ASTM International.

Fagernäs, L., Brammer, J., Wilén, C., Lauer, M., and Verhoeff, F. 2010. Drying of biomass for second generation synfuel production. *Biomass and Bioenergy* 34: 1267–1277.

Food, Nations, A.O.o.t.U. 1990. The potential use of wood residues for energy generation. *Energy Conservation in the Mechanical Forest Industries*.

Friedl, A., Padouvas, E., Rotter, H., and Varmuza, K. 2005. Prediction of heating values of biomass fuel from elemental composition. *Analytica Chimica Acta* 544: 191–198.

Goel, V., and Sharma, V. K. 2019. A brief review on renewable sources for biofuel. *Journal of Biofuels* 10: 97–100.

Gomez, L. D., Steele-King, C. G., and McQueen-Mason, S. J. 2008. Sustainable liquid biofuels from biomass: The writing's on the walls. *New Phytologist* 178: 473–485.

Gonçalves, M. A., Gonzaga, F. B., Fraga, I. C. S., de Matos Ribeiro, C., Sobral, S. P., Borges, P. P., and de Carvalho Rocha, W. F. 2011. Evaluation study of different glass electrodes by an interlaboratory comparison for determining the pH of fuel ethanol. *Sensors and Actuators B: Chemical* 158: 327–332.

Guo, M., Song, W., and Buhain, J. 2015. Bioenergy and biofuels: History, status, and perspective. *Renewable and Sustainable Energy Reviews* 42: 712–725.

Heo, H. S., Park, H. J., Park, Y.-K., Ryu, C., Suh, D. J., Suh, Y.-W., Yim, J.-H., and Kim, S.-S. 2010. Bio-oil production from fast pyrolysis of waste furniture sawdust in a fluidized bed. *Bioresource Technology* 101: S91–S96.

Hoekman, S. K., Broch, A., Robbins, C., Ceniceros, E., and Natarajan, M. 2012. Review of biodiesel composition, properties, and specifications. *Renewable and Sustainable Energy Reviews* 16: 143–169.

Hossain, M., Islam, M., Rahman, M., Kader, M., and Haniu, H. 2017. Biofuel from co-pyrolysis of solid tire waste and rice husk. *Energy Procedia* 110: 453–458.

Hou, S.-S., Rizal, F. M., Lin, T.-H., Yang, T.-Y., and Wan, H.-P. 2013. Microexplosion and ignition of droplets of fuel oil/bio-oil (derived from lauan wood) blends. *Fuel* 113: 31–42.

Huang, C.-W., Li, Y.-H., Xiao, K.-L., and Lasek, J. 2019. Cofiring characteristics of coal blended with torrefied Miscanthus biochar optimized with three Taguchi indexes. *Energy* 172: 566–579.

Huang, M., Ying, X., Shen, D., Feng, H., Li, N., Zhou, Y., and Long, Y. 2017. Evaluation of oil sludge as an alternative fuel in the production of Portland cement clinker. *Construction and Building Materials* 152: 226–231.

Huang, Y.-F., and Lo, S.-L. 2020. Predicting heating value of lignocellulosic biomass based on elemental analysis. *Energy* 191: 116501.

Hurt, R. H. 1998. Structure, properties, and reactivity of solid fuels. *Symposium (International) on Combustion* 27: 2887–2904.

Islam, M. R., Nabi, M. N., and Islam, M. N. 2001. Characterization of biomass solid waste for liquid fuel production. *4th International Conference on Mechanical Engineering*, pp. 77–82.

Jeyaseelan, S., and Qing, L. G. 1996. Development of adsorbent/catalyst from municipal wastewater sludge. *Water Science and Technology* 34: 499–505.

Jindarom, C., Meeyoo, V., Kitiyanan, B., Rirksomboon, T., and Rangsunvigit, P. 2007. Surface characterization and dye adsorptive capacities of char obtained from pyrolysis/gasification of sewage sludge. *Chemical Engineering Journal* 133: 239–246.

Junming, X., Jianchun, J., Yunjuan, S., and Yanju, L. 2008. Bio-oil upgrading by means of ethyl ester production in reactive distillation to remove water and to improve storage and fuel characteristics. *Biomass and Bioenergy* 32: 1056–1061.

Kaminsky, W., and Kummer, A. 1989. Fluidized bed pyrolysis of digested sewage sludge. *Journal of Analytical and Applied Pyrolysis* 16: 27–35.

Kang, S., Li, X., Fan, J., and Chang, J. 2012. Solid fuel production by hydrothermal carbonization of black liquor. *Bioresource Technology* 110: 715–718.

Karunanayake, J. 2018. Cinnamon firewood as a biofuel for electricity generation. *Engineer* 51: 31–38.

Klass, D. L. 1998. *Biomass for Renewable Energy, Fuels, and Chemicals*. Elsevier.

Koponen, K., and Hannula, I. 2017. GHG emission balances and prospects of hydrogen enhanced synthetic biofuels from solid biomass in the European context. *Applied Energy* 200: 106–118.

Koufopanos, C., Papayannakos, N., Maschio, G., and Lucchesi, A. 1991. Modelling of the pyrolysis of biomass particles. Studies on kinetics, thermal and heat transfer effects. *The Canadian Journal of Chemical Engineering* 69: 907–915.

Kuan, Y.-H., Wu, F.-H., Chen, G.-B., Lin, H.-T., and Lin, T.-H. 2020. Study of the combustion characteristics of sewage sludge pyrolysis oil, heavy fuel oil, and their blends. *Energy* 117559.

Lawal, A. I., Aladejare, A. E., Onifade, M., Bada, S., and Idris, M. A. 2020. Predictions of elemental composition of coal and biomass from their proximate analyses using ANFIS, ANN and MLR. *International Journal of Coal Science & Technology* 1–17.

Li, Y.-H., Lin, H.-T., Xiao, K.-L., and Lasek, J. 2018. Combustion behavior of coal pellets blended with Miscanthus biochar. *Energy* 163: 180–190.

Lloyd, W. G., and Davenport, D. A. 1980. Applying thermodynamics to fossil fuels: Heats of combustion from elemental compositions. *Journal of Chemical Education* 57: 56.

Long, H., Li, X., Wang, H., and Jia, J. 2013. Biomass resources and their bioenergy potential estimation: A review. *Renewable and Sustainable Energy Reviews* 26: 344–352.

Lu, G. M., and Lau, D. 1996. Characterisation of sewage sludge-derived adsorbents for H_2S removal. Part 2: Surface and pore structural evolution in chemical activation. *Gas separation & Purification* 10: 103–111.

Lu, J.-J., and Chen, W.-H. 2015. Investigation on the ignition and burnout temperatures of bamboo and sugarcane bagasse by thermogravimetric analysis. *Applied Energy* 160: 49–57.

Lundin, M., Olofsson, M., Pettersson, G., and Zetterlund, H. 2004. Environmental and economic assessment of sewage sludge handling options. *Resources, Conservation and Recycling* 41: 255–278.

Lyu, G., Wu, S., and Zhang, H. 2015. Estimation and comparison of bio-oil components from different pyrolysis conditions. *Frontiers in Energy Research* 3: 28.

Maryandyshev, P., Chernov, A., Lyubov, V., Trouvé, G., Brillard, A., and Brilhac, J.-F. 2015. Investigation of thermal degradation of different wood-based biofuels of the northwest region of the Russian Federation. *Journal of Thermal Analysis and Calorimetry* 122: 963–973.

Muchuweti, M., Birkett, J., Chinyanga, E., Zvauya, R., Scrimshaw, M. D., and Lester, J. 2006. Heavy metal content of vegetables irrigated with mixtures of wastewater and sewage sludge in Zimbabwe: Implications for human health. *Agriculture, Ecosystems & Environment* 112: 41–48.

Nizamuddin, S., Siddiqui, M. T. H., Baloch, H. A., Mubarak, N. M., Griffin, G., Madapusi, S., and Tanksale, A. 2018. Upgradation of chemical, fuel, thermal, and structural properties of rice husk through microwave-assisted hydrothermal carbonization. *Environmental Science and Pollution Research* 25: 17529–17539.

Nosek, R., Holubcik, M., and Jandacka, J. 2016. The impact of bark content of wood biomass on biofuel properties. *BioResources* 11: 44–53.

Ogle, R. A. 2016. *Dust Explosion Dynamics*. Butterworth-Heinemann.

Ohno, T., He, Z., Sleighter, R. L., Honeycutt, C. W., and Hatcher, P. G. 2010. Ultrahigh resolution mass spectrometry and indicator species analysis to identify marker components of soil-and plant biomass-derived organic matter fractions. *Environmental Science & Technology* 44: 8594–8600.

Parikh, J., Channiwala, S., and Ghosal, G. 2007. A correlation for calculating elemental composition from proximate analysis of biomass materials. *Fuel* 86: 1710–1719.

Parthasarathy, P., Narayanan, K. S., and Arockiam, L. 2013. Study on kinetic parameters of different biomass samples using thermo-gravimetric analysis. *Biomass and Bioenergy* 58: 58–66.

Paschalidou, A., Tsatiris, M., and Kitikidou, K. 2016. Energy crops for biofuel production or for food?-SWOT analysis (case study: Greece). *Renewable Energy* 93: 636–647.

Perea-Moreno, A.-J., Juaidi, A., and Manzano-Agugliaro, F. 2016. Solar greenhouse dryer system for wood chips improvement as biofuel. *Journal of Cleaner Production* 135: 1233–1241.

Pérez, J., Munoz-Dorado, J., De la Rubia, T., and Martinez, J. 2002. Biodegradation and biological treatments of cellulose, hemicellulose and lignin: An overview. *International Microbiology* 5: 53–63.

Pietrzak, R., and Bandosz, T. J. 2007. Reactive adsorption of NO_2 at dry conditions on sewage sludge-derived materials. *Environmental Science & Technology* 41: 7516–7522.

Pokorna, E., Postelmans, N., Jenicek, P., Schreurs, S., Carleer, R., and Yperman, J. 2009. Study of bio-oils and solids from flash pyrolysis of sewage sludges. *Fuel* 88: 1344–1350.

Ríos-Badrán, I. M., Luzardo-Ocampo, I., García-Trejo, J. F., Santos-Cruz, J., and Gutiérrez-Antonio, C. 2020. Production and characterization of fuel pellets from rice husk and wheat straw. *Renewable Energy* 145: 500–507.

Ruddy, D. A., Schaidle, J. A., Ferrell III, J. R., Wang, J., Moens, L., and Hensley, J. E. 2014. Recent advances in heterogeneous catalysts for bio-oil upgrading via "ex situ catalytic fast pyrolysis": Catalyst development through the study of model compounds. *Green Chemistry* 16: 454–490.

Saidur, R., Abdelaziz, E., Demirbas, A., Hossain, M., and Mekhilef, S. 2011. A review on biomass as a fuel for boilers. *Renewable and Sustainable Energy Reviews* 15: 2262–2289.

Sánchez, A. L., Urzay, J., and Liñán, A. 2015. The role of separation of scales in the description of spray combustion. *Proceedings of the Combustion Institute* 35: 1549–1577.

Seredych, M., and Bandosz, T. J. 2007. Sewage sludge as a single precursor for development of composite adsorbents/catalysts. *Chemical Engineering Journal* 128: 59–67.

Seyler, C., Hellweg, S., Monteil, M., and Hungerbühler, K. 2005. Life cycle inventory for use of waste solvent as fuel substitute in the cement industry – a multi-input allocation model (11 pp). *The International Journal of Life Cycle Assessment* 10: 120–130.

Shankar Tumuluru, J., Sokhansanj, S., Hess, J. R., Wright, C. T., and Boardman, R. D. 2011. A review on biomass torrefaction process and product properties for energy applications. *Industrial Biotechnology* 7: 384–401.

Shen, L., and Zhang, D.-K. 2003. An experimental study of oil recovery from sewage sludge by low-temperature pyrolysis in a fluidised-bed. *Fuel* 82: 465–472.

Sheng, C., and Azevedo, J. 2005. Estimating the higher heating value of biomass fuels from basic analysis data. *Biomass and Bioenergy* 28: 499–507.

Sher, F., Pans, M. A., Sun, C., Snape, C., and Liu, H. 2018. Oxy-fuel combustion study of biomass fuels in a 20 kWth fluidized bed combustor. *Fuel* 215: 778–786.

Shi, Z., Li, Y., Zhang, Y., Chen, Y., Li, X., Wu, D., Xu, T., Shan, C., and Du, G. 2017. High-efficiency and air-stable perovskite quantum dots light-emitting diodes with an all-inorganic heterostructure. *Nano Letters* 17: 313–321.

Sivaramakrishnan, K., and Ravikumar, P. 2011. Determination of higher heating value of biodiesels. *International Journal of Engineering Science and Technology* 3: 7981–7987.

Smith, K., Fowler, G., Pullket, S., and Graham, N. J. D. 2009. Sewage sludge-based adsorbents: A review of their production, properties and use in water treatment applications. *Water Research* 43: 2569–2594.

Stammbach, M. R., Kraaz, B., Hagenbucher, R., and Richarz, W. 1989. Pyrolysis of sewage sludge in a fluidized bed. *Energy & Fuels* 3: 255–259.

Syed-Hassan, S. S. A., Wang, Y., Hu, S., Su, S., and Xiang, J. 2017. Thermochemical processing of sewage sludge to energy and fuel: Fundamentals, challenges and considerations. *Renewable and Sustainable Energy Reviews* 80: 888–913.

Telmo, C., and Lousada, J. 2011. Heating values of wood pellets from different species. *Biomass and Bioenergy* 35: 2634–2639.

Tumuluru, J. S., Wright, C. T., Boardman, R. D., Yancey, N. A., and Sokhansanj, S. 2011. A review on biomass classification and composition, co-firing issues and pretreatment methods. *American Society of Agricultural and Biological Engineers*, Louisville, Kentucky, August 7–10, p. 1.

Tzanetakis, T., Ashgriz, N., James, D., and Thomson, M. 2008. Liquid fuel properties of a hardwood-derived bio-oil fraction. *Energy & Fuels* 22: 2725–2733.

USEIA. 2020. *Renewable Energy Explained*. Washington, DC: Office of Energy Statistics, U.S. Department of Energy.

Wasserman, J. C., Figueiredo, A. M. G., Pellegatti, F., and Silva-Filho, E. V. 2001. Elemental composition of sediment cores from a mangrove environment using neutron activation analysis. *Journal of Geochemical Exploration* 72: 129–146.

West, B. H., Sluder, S., Knoll, K., Orban, J., and Feng, J. 2012. *Intermediate Ethanol Blends Catalyst Durability Program*. Oak Ridge, TN: Oak Ridge National Lab (ORNL).

White, M. S., Curtis, M., Sarles, R., and Green, D. 1983. Effects of outside storage on the energy potential of hardwood particulate fuels: Part II. Higher and net heating values. *Forest Products Journal* 33: 61–65.

Wilk, M., and Magdziarz, A. 2017. Hydrothermal carbonization, torrefaction and slow pyrolysis of *Miscanthus giganteus*. *Energy* 140: 1292–1304.

Wilson, D. L. 1972. Prediction of heat of combustion of solid wastes from ultimate analysis. *Environmental Science & Technology* 6: 1119–1121.

Wu, S., Zhang, S., Wang, C., Mu, C., and Huang, X. 2018. High-strength charcoal briquette preparation from hydrothermal pretreated biomass wastes. *Fuel Processing Technology* 171: 293–300.

Yan, J., Yan, Y., Liu, S., Hu, J., and Wang, G. 2011. Preparation of cross-linked lipase-coated micro-crystals for biodiesel production from waste cooking oil. *Bioresource Technology* 102: 4755–4758.

Yang, Z., Wu, Y., Zhang, Z., Li, H., Li, X., Egorov, R. I., Strizhak, P. A., and Gao, X. 2019. Recent advances in co-thermochemical conversions of biomass with fossil fuels focusing on the synergistic effects. *Renewable and Sustainable Energy Reviews* 103: 384–398.

Yin, C.-Y. 2011. Prediction of higher heating values of biomass from proximate and ultimate analyses. *Fuel* 90: 1128–1132.

Yuan, W., and Bandosz, T. J. 2007. Removal of hydrogen sulfide from biogas on sludge-derived adsorbents. *Fuel* 86: 2736–2746.

Yusuff, A. S., Adeniyi, O. D., Olutoye, M. A., and Akpan, U. G. 2017. Performance and emission characteristics of diesel engine fuelled with waste frying oil derived biodiesel-petroleum diesel blend. *International Journal of Engineering Research in Africa*: 100–111. Trans Tech Publications.

Zhang, K., Zhou, L., Brady, M., Xu, F., Yu, J., and Wang, D. 2017. Fast analysis of high heating value and elemental compositions of sorghum biomass using near-infrared spectroscopy. *Energy* 118: 1353–1360.

Zhang, Y., Kang, L., Li, H., Huang, X., Liu, X., Guo, L., and Huang, L. 2019. Characterization of moxa floss combustion by TG/DSC, TG-FTIR and IR. *Bioresource Technology* 288: 121516.

3

Conversion of Municipal Solid Waste to Biofuels

Ravi Patel, Sonil Nanda and Ajay K. Dalai

TABLE OF CONTENTS

3.1 Introduction

A majority of the energy demand in the world is supplemented by fossil fuels. Fossil fuel burning has a detrimental impact on the ecosystem. The burning of fossil fuel hydrocarbons generates energy and massive amounts of greenhouse gases like CH_4, CO_2, SO_x, NO_x, and other volatile organic compounds (Nanda et al., 2016c). The developing and developed countries are stepping towards sustainable waste management due to adverse climate effects from fossil fuels. Greenhouse gases are also emitted through the microbial decomposition or burning of waste residues including municipal solid waste (MSW) and lignocellulosic biomass. Landfilling of such solid waste also does not restrict natural decomposition and the release of toxic greenhouse gases and emissions (Nanda and Berruti, 2021a). Some municipalities across the world have started to divert solid waste from landfills to other alternative methods because of the prevailing effects of landfilling. As a possible solution, to decrease the volume of solid waste and to convert them into clean energy, several waste-to-energy approaches are implemented (Nanda and Berruti, 2021b). As a sustainable alternative, biofuels are seen as a possible future of next-generation fuels.

Some of the naturally occurring and widely available raw materials can be used as feedstock for biofuels. First-generation biofuel feedstocks consist of food crops such as corn, sugarcane, soybeans, wheat, and sugar beet. However, such feedstocks are highly criticized since they threaten the food chain and compete with the food supply (Nanda et al., 2015). The second-generation feedstock includes non-consumable crops or waste materials such as wood, crop refuse, MSW, waste plastics, and food waste (Okolie et al., 2020; Okolie et al., 2021). Third-generation biofuel feedstocks include algae, which have no competition over edible food crops and arable lands.

With a high population, the per capita waste generation is increasing. Municipal solid waste refers to the everyday solid garbage and trash discarded by the public as waste materials. The main sources of MSW are households, but garbage collected from offices, institutions, and small-scale commercial enterprises located within a municipality are also accommodated. MSW is typically made of waste paper, trashed food, glasses, metals, rubber, leather, textiles, discarded wood, and yard trimmings (Nanda and Berruti, 2021a). Most of the MSW generated worldwide are landfilled or incinerated. Poorly managed solid wastes have tremendous effects on the local and global environment and public health. Landfill sites can be breeding grounds for disease vectors and carriers, thus resulting in respiratory diseases to the local communities through obnoxious foul odor. The improperly managed waste could have a higher downstream processing cost than the properly managed waste.

A landfill is a huge land area or excavated site landscaped to be separated from the populated areas. Landfilling is still considered to be the most common way to get rid of solid waste components. Continuous landfilling of MSW generates landfill gases and landfill leachate. Methane is a major constituent of landfill gas, which makes it combustible. Landfill leachate, a liquid that percolates via landfill, contains suspended and dissolved solids, metals, volatile compounds, toxic compounds, and microorganisms (Raghab et al., 2013). Leachate produced from MSW landfill is composed of many organic and inorganic contaminants, heavy metals, and hazardous components (Kettunen et al., 2009). The leachate composition varies depending on MSW composition, landfill operation, moisture content of solid waste, seasonal weather, and age of the landfill (Bohdziewicz et al., 2008). If the leachate is not processed efficiently, it can seep into the groundwater and water bodies, causing pollution and severe health hazards to animals and humans (Omar and Rohani, 2015).

MSW can be diverted from landfills to generate sustainable fuels through physical, thermochemical, and biological conversion processes. Some alternative technologies for handling waste materials are gasification, pyrolysis, liquefaction, anaerobic digestion, and other waste-to-energy techniques (Figure 3.1). After the physical segregation of inert materials such as metals, glasses, paints, batteries, the organic components of the MSW are diverted to waste-to-energy technologies to produce biochemicals and bioenergy. This chapter reviews such waste-to-energy techniques to convert MSW to biofuels. The challenges and prospects of the conversion technologies are systematically described.

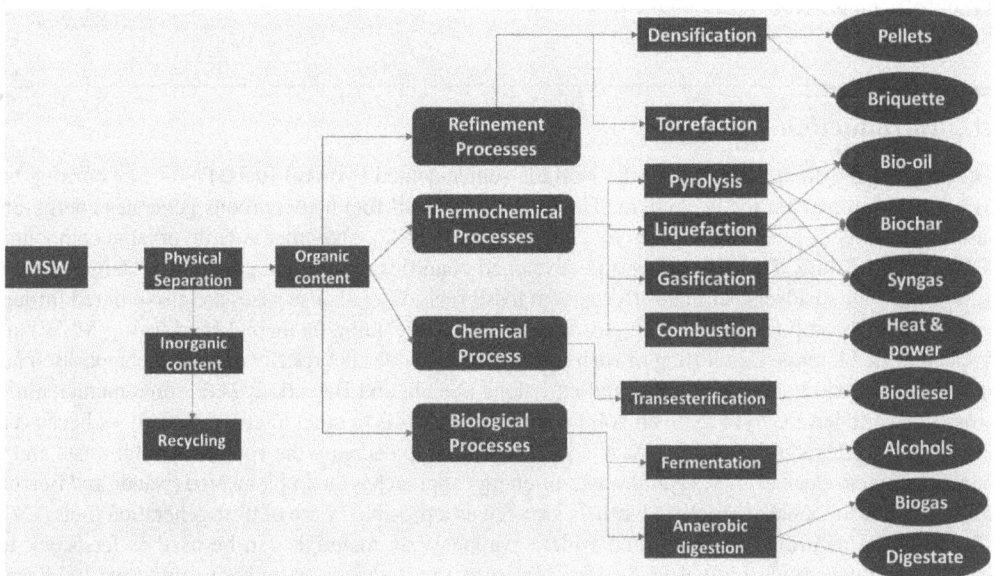

FIGURE 3.1   Conversion of municipal solid waste to biofuels.

3.2   Municipal Solid Waste Management in the Canadian Scenario

Canada sets a model example for its sustainable MSW management and efficient processing technologies. In Canada and most of North America, the households receive MSW collection services either by the municipality or through its private-sector waste management-contracting agency. The garbage pickup truck by the municipality collects the residential nonhazardous wastes such as garbage, used packaging, food waste, yard waste, electronic waste, wastepaper, and recyclables every week across the curbside of each urban residence. The drop-off recycling depots are available where curbside pickup of recyclables is not arranged. After collection, the MSW is self-hauled to transfer stations, material recycling facilities, and disposal sites (i.e., landfills or incinerators). Some municipalities also encourage operating composting at designated locations or household backyard to manage compostable organic materials. Backyard composting is a cost-effective method as 40% of residential MSW is compostable (Kelleher et al., 2000).

In Canada, the United States, Europe, and most other countries, the majority of municipalities have curbside MSW pickup scheduled every week. However, this process is labor-intensive and requires the use of fuels and other natural resources for collection and transportation to a resource recovery facility, waste-to-energy conversion plant, and/or landfill. Many cities are considering reducing the garbage pickup services so that they occur biweekly rather than every week. However, even with some incentives by the government, it is obvious that the pickup and disposal of MSW (especially food waste) is a growing issue for many cities. Fortunately, there are alternatives to the disposal of this food waste.

Canadian municipalities also accommodate nonresidential, nonhazardous solid wastes from institutions (e.g., offices, schools, colleges, universities, hospitals, government facilities, and senior homes), industries, shopping centers, and restaurants (Statistics Canada, 2020a). The nonresidential wastes also include construction, renovation, and demolition waste, which usually includes waste wood, drywall, scrap metals, cardboard, discarded doors and windows, wiring and pipes but excludes inert materials (e.g., asphalt, brick, cement, concrete, gravel, and sand). Environment Canada also allows the disposal of certain wastes at sea, which includes dredged materials, geological matter, fishery wastes, organic wastes, and decommissioned/retired vessels (Statistics Canada, 2012).

Figure 3.2 illustrates the current trend of residential and nonresidential wastes produced by the municipalities in some major Canadian provinces. According to Statistics Canada (2020a), Canada generated around 24,940,800 tons of MSW in 2016, of which residential wastes and non-residential wastes accounted for nearly 10,225,900 tons and 14,714,800 tons, respectively. Among all Canadian provinces, the maximum amounts of MSW were generated in Ontario (ca. 9,475,500 tons), Quebec (ca. 5,356,100 tons), Alberta (ca. 4,206,700 tons), and British Columbia (ca. 2,614,000 tons). As mentioned earlier, the production of MSW is directly proportional to the density of the urban population. Ontario is the highly populous province of Canada with a total human population of 14,711,827 followed by Quebec (8,537,674), British Columbia (5,110,917), and Alberta (4,413,146) as other significantly populated provinces (Statistics Canada, 2020b). Owing to their low population density, the Atlantic Provinces such as New Brunswick and Nova Scotia generate low amounts of residential and non-residential MSW every year.

Nearly 21% of the residential wastes in Canada are recycled while the residual 79% are disposed of through landfilling and incineration (Kelleher et al., 2000). More precisely, approximately 24 million tons of MSW is disposed of through landfilling and incineration across Canada and the United States. In Canada, less than 5% of MSW is incinerated (Statistics Canada, 2012). Instead of disposal, a portion of MSW is diverted to recycling and composting facilities, which includes glass, metal, plastic, paper, electronics, tires, organics as well as construction, renovation, and demolition. The per capita solid waste diversion amplified from 212 kg in 2002 to 254 kg in 2008. From the total MSW diverted for recycling and composting in 2008, about 51% accounted for residential wastes and 49% accounted for nonresidential waste (Statistics Canada, 2012).

The number of active and closed landfills in Ontario, Canada, is 2,382. Out of the total landfills, 1,206 are municipal landfills (4×10^4 to 19×10^6 m^3), 559 provincial landfills (4×10^4 m^3), 12

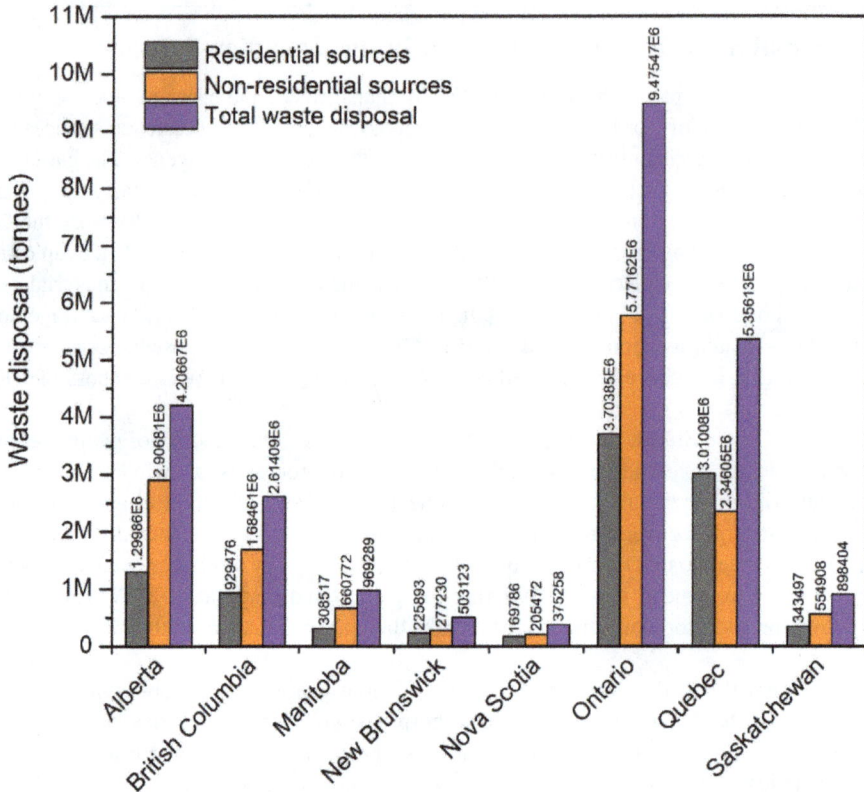

FIGURE 3.2 Sources of waste disposal in selected Canadian provinces (data source: Statistics Canada, 2020a). Note: The data presented are from the 2016 statistics year. 'M' is an abbreviation for million.

first nations landfills (4×10^4 m³), 2 federal landfills, 109 are private commercial landfills, 49 are private noncommercial landfills, and 445 are private industrial landfills (Ontario Waste Management Association, 2016). Ontario has nearly 805 active landfills that have the remaining capacity to dispose of 127.3 million tons of MSW. The landfills in Ontario received nearly 7.7 million tons of waste in 2014, excluding approximately 3.4 million tons of waste that was exported annually to the United States (Ontario Waste Management Association, 2016). Currently, approximately 11.1 million tons of MSW has been disposed of in Ontario landfills. Based on the remaining capacity of Ontario's active landfills (127.3 million tons), their remaining life is estimated to be between 11.4 and 16.5 years, if 30% of Ontario's MSW is exported to the United States. Toronto is the most populous city in Canada with more than 6 million people as of 2020. Solid Waste Management Services in Toronto collect wastes from more than 870,000 homes and non-residential establishments, street litter/recycling bins, parks' bins, annual special events as well as from schools, government offices, and corporations, and Industrial, Commercial and Institutional (IC&I) wastes are accepted at drop-off depots and landfills (The City of Toronto, 2018). The City of London in Western Ontario has a population of 404,000 and manages more than 159,640 tons of MSW per year of which 88,570 tons (55% share) are disposed of in landfills, and 71,070 tons (45% share) are diverted for composting, recycling, and other processing programs (The City of London, 2018a). In British Columbia, nearly 2.4 million tons of MSW is disposed of in landfills, which is equivalent to 520 kg/capita/annum (British Columbia Ministry of Environment, 2016).

In Canada, the expenditure for residential waste management services per household can typically range up to $360 (Gray, 2017). The total capital expenditures by the Canadian government for solid waste management amounted to $3.8 billion in 2016 as opposed to $5.3 billion in 2012 and $4.9 billion in 2008 (Statistics Canada, 2020b). The total number of full-time and part-time employees working in the MSW management activities in 2014 was 8177, and the total operating revenue was $2.6 billion. The major expenses of MSW management is related to the collection and transportation of MSW, which accounted for $1.3 billion in 2014 compared to $911 million in 2006. Nevertheless, according to the estimates by the United Nations Environment Programme (UNEP), the global market for waste management from collection to recycling is around $400 billion (Worldwatch Institute, 2012).

Table 3.1 summarizes the expenditures relating to the MSW management industry in some Canadian provinces. Following Ontario also are the major metropolitan and populous provinces such as Quebec, Alberta, and British Columbia that have high expenditures relating to waste management services. Among all the MSW management activities, waste collection and transportation as well as the operations of disposal facilities account for the mainstream expenditures. In some provinces such as Ontario, Quebec, British Columbia, and Alberta, the tipping fees also add significantly to the overall capital expenditures. The tipping fee or gate fee is the charge paid to the owner, lesser, or operator of a landfill or recycling facility for a specific amount of waste (by weight, volume, or item) for the right to its disposal or processing. The tipping fee also aids landfills to offset their revenues relating to operation and maintenance. The landfill tipping fees for residential waste disposal in the City of London, Ontario, can range from $8 (waste > 100 kg) to $75 (waste ~ 1000 kg) (The City of London, 2018b). In Ontario, landfill owners pay nearly $8.5 million to local communities in host agreements and $3.5 million in property taxes or payments-in-lieu of taxes (Ontario Waste Management Association, 2016). Furthermore, in 2014, the Ontario government held over $323 million in financial assurance for private landfills. It is recommended to restrict waste tipping into small areas and maintain good records of waste deliveries with a proper disposal plan (Kumar et al., 2011).

Alberta became the first Canadian province in 2004 for implementing an electronics stewardship program financed through the environmental handling fees (EHF) imposed on designated electronic items (Kelleher et al., 2000). All provinces in Canada enforce the EHF to fund their electronics recycling programs. The EHF fees vary as per the province and electronic device purchased (e.g., desktop/portable computer, peripherals, printer, television, audio devices, telephones, batteries). For example, the EHF per unit of desktop computers can vary from $1.00 in Quebec to $10.50 in Northwest Territories (The Source, 2020). A display device (television or monitor) of size 46 inches or larger is currently levied an EHF of $45 in New Brunswick to $10 in Alberta.

Alberta also implements a tire-recycling program by levying an advance disposal surcharge of $4 on new passenger and light truck vehicle tires and $200 for large off-road tires (Alberta Recycling, 2017). Similarly, Manitoba levies a steward-fee of all passenger tires purchased in Manitoba, for example, $3.75 for all-terrain vehicle tires to $135 for large off-the-road tires (Tire Stewardship Manitoba, 2016). In Ontario, the tire-stewardship fees can range from $3.30 for passenger and light truck tires to $1237.98 for heavy off-road pneumatic tires as well as solid and resilient tires (Rethink Tires, 2017). In Canada, oil fluid and oil containers (> 50 liters) are also subjected to environmental handling charges (EHC) as they are not consumed when in use and are available for collection and recycling. Lubricating oil is charged with an EHC of $0.05/liter in Alberta, British Columbia, Manitoba, New Brunswick, Prince Edward Island, Saskatchewan, and in Quebec it is $0.06/liter (Used Oil Management Associations of Canada, 2016). Moreover, EHC on oil filters (>8 inches) can vary from $0.85 in Quebec to $1.25 in British Columbia.

3.3 Physical Conversion of Municipal Solid Waste

The production of biofuels through sustainable sources is essential due to the tremendous increase in energy demand and environmental pressure. MSW can be used as a source of renewable material, but it comes with several challenges like heterogeneity, reduced heating value, and bulk density. MSW can be physically upgraded to solid biofuels to reduce its volume and enhance heating value. Bulk density of biomass can be improved from 40–200 kg/m^3 to 600–800 kg/m^3 via densification (Mani et al., 2003). Densification is an operation that characterizes fuels to make them more homogeneous and denser, which is achieved by either pelletizing or briquetting.

TABLE 3.1

Approximate expenditures (in Canadian dollars) Related to Waste Management Facilities in Some Canadian Provinces (2016 Statistics)

Activity	Alberta	British Columbia	Manitoba	New Brunswick	Newfoundland and Labrador	Nova Scotia	Ontario	Quebec	Saskatchewan	Canada
Collection and transportation	$231{,}789 \times 10^3$	$149{,}783 \times 10^3$	$27{,}913 \times 10^3$	$11{,}878 \times 10^3$	$13{,}223 \times 10^3$	$33{,}229 \times 10^3$	$557{,}031 \times 10^3$	$311{,}400 \times 10^3$	$29{,}246 \times 10^3$	$1{,}376{,}777 \times 10^3$
Tipping fees	$45{,}058 \times 10^3$	$58{,}723 \times 10^3$	$14{,}407 \times 10^3$	$10{,}857 \times 10^3$	$3{,}127 \times 10^3$	$14{,}411 \times 10^3$	$139{,}808 \times 10^3$	$148{,}307 \times 10^3$	$9{,}743 \times 10^3$	$447{,}597 \times 10^3$
Operation of disposal facilities	$150{,}751 \times 10^3$	$105{,}613 \times 10^3$	$15{,}694 \times 10^3$	$22{,}388 \times 10^3$	DS	$16{,}358 \times 10^3$	$193{,}475 \times 10^3$	$83{,}267 \times 10^3$	$15{,}921 \times 10^3$	$621{,}878 \times 10^3$
Operation of transfer stations	$34{,}565 \times 10^3$	$72{,}828 \times 10^3$	$1{,}058 \times 10^3$	DS	–	$7{,}247 \times 10^3$	$81{,}451 \times 10^3$	$4{,}705 \times 10^3$	207×10^3	$205{,}117 \times 10^3$
Operation of recycling facilities	$26{,}537 \times 10^3$	$17{,}011 \times 10^3$	$11{,}556 \times 10^3$	$8{,}383 \times 10^3$	DS	$9{,}302 \times 10^3$	$116{,}733 \times 10^3$	$35{,}966 \times 10^3$	$2{,}005 \times 10^3$	$229{,}201 \times 10^3$
Operation of organics processing facilities	$13{,}137 \times 10^3$	$9{,}092 \times 10^3$	$2{,}657 \times 10^3$	DS	DS	$10{,}590 \times 10^3$	$48{,}117 \times 10^3$	$4{,}886 \times 10^3$	740×10^3	$93{,}804 \times 10^3$
Contributions to landfills post closure and maintenance fund	$13{,}486 \times 10^3$	$16{,}801 \times 10^3$	$2{,}164 \times 10^3$	$1{,}043 \times 10^3$	DS	$1{,}548 \times 10^3$	$30{,}403 \times 10^3$	$4{,}620 \times 10^3$	$15{,}992 \times 10^3$	$89{,}569 \times 10^3$
Other expenditures	$14{,}461 \times 10^3$	$55{,}312 \times 10^3$	$2{,}208 \times 10^3$	$9{,}425 \times 10^3$	977×10^3	$10{,}562 \times 10^3$	$118{,}474 \times 10^3$	$52{,}333 \times 10^3$	$3{,}290 \times 10^3$	$270{,}027 \times 10^3$
Operating revenues	$506{,}446 \times 10^3$	$522{,}344 \times 10^3$	$60{,}797 \times 10^3$	$53{,}724 \times 10^3$	$18{,}576 \times 10^3$	$55{,}458 \times 10^3$	$999{,}938 \times 10^3$	$477{,}784 \times 10^3$	$70{,}877 \times 10^3$	$2{,}798{,}990 \times 10^3$
All current expenditures	$529{,}785 \times 10^3$	$485{,}163 \times 10^3$	$77{,}655 \times 10^3$	$67{,}585 \times 10^3$	$32{,}126 \times 10^3$	$103{,}247 \times 10^3$	$1{,}285{,}492 \times 10^3$	$645{,}483 \times 10^3$	$77{,}144 \times 10^3$	$3{,}333{,}970 \times 10^3$

Data source: Statistics Canada (2020b); abbreviation 'DS' denotes that the data is suppressed for confidentiality reasons by Statistics Canada.

3.3.1 Briquetting

Briquettes are a kind of refined solid fuels. MSW as well as agricultural and forestry residues have nonuniform physicochemical composition and are voluminous, creating challenges in their bulk long-distance transportation. This can be resolved through the densification method to make the biomass more compact, dense, and uniformly shaped. In this method, biomass is forced under high pressure using a screw or piston-press with lignin present in the woody material acting as a binder or external binding agents (Azargohar et al., 2019b). Densification can increase the net calorific value of a solid fuel per unit volume. For example, briquetting imparts good bulk density as compared to non-densified biomass. Moreover, the product formed is of homogeneous size and shape with low moisture content, which can be easily handled, transported, and stored.

3.3.2 Pelletization

Biomass fuel pellets typically have a cylindrical shape and are 6–8 mm in diameter, with varying length. Pelletization of agricultural biomass and wood chips as a renewable solid fuel is more popular. Biomass pellets are used in the household and industrial sector to produce heat and power. The conventional pellet production method includes raw materials to be pre-dried, which consume energy. During pelletization, the temperature increases that make lignin soften and act as a binder and hydrophobic agent (Azargohar et al., 2019a). As a result of pelletization, biomass pellets have lower water content. Due to homogeneity, variance in water content is less, which imparts superior combustion properties. The high bulk density and lower transportation and storage cost can be achieved for biomass pellets. The pellets are also easy to feed into the burner or furnace with high energy density and lower risk to ash or tar formation. The biomass pellets are also less prone to microbial decomposition.

3.4 Thermochemical Conversion of Municipal Solid Waste

3.4.1 Torrefaction

Torrefaction is a thermochemically driven process of carbonaceous compounds in which hemicellulose degradation prevails while cellulose and lignin fraction is partially altered (Ciolkosz and Wallace, 2011). Torrefaction is an upgrading process for solid fuels by lowering the moisture content and reducing the volatile organic fraction of the biomass (Rasanjani et al., 2019). Torrefaction increases fixed carbon and net calorific value as compared to non-torrified material. It is generally used as a thermochemical pretreatment process. Due to torrefaction, the lignocellulose content of MSW imparts brittle behavior, reduced physical strength, and improved grindability (Arias et al., 2008). Reduction in moisture content and hemicellulose content enhances the shelf life of the torrefied biomass to eliminate the chances of microbial decomposition during storage. Torrefaction characterizes biodegradable lignocellulosic materials into solid fuels by upgrading their fuel properties.

Torrefaction is considered a mild pyrolysis process. Torrefaction is often blended with densification to form an energy-dense fuel. Torrefaction involves treating feedstock typically at 200–350°C under an inert atmosphere. Considering the hydrophobicity of torrefied solids, they are easy to handle, access, ship, and store for longer periods (Iroba et al., 2017). Torrefaction can be used for processing MSW to enhance its energy density and lower the pretreatment cost for biorefining. Torrefied materials have a uniform shape than the original non-torrefied materials suggesting better flowability characteristics (Arias et al., 2008).

Food waste and wood subjected to torrefaction showed an increase in high heating value (HHV) from 17.5 MJ/kg to 19.3 MJ/kg and from 28.4 MJ/kg and 31.1 MJ/kg, respectively, when torrefied at 330°C (Samad et al., 2017). A study on 1 h of torrefaction of cotton gin waste using a batch system at 260°C showed a reduction in moisture content by 48.6% and increased HHV by 6.3% for torrefied biomass (Sadaka and Negi, 2009).

Torrefaction of landfill food waste at 300°C showed an improvement in the HHV of the torrefied solid from 19.8 MJ/kg to 27.1 MJ/kg (Pahla et al., 2018). The artificial waste sample resembling a typical Sri

Lankan MSW composition was torrefied in a lab-scale horizontal cylindrical batch reactor at 275°C for 20 min (Rasanjani et al., 2019). The resulting torrefied biomass showed an increase in the fixed carbon content from 37.4 wt.% to 83.4 wt.% and a lowering of moisture content from 54.9 wt.% to 4.1 wt.% when compared to the raw sample.

3.4.2 Gasification

Gasification is a thermochemical process that converts organic waste into synthesis gas. It is referred to as an incomplete oxidation of hydrocarbon materials in the presence of an oxygen amount less than its requirement for the stoichiometric combustion (Arena, 2011). Gasification is carried out at a temperature range of 550–900°C in the presence of air, steam, inert gas, or supercritical water (Arena, 2012; Parakh et al., 2020). Syngas from waste material can be of low, medium, or high calorific value depending upon its composition.

Gasification has many advantages over conventional incineration or combustion of solid wastes. Gasification produces syngas with no-to-negligible levels of dioxins, SO_x and NO_x. The produced syngas can be used to drive turbines or engines that convert fuel gas energy to electricity more efficiently than conventional incineration-based steam boilers. The syngas obtained from the gasification of solid waste is suited for wide-scale applications such as a chemical precursor or fuel gas to produce heat and power (Okolie et al., 2019).

Several gasification reactors are used to convert solid waste to syngas, which include a fixed bed, fluidized bed, and entrained flow gasifiers. Depending upon the type of gasifier and operating conditions, the tar content may vary. The operating temperature plays a critical role in tar formation during biomass gasification. Hence, temperatures of more than 800°C are recommended to minimize tar formation (Devi et al., 2003).

Plasma gasification utilizes an external power source to heat up and attain high reaction temperatures to break down the elementary structures of biomass except for radioactive materials (Sanlisoy and Carpinlioglu, 2017). Plasma gasification is useful for treating hazardous waste including MSW. On the other hand, it has demerits of low efficiency, large investment, and high power requirements (Heidenreich and Foscolo, 2015; Sikarwar et al., 2016).

MSW with raw wood was pretreated for 24 h at 105°C before feeding into a 10 kW plasmatron (plasma torch) (Shie et al., 2014). Plasmatron gasification at different reaction temperatures of 300°C, 500°C and 600°C resulted in syngas yields of 89.9, 90.2, 91.8 vol.%, respectively. Gasification of an MSW sample in a fixed-bed tubular reactor resulted in an increased syngas production as the temperature escalated from 550°C to 650°C (Gu et al., 2020). This suggests high syngas production at higher gasification temperatures.

Gasification of carbonaceous materials requires moisture content less than 10–20 wt.% (Lohri et al., 2017). However, feedstock with high-moisture content is suitable for hydrothermal gasification in the presence of subcritical and supercritical water (Nanda et al., 2017b). The biomass drying step and its associated costs can be eliminated for hydrothermal gasification of solid waste containing high-moisture content (Reddy et al., 2014). Water above its critical conditions (374°C and 22.1 MPa) is called supercritical water. At supercritical conditions, both gas and liquid phases exist homogeneously. Supercritical water shows different reactivity and solvent properties (Gong et al., 2017). A study on supercritical water gasification of food waste sample (i.e., orange peels) in a stainless steel tubular reactor at 5:1 water/biomass ratio under 23–25 MPa pressure and temperatures of 400, 500, and 600°C for 45 min resulted in total gas yields of 3.1, 4.1, and 5.5 mmol/g, respectively, suggesting higher gas yields at elevated temperatures (Nanda et al., 2016b).

3.4.3 Liquefaction

Liquefaction is a thermochemical process, which results in a liquid bio-crude oil and bioresidue from thermochemical, hydrothermal, and catalytic cracking of waste biomass. Subcritical water is used as an ecofriendly reaction medium for hydrothermal liquefaction (HTL) of organic compounds (Nanda et al., 2014c). Liquefaction breaks down the long-chain bio-polymeric compounds in biomass to liquid products. In contrast to pyrolysis, lower reaction temperature, enhanced energy recovery, and lower tar yields are characteristic of the liquefaction process (Gollakota et al., 2018).

HTL produces a binary-phase composite of processed water and bio-oil with suspended solids (char) and a trace amount of syngas (Lohri et al., 2017; Arturi et al., 2016). To make bio-crude oil less viscous, solvents such as alcohols, ethers, and ketones are added to the reaction system (Lohri et al., 2017). The energy density of bio-crude oil produced from HTL varies from 30 to 37 MJ/kg with potential applications as a bunker or residual fuel oil (Toor et al., 2011; Lohri et al., 2017). The bio-crude oil also contains less oxygen and moisture content when compared to the bio-oil obtained from pyrolysis. This makes bio-crude oil thermally stable with greater calorific value and less corrosive properties (Lohri et al., 2017). However, the moisture content from the bio-crude oil can be further removed through catalytic hydroprocessing and upgrading technologies (Toor et al., 2011).

3.4.4 Pyrolysis

Pyrolysis is a thermochemical conversion technology that converts carbonaceous materials through controlled heating under an oxygen-free atmosphere to produce biochar, bio-oil, and gases. Pyrolysis is generally processed in the presence of inert gas at a temperature range of 400–800°C (Kumar and Samadder, 2017). The yields and composition of pyrolysis products, that is, biochar, bio-oil, and gases depend upon the operating conditions such as temperature, heating rates, and vapor residence time (Nanda et al., 2016a). In slow pyrolysis, the primary product is biochar, whereas fast pyrolysis results in a higher yield of bio-oil (Lohri et al., 2017). A typical reaction representing the pyrolysis process is shown in Eq. (3.1):

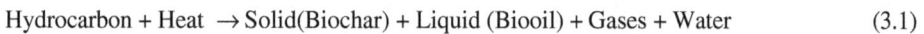

$$\text{Hydrocarbon} + \text{Heat} \rightarrow \text{Solid(Biochar)} + \text{Liquid (Biooil)} + \text{Gases} + \text{Water} \tag{3.1}$$

In slow pyrolysis, the main solid product is biochar followed by bio-oil and producer gas. Biochar is a co-product of pyrolysis and potentially applicable as an adsorbent, solid fuel, and soil-amending agent (Nanda et al., 2016a). Biochar is a porous carbon-enriched material that also contains alkali and alkaline earth metals (Mohanty et al., 2013). The liquid product of pyrolysis is called pyrolysis oil, bio-oil, pyrolytic oil, wood-oil, wood-distillate, pyroligneous acid, liquid-wood, and liquid-smoke (Lohri et al., 2017; Mohan et al., 2006). Bio-oil produced through pyrolysis has physical characteristics of being dark brownish-red to black in color and of distinct acidic, smoky aroma, and which can create possible irritation to eyes upon exposure (Venderbosch and Prins, 2010).

The liquid product obtained from pyrolysis typically contains an aqueous and organic phase (Nanda et al., 2014c). The aqueous phase mainly contains acids, water, and functionally grouped hydrocarbons such as ether, ester, alcohols, aldehyde, and ketone (Boucher et al., 2000). On the other hand, the organic phase is predominantly the bio-oil fraction mainly composed of phenolics and carbonyls. Because of relatively higher water and acid content, the crude bio-oil is unstable, caustic, and highly viscous with lower energy value (Venderbosch and Prins, 2010; Guo et al., 2015; Kan et al., 2016).

The gaseous product of pyrolysis is composed mainly of CO_2, CO, CH_4, H_2, C_2H_6, C_2H_4, and a small quantity of higher molecular hydrocarbon gases and water vapor (Nanda et al., 2014a). The average lower heating value (LHV) of pyrolysis gases falls in the range of 10–20 MJ/Nm^3 (Kan et al., 2016; Lohri et al., 2017). The pyrolysis gas is either directly applied for electricity generation or as a precursor for the synthesis of hydrocarbons and chemicals through the Fischer–Tropsch process.

Pyrolysis of waste high-density polyethylene plastic bags (a component of MSW) at 440°C produced around 74 wt.% of pyrolysis oil (Sharma et al., 2014). Pyrolysis of polystyrene-disposable plastics plates at 450°C with a heating rate of 10°C/min for 75 min resulted in 81 wt.%, 13 wt.%, and 6.2 wt.% of bio-oil, gases, and biochar, respectively (Rehan et al., 2017). Pyrolysis of garbage bags collected from an MSW landfill site in Padang Siding, Malaysia, was performed at 200–750°C (Tursunov, 2014). The yields of gases, bio-oil, and biochar were 39.9 wt.%, 21.7 wt.%, and 38.4 wt.% without a catalyst. However, the yields of gases, bio-oil, and biochar increased to 56.7 wt.%, 10.9 wt.%, and 32.4 wt.%, respectively, in the presence of calcined dolomite catalyst. Pyrolysis in the presence of zeolite catalysts resulted in 25 wt.% gases, 36.4 wt.% bio-oil, and 38.7 wt.% biochar (Tursunov, 2014).

Microwave-assisted pyrolysis is an emerging technology that uses dielectric heating via microwaves. In microwave pyrolysis, heating is carried out due to a change in the arrangement of electric dipoles in the materials being heated (Huang et al., 2016). The main advantage of microwave heating is that it offers a uniform distribution and steady, intense, and relatively cheaper source of heating (Osepchuk, 2002). Microwave pyrolysis is gaining attention due to its heating efficiency. However, some challenges associated with this technology that needs to be addressed are controlling dielectric properties of feedstock, maintaining an inert atmosphere, and avoiding the combustion of pyrolysis products (Bu et al., 2016).

Food waste collected from a residential quarter in Guangzhou, China, containing 32.7 wt.% white rice, 44.2 wt.% vegetable leaves, and 23.1 wt.% meat/bones was predried to 105°C for 24 h before microwave pyrolysis in a 2450 MHz microwave reactor at 300–600 W microwave power (Liu et al., 2014). Microwave pyrolysis at 300, 400, 500, and 600 W microwave power produced bio-oil with production energy of 13.9, 165.8, 148.7, and 125.9 kJ, respectively.

3.4.5 Transesterification

MSW contains a significant amount of food waste. Food waste contains oil from vegetables, waste cooking oil and grease, and oilseeds and fats from animal products that can be used for transesterification to produce biodiesel. Food waste can be diverted into a more sustainable way to produce bioenergy. Urban biowaste such as waste cooking oil, animal waste fats from slaughterhouses, and grease typically accumulate in the garbage bins of restaurants and households, which can be potential raw materials for transesterification reaction (Canakci, 2007; Wang et al., 2008; Park et al., 2010; Nanda et al., 2019).

Transesterification is a chemical reaction that occurs in the presence of fatty acid, alcohol, and a catalyst (Reddy et al., 2016). The catalyst can be either a strong acid or base. This process is also termed as alcoholysis. In this process, organic alkyl groups of vegetable or plant oils are replaced by methyl or ethyl group from methanol or ethanol, respectively. Transesterification aims to decrease the viscosity of animal fat or vegetable oil to make it suitable for diesel-based engines. The typical transesterification reaction is represented as shown in Eq. (3.2):

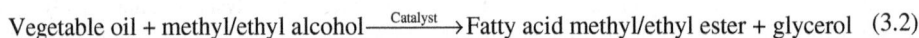

$$\text{Vegetable oil} + \text{methyl/ethyl alcohol} \xrightarrow{\text{Catalyst}} \text{Fatty acid methyl/ethyl ester} + \text{glycerol} \quad (3.2)$$

The end products of transesterification reaction are crude biodiesel, glycerol, and waste alkaline effluents (Reddy et al., 2016). Biodiesel is a yellowish liquid product of transesterification reaction having an energy potential of 38–45 MJ/kg (Lohri et al., 2017; Guo et al., 2015). The purified glycerol is potentially applicable in oleochemistry, cosmetics, pharmaceuticals, and food industries. Glycerol can be used as animal feed (Yang et al., 2012), the biochemical precursor (Li et al., 2013), a substrate for the anaerobic digestion (Larsen et al., 2013; Hutnan et al., 2013), feedstock for ethanol production (Liu et al., 2012), or electricity generation in microbial fuel cells (Reiche and Kirkwood, 2012).

Waste cooking oil can be considered as a potential raw material for biodiesel production since it is several magnitudes cheaper than virgin vegetable oil. On the other hand, waste cooking oil comes with a few disadvantages, for instance, high free fatty acid and high-moisture content. To lower the free fatty acid content, chemical pretreatments are required that subsequently increase the cost of biodiesel production (Yaakob et al., 2013). Several researchers have used many homogeneous or heterogeneous catalysts for transesterification reaction to produce biodiesel. Table 3.2 summarizes a few such notable studies on the transesterification of waste cooking oil using different catalysts.

3.4.6 Combustion

Combustion or incineration is a thermally driven process where solid fuels are burnt in the presence of air or oxygen to generate heat energy that can be further converted to electricity. It is an alternative pathway to convert chemical energy stored in organic wastes into heat and electricity. Combustion is carried out in huge furnaces, boilers, steam turbines, and turbo-generators. The typical temperature range for the combustion process to produce hot gases is around 800–1000°C or higher. Most of the MSW diverted

TABLE 3.2

Transesterification of Waste Cooking Oil

Catalyst	Optimum Conditions	Ester Yield and Conversion (%)	Reference
0.5 wt.% NaOH	Sonic dismembrator with 500W and 20 kHz; methanol/oil ratio of 9:1	Yield: 90%	Gude and Grant (2013)
1 wt.% NaOH	A dry reaction flask with reflux condenser; methanol/oil ratio of 9:1; 90 min; 50°C	Conversion: 85.8%	Meng et al. (2008)
10 wt.% Cao-ZrO_2 (Ca:Zr molar ratio of 0.5)	Methanol/oil ratio of 30:1; 2 h; 65°C	Biodiesel yield: 92.1%	Dehkordi and Ghasemi (2012)
2.7 wt.% Sr/ZrO_2	Methanol/oil ratio of 29:1; 2.8 h; 115.5°C	Yield: 79.7%	Omar and Amin (2011)
3 wt.% KBr impregnated CaO	Methanol/oil ratio of 12:1; 1.8 h; 65°C	Yield: 78.9%	Mahesh et al. (2015)
5 wt.% calcined (900°C, 2 h), CaO sourced from waste mud crab shells and cockleshells (1:1 mixed mass ratio)	Methanol/oil ratio of 13:1; 3 h; refluxing temperature	Yield: 98%	Boey et al. (2012)

FIGURE 3.3 Incineration route of MSW to energy.

from landfills is incinerated to clear the space required at the dumping sites and generate energy. It is by far the most popular waste-to-energy technology practiced worldwide. Incineration allows a huge reduction of the volume (about 90 vol%) and mass (about 70 wt.%) of MSW (Saikia et al., 2007).

Efficient combustion of organic compounds produces CO_2 and water vapor, but if the feedstock is contaminated with chlorine and sulfur and nitrogen-containing compounds, then there is a chance to produce dioxins and other toxic gases such as SO_x and NO_x. MSW is hazardous due to the containment of plastics, demolition waste, metals, oils, paints, cleaners, pesticides, batteries, etc. Hence, it cannot be directly subjected to combustion. As a prior process, hazardous components and noncombustible materials are removed. The explosive materials such as batteries and e-wastes are also separated from the MSW stream before incineration. MSW to energy is typically implemented via the following route as shown in Figure 3.3.

The major advantage of MSW combustion or incineration is the destruction of biohazardous microorganisms and biomedical wastes along with the conversion of carbonaceous materials into less harmful products and bioenergy (Brunner and Rechberger, 2015; Kumar and Samadder, 2017). Combustion is viable for low moisture-containing nonbiodegradable and combustible MSW components (Tan et al., 2014). The average heating value of MSW must be a minimum of 1700–1900 kcal/kg for efficient combustion with energy recovery option (Melikoglu, 2013; Kumar and Samadder, 2017). One ton of MSW could produce an average of 544 kWh of energy and 180 kg of solid residuals upon combustion (Zaman, 2010; Kumar and Samadder, 2017). Direct combustion is not only a widely accepted practice to handle MSW, but it is also one of the most polluting approaches (Mittal et al., 2017). Direct combustion of 1000 kg of MSW could generate emissions equivalent to 1090 kg CO_2 (Botello-Alvareza et al., 2018).

Two main solid by-products of MSW incineration are bottom ash and fly ash of which nearly 80% account for the former. MSW incineration bottom ash can be considered as an important precursor in the cement industry. Concrete blocks produced from cement aggregates could completely be replaced with MSW bottom ash (Silva et al., 2017). Saffarzadeh et al. (2016) suggested that bottom ash residues generated from incinerators could help characterize metal aluminum and aluminum alloy with potential use in hydrogen production.

3.5 Biological Conversion of Municipal Solid Waste

3.5.1 Anaerobic Digestion

Anaerobic digestion is a mature technique that biochemically degrades high-moisture containing solid and slurry-based organic wastes through microbial activity under anaerobic conditions. Anaerobic digestion consists of sequential chemical reactions that are catalyzed by various microorganisms to convert carbonaceous materials into biogas (predominantly CH_4) and a slurry with a high organic load. Long-chain polymers (e.g., cellulose, hemicellulose, pectin, and starch) are hydrolyzed and converted into smaller units like oligomers and monomers for further metabolism by fermentative methanogenic bacteria to produce biogas and volatile organics like acetate, propionate, and butyrate (Hutnan et al., 2013). The volatile organic acids are converted to methanogenic precursors (e.g., H_2, CO_2, and acetate) by syntrophic acetogens except for acetates. Methanogenic bacteria produce methane from acetate and other methanogenic precursors.

Apart from direct composting or animal feeding, anaerobic digestion is a promising way to degrade organic fraction to produce biogas. Since this process mainly produces methane, it is also called biomethanation. For many decades, the significant applications of anaerobic digestion have been found to valorize the organic fraction of MSW, agriculture residues, cattle manure, and sewage sludge (Jimenez et al., 2015). In rural areas, biogas is used for cooking that provides the advantage of reducing the usage of fossil fuel and firewood. Consequently, the reliance on renewable sources for energy production could increase while mitigating the air pollution caused by the burning of raw waste biomass and fossil fuels. The slurry obtained as a co-product of anaerobic digestion can be used as compost in agriculture to enhance soil fertility. The yield and composition of biomass depend on the biomass type, process condition of anaerobic digestion, and microorganisms used. Biomass typically consists of 50–75 vol% CH_4, 25–40 vol% CO_2, and 1–15 vol% of other gases such as water vapor, ammonia, and hydrogen sulfide. (Surendra et al., 2014; Kumar and Samadder, 2017). Biogas has a typical LHV of 21–24 MJ/m³ mainly attributed to CH_4 content (Bond and Templeton, 2011; Lohri et al., 2017). The overview of anaerobic digestion technology applied to MSW can be given by the flowchart in given Figure 3.4.

A wide range of feedstocks such as sewage waste, animal manure, food waste, agricultural crop residues, kitchen waste, and organic fraction of MSW can be used for the anaerobic digestion (Romero-Guiza et al., 2016). Anaerobic digestion is typically favorable in a moist environment even if the feedstock contains greater than 60% moisture (Appels et al., 2011). Raw materials with high lignin fraction such as woody biomass are not appropriate for anaerobic digestion because microorganisms are inefficient in their biological degradation. The biodegradable substances degrade through sequential processes of

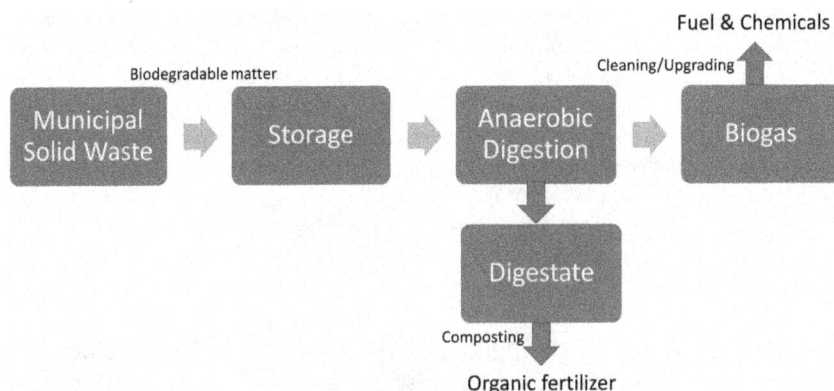

FIGURE 3.4 Representation of anaerobic digestion of MSW.

hydrolysis followed by acetogenesis and methanogenesis to produce CH_4 and CO_2. The anaerobic digestion technique can be further differentiated by the operating temperature of the reactor, microorganisms used (i.e., thermophilic or mesophilic), total solids present (i.e., low or high solid concentration), mode of feeding (i.e., batch, fed-batch, or continuous), and processing steps (i.e., single or multi-step process) (Hartmann and Ahring, 2006; Kothari et al., 2014; Mao et al., 2015; Lohri et al., 2017).

Some notable advantages of anaerobic digestion of MSW are proper handling of waste materials, ecological security by the sanitation of pathogenic organisms, prevention of air pollution, mitigating the breeding of disease vectors due to anaerobic conditions, and efficient utilization of CH_4 as a greenhouse gas (Mao et al., 2015). MSW contains highly heterogeneous, toxic, and nonbiodegradable materials, which require proper separation to prevent inhibition of methanogenic bacteria. Process optimization and energy density of biogas are other significant challenges for anaerobic digestion (Fan et al., 2018).

The average CH_4 yield from carbonaceous solid waste is between 0.36 and 0.56 m^3/kg of volatile solids (Bouallagui et al., 2005; Khalid et al., 2011; Lohri et al., 2017). A feedstock containing food waste and straw was investigated for anaerobic digestion in a 1 L batch reactor to produce CH_4 at 0.39 m^3/kg of volatile solids under mesophilic temperatures, that is, 35°C (Yong et al., 2015).

3.5.2 Fermentation

Fermentation is mainly driven by enzymes that break the polysaccharide sugar component of waste biomass into monomeric components through a series of chemical reactions assisted by yeast, bacteria, or their enzymes. Fermentation is a biological technique used to convert starch and sugar-based feedstocks to bioethanol, biobutanol, biohydrogen, and other biofuels and biochemicals (Nanda et al., 2014b; Nanda et al., 2017a; Sarangi and Nanda, 2020). Fermentation is different from anaerobic digestion as its main products are organic acids and alcohols as compared to biogas in anaerobic digestion. Moreover, anaerobic digestion is performed under an oxygen-free environment, whereas fermentation can be performed under both aerobic and anaerobic conditions.

Ethanol can be prepared through fermentative consumption of pyruvate from glucose (Kang and Lee, 2015). Glycolysis is a process that converts glucose components to partially oxidizable products (e.g., pyruvate) through microbial metabolic activities. Pyruvate undergoes fermentation to produce ethanol by subsequent reactions of pyruvate-decarboxylase followed by alcohol-dehydrogenase (Kang and Lee, 2015). Ethanol fermentation is thoroughly investigated using Saccharomyces cerevisiae and Escherichia coli (Bai et al., 2008; Geddes et al., 2011; Kang and Lee, 2015). Other microorganisms have also been investigated due to the merits of their native enzymes and pathways as a production host such as Zymomonas mobilis. Z. mobilis is considered a substitute host to yeast because of its ethanol production yield and pathway for glycolysis (Sprenger, 1996; Kang and Lee, 2015).

Fermentation is a crucial step in the production of bioethanol, which is one of the emerging alcohol-based biofuels in the world. Bioethanol can be used in a blend with gasoline with the most popular ones being E85 (85% bioethanol and 15% gasoline), E20 (20% bioethanol and 80% gasoline), and E10 (10% bioethanol and 90% gasoline) (Nanda et al., 2014c). Ethanol is also applicable in the transesterification reaction of cooking oil to produce biodiesel (Sarris and Papanikolaou, 2016).

3.6 Conclusions

The generation of MSW is considered a global issue for solid waste management. As a potential solution, landfilling or direct incineration is being widely practiced to manage MSW. A larger portion of MSW is organic waste, which can be converted to biofuels through different waste-to-energy technologies. A major challenge of these alternative techniques is the requirement of biomass pretreatment and specific feedstock properties. Therefore, it can be deduced that most of the organic biodegradable MSW is suitable for biological waste-to-energy processes such as anaerobic digestion and fermentation. On the other hand, recalcitrant organic components in MSW (e.g., plastics, waste paper, and woody biomass) can be processed through thermochemical processes such as liquefaction, gasification, and pyrolysis. The remaining inert waste components of MSW end up in landfills.

Each technology has its merits and drawbacks, but they are capable of producing specific biofuels to support sustainable waste management. Bioenergy and biofuels can be used for domestic and industrial sectors as well to supplement the worldwide energy demand. Therefore, the dependency on nonrenewable conventional fossil fuels can be greatly reduced while simultaneously lowering greenhouse gas emissions. To make waste-to-energy technologies efficient and more sustainable, integrated or hybrid technologies must be implemented, but at the same time, it should be economical and commercially viable.

Acknowledgments

The financial support from the Natural Sciences and Engineering Research Council of Canada (NSERC), Canada Research Chairs (CRC) program, and Agriculture and Agri-Food Canada (AAFC) is greatly acknowledged.

REFERENCES

Alberta Recycling. 2017. Tire recycling program. www.albertarecycling.ca/tire-recycling-program/eligible-tires (accessed 19 February 2018)

Appels, L., Lauwers, J., Degreve, J., Helsen, L., Lievens, B., Willems, K., Van Impe, J., and Dewil, R. 2011. Anaerobic digestion in global bio-energy production: Potential and research challenges. *Renewable Sustainable Energy Reviews* 15: 4295–4301.

Arena, U. 2011. Editorial. Gasification: An alternative solution for waste treatment with energy recovery. *Waste Management* 31: 405–406.

Arena, U. 2012. Process and technological aspects of municipal solid waste gasification. A review. *Waste Management* 32: 625–639.

Arias, B., Pevida, C., Fermoso, J., Plaza, M., Rubiera, F., and Pis, J. 2008. Influence of torrefaction on the grindability and reactivity of woody biomass. *Fuel Processing Technology* 89: 169–175.

Arturi, K. R., Toft, K. R., Nielsen, R. P., Rosendahl, L. A., and Søgaard, E. G. 2016. Characterization of liquid products from hydrothermal liquefaction (HTL) of biomass via solid phase microextraction (SPME). *Biomass and Bioenergy* 88: 116–125.

Azargohar, R., Nanda, S., Dalai, A. K., and Kozinski, J. A. 2019a. Physico-chemistry of biochars produced through steam gasification and hydro-thermal gasification of canola hull and canola meal pellets. *Biomass & Bioenergy* 120: 458–470.

Azargohar, R., Nanda, S., Kang, K., Bond, T., Karunakaran, C., Dalai, A. K., and Kozinski, J. A. 2019b. Effects of bio-additives on the physicochemical properties and mechanical behavior of canola hull fuel pellets. *Renewable Energy* 132: 296–307.

Bai, F. W., Anderson, W. A., and Moo-Young, M. 2008. Ethanol fermentation technologies from sugar and starch feedstocks. *Biotechnology Advances* 26: 89–105.

Boey, P.-L., Ganesan, S., Maniam, G. P., and Khairuddean, M. 2012. Catalysts derived from waste sources in the production of biodiesel using waste cooking oil. *Catalysis Today* 190: 117–121.

Bohdziewicz, J., Neczaj, E., and Kwarciak, A. 2008. Landfill leachate treatment by means of anaerobic membrane bioreactor. *Desalination* 221: 559–565.

Bond, T., and Templeton, M. R. 2011. History and future of domestic biogas plants in the developing world. *Energy for Sustainable Development* 15: 347–354.

Botello-Alvareza, J., Rivas-Garciab, P., Fausto-Castro, L., Estrada-Baltazar, A., and Gomez-Gonzalez, R. 2018. Informal collection, recycling and export of valuable waste as transcendent factor in the municipal solid waste management: A Latin-American reality. *Journal of Cleaner Production* 182: 485–495.

Bouallagui, H., Touhami, Y., Ben Cheikh, R., and Hamdi, M. 2005. Bioreactor performance in anaerobic digestion of fruit and vegetable wastes. *Process Biochemistry* 40: 989–995.

Boucher, M. E., Chaala, A., and Roy, C. 2000. Bio-oils obtained by vacuum pyrolysis of softwood bark as a liquid fuel for gas turbines. Part I: Properties of bio-oil and its blends with methanol and a pyrolytic aqueous phase. *Biomass and Bioenergy* 19: 337–350.

British Columbia Ministry of Environment. 2016. *A Guide to Solid Waste Management Planning.* Version 1. Government of British Columbia, Canada.

Brunner, P., and Rechberger, H. 2015. Waste to energy – key element for sustainable waste management. *Waste Management* 37: 3–12.

Bu, Q., Morgan Jr., H., Liang, J., Lei, H., and Ruan, R. 2016. Catalytic microwave pyrolysis of lignocellulosic biomass for fuels and chemicals. *Advances in Bioenergy* 1: 69–123.

Canakci, M. 2007. The potential of restaurant waste lipids as biodiesel feedstocks. *Bioresource Technology* 98: 183–190.

Ciolkosz, D., and Wallace, R. 2011. A review of torrefaction for bioenergy feedstock production. *Biofuels, Biofuels Bioproducts & Biorefining* 5: 317–329.

Dehkordi, A. M., and Ghasemi, M. 2012. Transesterification of waste cooking oil to biodiesel using Ca and Zr mixed oxides as heterogeneous base catalysts. *Fuel Processing Technology* 97: 45–51.

Devi, L., Ptasinski, K. J., and Janssen, F. J. J. G. 2003. A review of the primary measures for tar elimination in biomass gasification processes. *Biomass and Bioenergy* 24: 125–140.

Fan, Y., Klemes, J., Lee, C., and Perry, S. 2018. Anaerobic digestion of municipal solid waste: Energy and carbon emission footprint. *Journal of Environmental Management* 223: 888–897.

Geddes, C. C., Nieves, I. U., and Ingram, L. O. 2011. Advances in ethanol production. *Current Opinion in Biotechnology* 22: 312–319.

Gollakota, A., Kishore, N., and Gu, S. 2018. A review on hydrothermal liquefaction of biomass. *Renewable and Sustainable Energy Reviews* 81: 1378–1392.

Gong, M., Nanda, S., Romero, M. J., Zhu, W., and Kozinski, J. A. 2017. Subcritical and supercritical water gasification of humic acid as a model compound of humic substances in sewage sludge. *The Journal of Supercritical Fluids* 119: 130–138.

Gray, J. 2017. Pay-as-you-throw' pegged at $62. *The Globe and Mail*. www.theglobeandmail.com/news/national/pay-as-you-throw-pegged-at-62/article686153/ (accessed 15 February 2018)

Gu, Q., Wu, W., Jin, B., and Zhou, Z. 2020. Analyses for synthesis gas from municipal solid waste gasification under medium temperatures. *Processes* 8: 84.

Gude, V. G., and Grant, G. E. 2013. Biodiesel from waste cooking oils via direct sonication. *Applied Energy* 109: 135–144.

Guo, M., Song, W., and Buhain, J. 2015. Bioenergy and biofuels: History, status, and perspective. *Renewable and Sustainable Energy Reviews* 42: 712–725.

Hartmann, H., and Ahring, B. K. 2006. Strategies for the anaerobic digestion of the organic fraction of municipal solid waste: An overview. *Water Science & Technology* 53: 7–22.

Heidenreich, S., and Foscolo, P. 2015. New concepts in biomass gasification. *Progress in Energy and Combustion Science* 46: 72–95.

Huang, Y. F., Chiueh, P. T., and Lo, S. L. 2016. A review on microwave pyrolysis of lignocellulosic biomass. *Sustainable Environment Research* 26: 103–109.

Hutnan, M., Kolesarova, N., and Bodík, I. 2013. Anaerobic digestion of crude glycerol as sole substrate in mixed reactor. *Environmental Technology* 34: 2179–2187.

Iroba, K., Baik, O., and Tabil, L. 2017. Torrefaction of biomass from municipal solid waste fractions II: Grindability characteristics, higher heating value, pelletability and moisture adsorption, *Biomass and Bioenergy* 106: 8–20.

Jimenez, J., Latrille, E., Harmand, J., Robles, A., Ferrer, J., Gaida, D., Wolf, C., Mairet, F., Bernard, O., Alcaraz-Gonzalez, V., Mendez-Acosta, H., Zitomer, D., Totzke, D., Spanjers, H., Jacobi, F., Guwy, A., Dinsdale, R., Premier, G., Mazhegrane, S., Ruiz-Filippi, G., Seco, A., Ribeiro, T., Pauss, A., and Steyer, J. P. 2015. Instrumentation and control of anaerobic digestion processes: A review and some research challenges. *Reviews in Environmental Science and Bio/Technology* 14: 615–648.

Kan, T., Strezov, V., and Evans, T. J. 2016. Lignocellulosic biomass pyrolysis: A review of product properties and effects of pyrolysis parameters. *Renewable and Sustainable Energy Reviews* 57: 1126–1140.

Kang, A., and Lee, T. S. 2015. Review: Converting sugars to biofuels: Ethanol and beyond. *Bioengineering* 2: 184–203.

Kelleher, M., Robins, J., and Dixie, J. 2000. *Taking Out the Trash: How to Allocate the Costs Fairly*. C.D. Howe Institute Commentary, Toronto, Canada.

Kettunen, R. H., Hoilijoki, T. H., and Rintala, J. A. 2009. Anaerobic and sequential anaerobic-aerobic treatments of municipal landfill leachate at low temperatures. *Bioresource Technology* 58: 40–41.

Khalid, A., Arshad, M., Anjum, M., Mahmood, T., and Dawson, L. 2011. The anaerobic digestion of solid organic waste. *Waste Management* 31: 1737–1744.

Kothari, R., Pandey, A. K., Kumar, S., Tyagi, V. V., and Tyagi, S. K. 2014. Different aspects of dry anaerobic digestion for bio-energy: An overview. *Renewable and Sustainable Energy Reviews* 39: 174–195.

Kumar, A., and Samadder, S. 2017. A review on technological options of waste to energy for effective management of municipal solid waste. *Waste Management* 69: 407–422.

Kumar, S., Chiemchaisri, C., and Mudhoo, A. 2011. Bioreactor landfill technology in municipal solid waste treatment: An overview. *Critical Reviews in Biotechnology* 31: 77–97.

Larsen, A. C., Gomes, B. M., Gomes, S. D., Zenatti, D. C., and Torres, D. G. B. 2013. Anaerobic co-digestion of crude glycerin and starch industry effluent. *Engenharia Agricola* 33:341–352.

Li, C., Lesnik, K., and Liu, H. 2013. Microbial conversion of waste glycerol from biodiesel production into value-added products. *Energies* 6:4739–4768.

Liu, H., Ma, X., Li, L., Hu, Z., Guo, P., and Jiang, Y. 2014. The catalytic pyrolysis of food waste by microwave heating. *Bioresource Technology* 166:45–50.

Liu, X., Jensen, P. R., and Workman, M. 2012. Bioconversion of crude glycerol feedstocks into ethanol by *Pachysolen tannophilus*. *Bioresource Technology* 104: 579–586.

Lohri, C. R., Diener, S., Zabaleta, T., Mertenat, A., and Zurbrugg, C. 2017. Treatment technologies for urban solid biowaste to create value products: A review with focus on low- and middle-income settings. *Reviews in Environmental Science and Bio/Technology* 16: 81–130.

Mahesh, S. E., Ramanathan, A., Begum, K. M. M. S., and Narayanan, A. 2015. Biodiesel production from waste cooking oil using KBr impregnated CaO as catalyst. *Energy Conversion and Management* 91: 442–450.

Mani, S., Tabil, L., and Sokhansanj, S. 2003. Compaction of biomass grinds-an overview of compaction of biomass grinds. *Powder Handling & Processing* 15: 160–168.

Mao, C., Feng, Y., Wang, X., and Ren, G. 2015. Review on research achievements of biogas from anaerobic digestion. *Renewable and Sustainable Energy Reviews* 45: 540–555.

Melikoglu, M. 2013. Vision 2023: Assessing the feasibility of electricity and biogas production from municipal solid waste in Turkey. *Renewable Sustainable Energy Reviews* 19: 52–63.

Meng, X., Chen, G., and Wang, Y. 2008. Biodiesel production from waste cooking oil via alkali catalyst and its engine test. *Fuel Processing Technology* 89: 851–857.

Mittal, S., Pathak, M., Shukla, P., and Ahlgren, E. 2017. GHG mitigation sustainability co-benefits of urban solid waste management strategies: A case study of Ahmedabad, India. *Chemical Engineering Transactions* 56: 457–462.

Mohan, D., Pittman, C. U., and Steele, P. H. 2006. Pyrolysis of wood/biomass for bio-oil: A critical review. *Energy Fuels* 20: 848–889.

Mohanty, P., Nanda, S., Pant, K. K., Naik, S., Kozinski, J. A., and Dalai, A. K. 2013. Evaluation of the physio-chemical development of biochars obtained from pyrolysis of wheat straw, timothy grass and pinewood: Effects of heating rate. *Journal of Analytical and Applied Pyrolysis* 104: 485–493.

Nanda, S., Azargohar, R., Dalai, A. K., and Kozinski, J. A. 2015. An assessment on the sustainability of lignocellulosic biomass for biorefining. *Renewable and Sustainable Energy Reviews* 50: 925–941.

Nanda, S., Azargohar, R., Kozinski, J. A., and Dalai, A. K. 2014a. Characteristic studies on the pyrolysis products from hydrolyzed Canadian lignocellulosic feedstocks. *Bioenergy Research* 7: 174–191.

Nanda, S., and Berruti, F. 2021a. A technical review of bioenergy and resource recovery from municipal solid waste. *Journal of Hazardous Materials* 403: 123970.

Nanda, S., Dalai, A. K., Berruti, F., and Kozinski, J. A. 2016a. Biochar as an exceptional bioresource for energy, agronomy, carbon sequestration, activated carbon and specialty materials. *Waste and Biomass Valorization* 7: 201–235

Nanda, S., Dalai, A. K., and Kozinski, J. A. 2014b. Butanol and ethanol production from lignocellulosic feedstock: Biomass pretreatment and bioconversion. *Energy Science & Engineering* 2: 138–148.

Nanda, S., Golemi-Kotra, D., McDermott, J. C., Dalai, A. K., Gökalp, I., and Kozinski, J. A. 2017a. Fermentative production of butanol: Perspectives on synthetic biology. *New Biotechnology* 37: 210–221.

Nanda, S., Gong, M., Hunter, H. N., Dalai, A. K., Gökalp, I., and Kozinski, J. A. 2017b. An assessment of pinecone gasification in subcritical, near-critical and supercritical water. *Fuel Processing Technology* 168: 84–96.

Nanda, S., Isen, J., Dalai, A. J., and Kozinski, J. A. 2016b. Gasification of fruit wastes and agro-food residues in supercritical water. *Energy Conversion and Management* 110: 296–306.

Nanda, S., Mohammad, J., Reddy, S. N., Kozinski, J. A., and Dalai, A. K. 2014c. Pathways of lignocellulosic biomass conversion to renewable fuels. *Biomass Conversion and Biorefinery* 4: 157–191.

Nanda, S., Rana, R., Hunter, H. N., Fang, Z., Dalai, A. K., and Kozinski, J. A. 2019. Hydrothermal catalytic processing of waste cooking oil for hydrogen-rich syngas production. *Chemical Engineering Science* 195: 935–945.

Nanda, S., Reddy, S. N., Mitra, S. K., and Kozinski, J. A. 2016c. The progressive routes for carbon capture and sequestration. *Energy Science & Engineering* 4: 99–122.

Okolie, J. A., Nanda, S., Dalai, A. K., Berruti, F., and Kozinski, J. A. 2020. A review on subcritical and super-critical water gasification of biogenic, polymeric and petroleum wastes to hydrogen-rich synthesis gas. *Renewable and Sustainable Energy Reviews* 119: 109546.

Okolie, J. A., Nanda, S., Dalai, A. K., and Kozinski, J. A. 2021. Chemistry and specialty industrial applications of lignocellulosic biomass. *Waste and Biomass Valorization* 12: 2145–2169.

Okolie, J. A., Rana, R., Nanda, S., Dalai, A. K., and Kozinski, J. A. 2019. Supercritical water gasification of bio-mass: A state-of-the-art review of process parameters, reaction mechanisms and catalysis. *Sustainable Energy & Fuels* 3: 578–598.

Omar, H., and Rohani, S. 2015. Treatment of landfill waste, leachate and landfill gas: A review. *Frontiers of Chemical Science and Engineering* 9: 15–32.

Omar, W. N. N. W., and Amin, N. A. S. 2011. Biodiesel production from waste cooking oil over alkaline modi-fied zirconia catalyst. *Fuel Processing Technology* 92: 2397–2405.

Ontario Waste Management Association. 2016. *State of Waste in Ontario: Landfill Report.* First Annual Report.

Osepchuk, J. M. 2002. Microwave power applications. *IEEE Transactions on Microwave Theory and Techniques* 50: 975–985.

Pahla, G., Ntuli, F., and Muzenda, E. 2018. Torrefaction of landfill food waste for possible application in bio-mass co-firing. *Waste Management* 71: 512–520.

Parakh, P. D., Nanda, S., and Kozinski, J. A. 2020. Eco-friendly transformation of waste biomass to biofuels. *Current Biochemical Engineering* 6: 120–134.

Park, J.-Y., Lee, J.-S., Wang, Z.-M., and Kim, D.-K. 2010. Production and characterization of biodiesel from trap grease. *Korean Journal of Chemical Engineering* 27: 1791–1795.

Raghab, S. M., Abd El Meguid, A. M., and Hegazi, H. A. 2013. Treatment of leachate from municipal solid waste landfill. *HBRC Journal* 9: 187–192.

Rasanjani, C., Gunathilaka, T., Pieris., Bandara, H., and Narayana, M. 2019. Torrefaction of urban bio waste in Sri Lanka. *Moratuwa Engineering Research Conference (MERCon),* 18957817.

Reddy, S. N., Nanda, S., Dalai, A. K., and Kozinski, J. A. 2014. Supercritical water gasification of biomass for hydrogen production. *International Journal of Hydrogen Energy* 39: 6912–6926.

Reddy, S. N., Nanda, S., and Kozinski, J. A. 2016. Supercritical water gasification of glycerol and methanol mixtures as model waste residues from biodiesel refinery. *Chemical Engineering Research and Design* 113: 17–27.

Rehan, M., Miandad, R., Barakat, M., Ismail, I., Almeelbi, T., Gardy, J., Hassanpour, A., Khan, M., Demirbas, A., and Nizami, A. 2017. Effect of zeolite catalysts on pyrolysis liquid oil. *International Biodeterioration & Biodegradation* 119: 162–175.

Reiche, A., and Kirkwood, K. M. 2012. Comparison of *Escherichia coli* and anaerobic consortia derived from compost as anodic biocatalysts in a glycerol-oxidizing microbial fuel cell. *Bioresource Technology* 123: 318–323.

Rethink Tires. 2017. Tire Stewardship fee (TSF) chart. http://rethinktires.ca/program-participants/stewards/tsf-fee-chart/#sthash.NP920w0K.dpbs (accessed 19 February 2018)

Romero-Guiza, M. S., Vila, J., Mata-Alvarez, J., Chimenos, J. M., and Astals, S. 2016. The role of additives on anaerobic digestion: A review. *Renewable Sustainable Energy Reviews* 58: 1486–1499.

Sadaka, S., and Negi, S. 2009. Improvements of biomass physical and thermochemical characteristics via tor-refaction process. *Environmental Progress & Sustainable Energy* 28: 427–434.

Saffarzadeh, A., Arumugam, N., and Shimaoka, T. 2016. Aluminum and aluminum alloys in municipal solid waste incineration (MSWI) bottom ash: A potential source for the production of hydrogen gas. *International Journal of Hydrogen Energy* 41: 820–831.

Saikia, N., Kato, S., and Kojima, T. 2007. Production of cement clinkers from municipal solid waste incinera-tion (MSWI) fly ash. *Waste Management* 27: 1178–1189.

Samad, N., Jamin, N., and Saleh, S. 2017. Torrefaction of municipal solid waste in Malaysia. *Energy Procedia* 138: 313–318.

Sanlisoy, A., and Carpinlioglu, M. 2017. A review on plasma gasification for solid waste disposal. *International Journal of Hydrogen Energy* 42: 1361–1365.

Sarangi, P. K., and Nanda, S. 2020. Biohydrogen production through dark fermentation. *Chemical Engineering & Technology* 43: 601–612.

Sarris, D., and Papanikolaou, S. 2016. Biotechnological production of ethanol: Biochemistry, processes and technologies. *Engineering in Life Sciences* 16: 307–329.

Sharma, B. K., Moser, B. R., Vermillion, K. E., Doll, K. M., and Rajagopalan, N. 2014. Production, characterization and fuel properties of alternative diesel fuel from pyrolysis of waste plastic grocery bags. *Fuel Processing Technology* 122: 79–90.

Shie, J., Chen, L., Lin, K., and Chang, C. 2014. Plasmatron gasification of biomass lignocellulosic waste materials derived from municipal solid waste. *Energy* 64: 82–89.

Sikarwar, V., Zhao, M., Clough, P., Yao, J., Zhong, X., Memon, M., Shah, N., Anthony, E., and Fennell, P. 2016. An overview of advances in biomass gasification. *Energy & Environmental Science* 9: 2939–2977.

Silva, R. V., Brito, J., Lynn, C. J., and Dhir, R. K. 2017. Use of municipal solid waste incineration bottom ashes in alkali-activated materials, ceramics and granular applications: A review. *Waste Management* 68: 207–220.

Sprenger, G. 1996. Carbohydrate metabolism in *Zymomonas mobilis*: A catabolic highway with some scenic routes. *FEMS Microbiology Letters* 145: 301–307.

Statistics Canada. 2012. *Human Activity and the Environment: Waste management in Canada*. Ottawa, Canada: Environment Accounts and Statistics Division. Minister of Industry.

Statistics Canada. 2020a. Disposal of waste, by source. Table: 38-10-0032-01. https://www150.statcan.gc.ca/t1/tbl1/en/tv.action?pid=3810003201 (accessed 22 April 2020)

Statistics Canada. 2020b. Local government characteristics of the waste management industry. Table: 38-10-0036-01. https://www150.statcan.gc.ca/t1/tbl1/en/tv.action?pid=3810003601 (accessed 22 April 2020)

Surendra, K., Takara, D., Hashimoto, A., and Khanal, S. 2014. Biogas as a sustainable energy source for developing countries: Opportunities and challenges. *Renewable Sustainable Energy Reviews* 31: 846–859.

Tan, S. T., Hashim, H., Lim, J. S., Ho, W. S., Lee, C. T., and Yan, J. 2014. Energy and emissions benefits of renewable energy derived from municipal solid waste: Analysis of a low carbon scenario in Malaysia. *Applied Energy* 136: 797–804.

The City of London. 2018a. Too good to waste. Waste reduction week. www.london.ca/residents/Garbage-Recycling/Recycling/Pages/Waste-Reduction-Week.aspx (accessed 20 February 2018)

The City of London. 2018b. Waste disposal fees. www.london.ca/residents/Garbage-Recycling/Garbage/Pages/Waste-Disposal-Fees.aspx (accessed 20 February 2018)

The City of Toronto. 2018. Solid waste management services. www.toronto.ca/city-government/accountability-operations-customer-service/city-administration/staff-directory-divisions-and-customer-service/solid-waste-management-services/ (accessed 20 February 2018)

The Source. 2018. Environmental handling fees. www.thesource.ca/en-ca/envHandlingFee (accessed 1 December 2020)

Tire Stewardship Manitoba. 2016. Steward-fees. www.tirestewardshipmb.ca/retailers/steward-fees (accessed 19 February 2018)

Toor, S. S., Rosendahl, L., and Rudolf, A. 2011. Hydrothermal liquefaction of biomass: A review of subcritical water technologies. *Energy* 36: 2328–2342.

Tursunov, O. 2014. A comparison of catalysts zeolite and calcined dolomite for gas production from pyrolysis of municipal solid waste (MSW). *Ecological Engineering* 69: 237–243.

Used Oil Management Associations of Canada. Environmental handling charges. http://usedoilrecycling.com/wordpress/wp-content/uploads/2016/09/EHC-Applicable-Product-List-2016-10-01.pdf (accessed 19 February 2018)

Venderbosch, R. H., and Prins, W. 2010. Fast pyrolysis technology development. *Biofuels, Bioproducts and Biorefining* 4: 178–208.

Wang, Z.-M., Lee, J.-S., Park, J.-Y., Wu, C.-Z., and Yuan, Z.-H. 2008. Optimization of biodiesel production from trap grease via acid catalysis. *Korean Journal of Chemical Engineering* 25: 670–674.

Worldwatch Institute. 2012. Global municipal solid waste continues to grow. www.worldwatch.org/global-municipal-solid-waste-continues-grow (accessed 31 January 2019)

Yaakob, Z., Mohammad, M., Alherbawi, M., Alam, Z., and Sopian, K. 2013. Overview of the production of biodiesel from waste cooking oil. *Renewable and Sustainable Energy Reviews* 18: 184–193.

Yang, F., Hanna, M. A., and Sun, R. 2012. Value-added uses for crude glycerol – a byproduct of biodiesel production. *Biotechnology for Biofuels* 5: 13.

Yong, Z., Dong, Y., Zhang, X., and Tan, T. 2015. Anaerobic co-digestion of food waste and straw for biogas production. *Renewable Energy* 78: 527–530.

Zaman, A. U. 2010. Comparative study of municipal solid waste treatment technologies using life cycle assessment method. *International Journal of Environmental Science Technology* 7: 225–234.

4

Conversion of Plastic Waste to Fuels and Chemicals

Ravi Patel, Sonil Nanda and Ajay K. Dalai

CONTENTS

4.1 Introduction

The development of plastic-based polymers is a significant invention that has upgraded the standard of living for human beings. Plastics are used widely in many applications such as transportation, health, aviation, engineering, and construction. Plastics have replaced the usage of metals and wood in several commercial and industrial products due to their specific characteristics of being lightweight, durable, having resistance to chemicals and corrosion, flexible, and having low production cost (Wong et al., 2015). There is a significant increase in plastic production of approximately 10% globally each year since 1950 (Singh and Ruj, 2016). Figure 4.1 illustrates the worldwide production of waste plastics over the years. The global manufacturing of plastics has dramatically increased in 2018 (359 million metric tons, MMT) when compared to the previous years (Statista, 2020). The data for 2010 (270 MMT), 1989 (100 MMT), and 1950 (1.5 MMT) can be seen in Figure 4.1.

Waste plastics are a significant portion of municipal solid waste (MSW) (Nanda and Berruti, 2021a, 2021b). According to the World Bank, more than 242 million tons (MT) of plastic waste were generated globally in 2016, which comprises a substantial portion (approximately 12%) of MSW (The World Bank, 2020). From the total plastic waste in the world, 35 MT was generated from North America, 45 MT from Europe and Central Asia, and 57 MT from East Asia and the Asia-Pacific region (The World Bank, 2020). The large-scale production of plastics imparts a substantial environmental burden because of the accumulation of plastic waste in landfills and oceans. Moreover, plastics are nonbiodegradable, which allows them to linger in landfills and water bodies for several hundred years. Therefore, more consistent and sustainable methods should be devised to manage plastic wastes.

Not all plastics are recyclable. Plastics that can be recycled are turned back in a closed-loop system and converted to other forms for reuse. However, nonrecyclable plastics are difficult to process and hence end up in landfills and thereby increase environmental constraints. As an alternative sustainable solution,

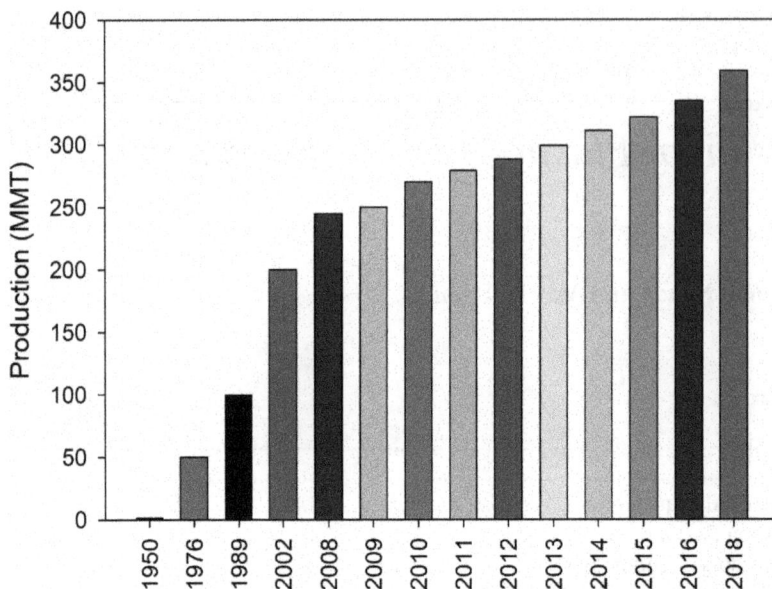

FIGURE 4.1   Global production of plastics (data source: Statista, 2020).

energy recovery through thermochemical pathways can be possible. The thermochemical waste-to-energy conversion technologies are gaining interest because of several advantages such as effective waste minimization, valorization, and sustainable energy generation (Nanda and Berruti, 2021c). High energy-dense products can be obtained by treating nonrecyclable mixed plastic waste under specific operating conditions. The Society of the Plastics Industry (SPI) has developed a resin-coding system to recognize different resins that are found in plastic materials to categorize and recycle plastics (Table 4.1).

Thermal and catalytic processes are used to convert waste plastic into fuels. Such waste-to-energy technologies are promising approaches to decrease the volume of waste plastic, which could otherwise create environmental problems and health hazards. This chapter reviews some thermochemical processes used for valorizing plastics such as pyrolysis, gasification, and liquefaction. Incineration of waste plastic is not recommended as it generates toxic emissions and greenhouse gases. The possible pathways of mixed plastic thermochemical treatment are shown in Figure 4.2.

4.2 Pyrolysis

Pyrolysis results in thermal degradation of long-chain complex carbonaceous compounds under an inert environment and controlled temperature, heating rate, and vapor residence time (Mohanty et al., 2013; Nanda et al., 2014a). Pyrolysis is favored over other technologies because it operates under an oxygen-free atmosphere. Therefore, dioxin formation from reaction products with oxygen can be eliminated. Depending on the reaction temperature, heating rate, and vapor residence time, pyrolysis of biomass can be classified into slow, fast, intermediate, and flash pyrolysis. While slow pyrolysis results in greater yields of biochar, fast and flash pyrolysis processes produce higher yields of bio-oil (Nanda et al., 2016a).

Pyrolysis can convert waste plastics and polymers into pyrolytic oil, char, and gases at relatively high temperatures. However, pyrolysis of polyethylene-type plastics such as low-density polyethylene (LDPE), high-density polyethylene (HDPE), and polypropylene (PP) could be challenging because of their cross-linked and long-chain hydrocarbon structures (Miandad et al., 2016; Achilias et al., 2017). In such a scenario, catalytic pyrolysis could help in eliminating some of such challenges. The differences between non-catalytic and catalytic pyrolysis are shown in Table 4.2.

TABLE 4.1

Description of Types of Polymers

Type of Polymers	Plastic Resin Code and Abbreviation	Recyclability	Percentage Recycled Annually	Commercial Usage
Polyethylene terephthalate	1 (PET)	Commonly recycled	36%	Soda bottles, rope, comb, medicine jars, tote bags, etc.
High-density polyethylene	2 (HDPE)	Commonly recycled	30–35%	Milk jugs, toys, soap and detergent bottles, buckets, crates, etc.
Polyvinyl chloride	3 (PVC)	Sometimes recycled	<1%	Plumbing pipes, credit cards, floor coverings, cooking oil bottles, etc.
Low-density polyethylene	4 (LDPE)	Sometimes recycled	6%	Plastic wraps, garbage bags, wire and cable coverings, etc.
Polypropylene	5 (PP)	Occasionally recycled	3%	Most bottle tops, drinking straws, car parts, hot food containers, heavy-duty bags, etc.
Polystyrene	6 (PS)	Difficult but commonly recycled	34%	Disposable foam cups, plastic cutlery, coat hangers, egg cartons, etc.
Other plastics (acrylic, fiberglass, and polycarbonate)	7 (others)	Difficult to recycle	Low	Baby bottles, eyeglasses, CDs and DVDs, dental sealants, etc.

Reference: Wallace (2020).

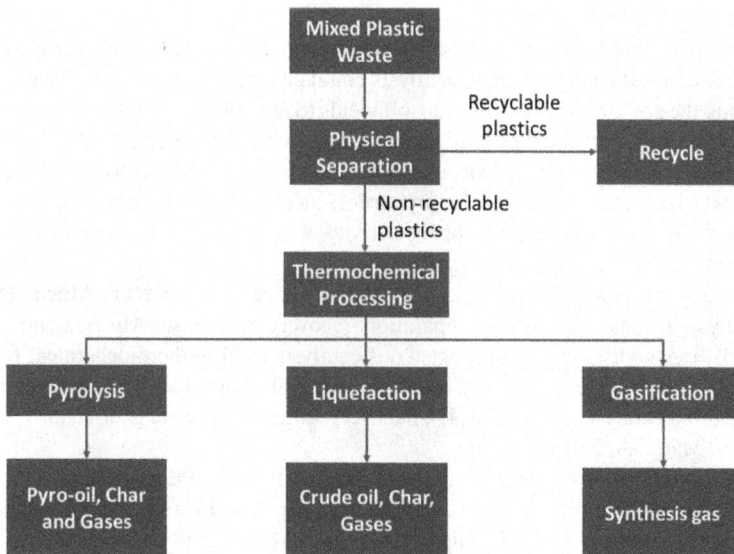

FIGURE 4.2 Thermochemical pathways to convert mixed plastic waste to fuels.

TABLE 4.2

Comparison between Thermal and Catalytic Pyrolysis of Biomass

Attribute	Non-Catalytic Pyrolysis	Catalytic Pyrolysis
Catalyst involvement	No	Yes
Temperature	High (400–600°C)	Medium (350–500°C)
Impurities in liquid oil	Yes	Comparatively less
Bio-oil	• Bio-oil could have a low octane number • High water content • Higher aqueous phase and lower organic phase in the bio-oil	• Bio-oil could have a high octane number • Less water content • Lower aqueous phase and higher organic phase in the bio-oil
Gases	• Less quantity of gas fraction • High yields of CO and CO_2	• High quantity of gas fraction • High yields of C_3 and C_4 components
Char	• A high amount of char is obtained at low temperatures	• Less char is produced for catalysts having a high surface area

References: Miandad et al. (2016); Chen et al. (2014).

Catalysts are generally used to enhance the quality of crude oil produced from plastics. The role of the catalyst is crucial in selective cracking reactions. Catalysts increase the number of lighter-fraction hydrocarbons in the liquid oil and reduce the overall process energy requirements (Miandad et al., 2016; Lerici et al., 2015; Lopez et al., 2011). Some catalysts reduce the concentration of contaminants like phosphorous, nitrogen, chlorine, and sulfur in the liquid fraction.

Catalysts such as natural zeolite, zeolite-β, Cu-Al$_2$O$_3$, red mud, and ZSM-5 have been studied to improve the mechanism and yield of the catalytic pyrolysis of plastic wastes (Miandad et al., 2016). A study by Syamsiro et al. (2014) suggested that by using a higher BET surface area catalyst, more contact can be established between the catalyst and reactant components. Consequently, it accelerates thermal cracking to yield more gases than the liquid fraction by increasing the reaction rate. The highly porous internal catalyst structure acts as a medium to selectively pass and further break the larger plastic components into smaller units (Lee, 2009). The gas formation primarily occurs in the pores of the catalyst support, while waxes form on the external sites of the catalyst by polymerization reaction (Lin, 2000; Mastral et al., 2002). Degradation of high olefins proceeds on the exterior catalyst surface but further decomposition and product selectivity take place in the inner catalyst porous structure (You et al., 2000).

Catalysts improve the quality of the pyrolysis products and reduce the required temperature and reaction time. In some cases, the application of catalysts could enhance the rate of cracking reaction, consequently increasing the gas yields and decreasing oil yields (Syamsiro et al., 2014). The product oil quality can be enhanced by further cracking of long-chain organic compounds that are adsorbed on the catalyst surface to smaller components. The physicochemical and structural characteristics of catalysts such as pore volume, pore size, surface area, acidic properties, and functional groups play important roles in affecting their activity, performance, product selectivity, and product yield during pyrolysis (Miandad et al., 2016).

Homogeneous and heterogeneous catalysts are generally used in polymer cracking reactions, but heterogeneous catalysts are more suitable for separation, recovery, and reuse (Almeida and Marques, 2016; Okolie et al., 2019). A wide range of heterogeneous catalysts used in thermochemical biorefining processes is solid-acid catalysts (e.g., zeolites, alumina, silica–alumina, and fluid catalytic cracking catalysts), mesostructured catalyst (e.g., MCM-41), nano-crystalline zeolites (e.g., n-HZSM-5), Lewis acids, etc. (Almeida and Marques, 2016).

Nair et al. (2016) studied thermal and catalytic pyrolysis of 100 g LDPE at 400°C and 0.1 MPa pressure in non-catalytic and catalytic systems. The catalysts used were zeolite and calcium bentonite. Catalytic pyrolysis resulted in higher yields of oil compared to non-catalytic pyrolysis. The catalysts were found to significantly reduce the cracking temperature during pyrolysis of LDPE. Marcilla et al. (2009) comparatively studied the pyrolysis of HDPE and LDPE in thermal and catalytic cracking at 550°C. The product yields are shown in Table 4.3.

TABLE 4.3

Catalytic and Thermal Pyrolysis of HDPE and LDPE by HZSM-5 Catalyst

Attribute	Thermal Pyrolysis		Catalytic Pyrolysis	
Yield (mg/100 mg of Polyethylene)	HDPE	LDPE	HDPE + HZSM-5	LDPE + HZSM-5
Gases	16.3	14.6	72.6	70.7
Liquids/wax	84.7	93.1	17.3	18.3
Char	–	–	0.7	0.5

Reference: Marcilla et al. (2009).

4.3 Hydrothermal Liquefaction

Hydrothermal liquefaction (HTL) of biomass uses subcritical water to convert organic feedstocks into liquids biofuels, that is, bio-crude oil. Liquefaction is defined as catalytic or non-catalytic thermal degradation of long-chain molecules into liquid hydrocarbon fuels. HTL is typically carried out at relatively moderate-to-high pressures (7–30 MPa) and moderate temperatures (280–380°C) (Chen et al., 2019; Peterson et al., 2008). In hydrothermal liquefaction, subcritical or supercritical water serves as a catalyst, reactant, and solvent. The critical temperature (374°C) and critical pressure (22.1 MPa) of water determine its subcritical phase (temperature < 374°C and pressure < 22.1 MPa) or supercritical phase (temperature > 374°C and pressure > 22.1 MPa) (Reddy et al., 2014; Nanda et al., 2016b). The significant changes in the properties of water at the subcritical and supercritical conditions make the reaction faster, selective, and more efficient to transform carbonaceous organic compounds into bio-crude oil when compared to other biomass-to-liquid conversion technologies such as pyrolysis.

HTL technology is compatible with high-moisture-containing feedstock such as algae, sewage sludge, and food waste. Hence, the need for biomass drying can be eliminated, which can reduce the associated cost. Chen et al. (2019) suggested that supercritical water liquefaction emits 5 to 16 times lower greenhouse gas emissions than incineration. Rangarajan et al. (1998) proposed possible reaction pathways for the liquefaction of polypropylene. The three main processes involved are the following:

(i)   depolymerization reaction at the chain end to form basic monomer compound,

(ii) intramolecular transfer at the chain terminal to generate lighter fractions, and

(iii) intermolecular transfer resulting in chain scission reaction to produce smaller chain fragments.

Pedersen and Conti (2017) investigated the processing of different high-density waste plastics in a supercritical water medium. They found that all these polymers have the potential to be treated hydrothermally and transformed into crude oil with a lower fraction of gases and solids. Williams and Slaney (2007) studied the liquefaction of single plastic waste and plastic waste mixture at 18 MPa and 500°C for 60 min. At these operating conditions, a higher conversion rate of plastics to crude oil was reported with less gas fraction and solid residues. Moreover, the authors concluded that the production of liquid crude oil was higher in liquefaction compared to pyrolysis.

Chen et al. (2019) studied supercritical water liquefaction of polypropylene at a temperature range of 380–500°C at 23 MPa in a batch reactor. The results showed the highest oil yield of 91% at 425°C for 2–4 h of reaction time or at 450°C for 0.5–1 h. The authors also suggested that with an increase in temperature, oil yield increased. The produced oils had HHV values of 46.3–49.3 MJ/kg, almost comparable to that of gasoline (44–46 MJ/kg). Additionally, the oil derived from the liquefaction of polypropylene had a greater HHV when compared to the bio-oil derived from the liquefaction of biomass (30–36 MJ/kg) mainly due to lower oxygen content in polypropylene-derived oil via liquefaction.

Another study on liquefaction of plastic was carried out by Mansur et al. (2018) at 200–350°C under 1–10 kg/cm^2 pressure for 0–60 min holding time with volcanic ash as a catalyst. They found that the plastic liquefaction oil yield reached 87% at 350°C for 10 min over the natural volcanic ash catalyst. The obtained oil had an HHV of 48.6 MJ/kg, which is comparable with commercial diesel (42–46 MJ/kg). The metal oxides present in the volcanic ash acted as catalytic particles to enhance the liquefaction of plastic wastes.

4.4 Gasification

Gasification is an attractive biomass-to-gas conversion technology that converts solid waste materials into energy-dense synthesis gas, which has a high-value as a gaseous fuel and precursor for the synthesis of industrial chemicals and solvents (Nanda et al., 2014b). The product yield and composition depend on the gasification agent (i.e., steam, air, inert gas, subcritical water, or supercritical water), temperature, pressure, reaction time, feedstock concentration, and catalyst (Nanda et al., 2016c; Okolie et al., 2020). Among the several catalysts used in gasification, nickel is advantageous because of its high efficiency in maximizing gas production and eliminating tar in the product stream (Nanda et al., 2016d; Okolie et al., 2019).

Catalytic gasification technology is developed to be the most promised path for hydrogen-rich syngas production that can address sustainability and fuel demands. Gasification of waste plastics using air as a gasifying agent could produce syngas with an average heating value of 6–8 MJ/m^3 (Lopez et al., 2018). Using steam as a gasifying agent, syngas with an average heating value of more than 15 MJ/m^3 could be obtained (Lopez et al., 2018). Syngas can be used as a source of hydrogen or converted into motor fuels such as green diesel and gasoline-grade hydrocarbons through Fischer–Tropsch synthesis (Nanda et al., 2017). Additionally, syngas can be used in gas turbines or fuel cells for power generation.

He et al. (2009) and Melo and Morlanes (2008) proposed mechanisms for steam catalytic gasification of waste plastics to describe the selective adsorption of hydrocarbon molecules on the dual sites of the catalyst surface where nickel particles attack the chain-end carbon of the catalyst via α-scission. Water molecules are selectively adsorbed on the catalyst support, which allows the mobility for oxygen atoms to be adsorbed on the catalyst surface. Furthermore, the steam reforming reaction takes place on the metal–support interface. The species produced (C_1) can react with oxygen species from water adsorption–dissociation process for the active sites of nickel. They could also remain adsorbed on the active site for possible deactivation of the catalyst.

He et al. (2009) studied the catalytic steam gasification of waste polyethylene from MSW in a bench-scale fixed bed downstream reactor to produce syngas using NiO/γ-Al$_2$O$_3$ catalyst at 700–900°C. They found that steam increased gas production and improved the conversion of waste polyethylene. Additionally, with an increase in the temperature from 700°C to 900°C, the tar and char yields significantly decreased. Moreover, by using steam and NiO/γ-Al$_2$O$_3$ catalyst, higher yields of hydrogen-enriched syngas were obtained with lower production of tar and char.

Sancho et al. (2008) investigated catalytic air gasification of 100 wt.% polypropylene in a bubbling fluidized bed reactor using dolomite or olivine as gasifier bed additives at 850°C. They found that dolomite was more efficient than olivine in tar removal. However, the char particles generated during gasification caused plugging of pipes downstream of the gasifier. However, this problem was not evident in the case of gasification of 100 wt.% olivine.

4.5 Co-Processing Technologies

Co-processing involves thermochemical processing of two or more feedstocks at the same time, which could help supplement the organic loading, catalytic properties, and product quality. Co-processing also reduces production costs, increases waste disposal options, and reduces environmental impacts (Zhang et al., 2018). The life-cycle assessment studies of co-processing suggest better environmental performance than landfilling or direct incineration (Geocycle, 2020).

Karagoz et al. (2003) studied the effects of the type of catalyst used and the temperature range on the yield and quality of the crude oil produced from co-cracking. In this investigation, waste plastic was co-processed with vacuum gas oil (VGO) through liquefaction at 425–450°C using catalysts such as DHC-8, HZSM-5, and cobalt-loaded activated carbon (Co-AC). Considering the yield and quality of the liquid product obtained, Co-AC showed the best results in the cracking of municipal waste plastic and VGO.

Ahmaruzzaman and Sharma (2007) studied co-cracking of polypropylene (PP), petroleum vacuum residue (XVR), and Calotropis procera (CL) over the temperature range of 380–460°C in a batch reactor under isothermal conditions. They found that the decomposition of a single feedstock (XVR) followed the reaction of the order 1.3 with an activation energy of 126 KJ/mol. However, co-cracking of XVR + PP + CL in the same reactor followed the reaction of the first order with an activation energy of 66 KJ/mol. They concluded that synergistic interactions occur because of the early production of free radicals from the cracking of PP polymer, which further reacted with the intermediate decomposition products from CL and XVR.

4.6 Industrial Perspectives on Plastic Recycling: Special Focus on the Canadian Scenario

Canada usually produces around 3.3 million tons of waste plastic each year (Young, 2019). According to recent estimates, Canadians use approximately 15 billion single-use plastic bags each year and 57 million straws every day. In addition, 33% of plastic manufactures are used as single-used products or packing (Young, 2019). In Canada, 86% of its total plastic waste is landfilled, and only 9% is recycled without disregarding the high costs associated with the collection, hauling, sorting, storing, and processing of the waste plastics (Young, 2019). Of the 9% of the recyclable plastics generated in Canada, nearly 12% are exported outside the country for recycling purposes.

A few industries in Canada aim towards closing the loop of waste plastic generation. Pyrowave™ situated in Ontario and Quebec, Canada, uses catalytic microwave depolymerization (CMD) technology to depolymerize mixed plastics to predominantly wax, oil, and styrene monomers. These recycled end products serve as the virgin raw materials for plastic manufacturers for many FDA (Food and Drug Administration) compliant applications, thus closing the loop of polymer lifecycle. Pyrowave™'s patented microwave technology uses small-scale modular units that can process 50–100 kg of plastics per cycle every 30 min, thereby treating 400–1,200 tons of plastics annually (Pyrowave™, 2018).

GreenMantra located in Brantford, Ontario, Canada, recycles post-consumer and post-industrial plastics using its proprietary thermo-catalytic depolymerization system to produce industrially relevant synthetic waxes, polymer additives, and fine chemicals (GreenMantra Technologies®, 2018). These end products meet specific performance requirements for diverse applications in asphalt roofing and paving, polymer processing, adhesives, inks, coatings, and plastic composites. The company intends toward a circular economy to efficiently reuse the waste plastics rather than landfilling.

ReVital Polymers headquartered in Sarnia, Ontario, operates a 180,000 sq. ft. recovery facility to recycle post-consumer plastic wastes into virgin plastic pellets and other polymer products for manufacturing new sustainable commodities (ReVital Polymers, 2018). The company aims at the end-of-life management of plastics by producing high-quality resin products for greater economic and environmental returns. Canada Fibers, headquartered in Toronto, along with its affiliates (e.g., Urban Polymers, Urban Waste Recycling, Urban Biofuels, Urban Garden Products and YESS Management) is committed to making many positive impacts to the plastic-recycling industries by reducing greenhouse gas emissions and saving landfill capacity (Canada Fibers, 2018).

Globally, there are also a few notable companies involved in a circular economy and end-of-the-life plastic recycling. SABIC, a petrochemical manufacturing company, headquartered in Riyadh, Saudi Arabia, has its key ambition to decouple plastic from fossil resources and reduce their landfilling and littering. The company strives in its research, development, and innovation at its global branches to scale up its high-quality chemical recycling processes of mixed plastic wastes to produce the original polymer and other industrially relevant commodities (SABIC, 2018).

Agilyx situated in Tigard, Oregon, the United States, performs differential pyrolysis to thermochemically recycle waste plastics into value-added synthetic oils, chemicals, and monomers to be used to remanufacture plastic products (Agilyx, 2018). The facility's chemical recycling processes can be performed at a significantly improved environmental profile than traditional manufacturing. Like other counterparts, the Dow Chemical Company also assures responsible production, disposal, and recycling of plastic products. Recently, in Mumbai, India, the company has announced an innovative formulation of polyethylene resins for sustainable polyethylene laminate solution with flexible packaging applications (DOW®, 2018). Unlike the current flexible packaging materials that are made of multi-layered polymeric materials, DOW has developed mono-layered polyethylene laminate packaging that is 100% recyclable at the end of its lifecycle.

4.7   Challenges and Prospects

Refineries related to plastic wastes are a plausible solution for the environmental, societal, and geopolitical concerns associated with fossil fuels. Fuels and chemicals from waste plastics can play an active role in reducing greenhouse gas emissions and environmental pollution. Additionally, they can generate economic benefits to the circular economy in a country. These refineries can be the potential option to lower down increasing energy demand, but there are several challenges associated with their large-scale implementation.

Plastic recycling is a promising option for reducing the accumulation of waste plastic in the environment. The recycling of plastics mainly consists of four categories such as (Hopewell et al., 2009):

(i)   primary or preliminary (i.e., mechanical treatment to obtain a product of equivalent or same properties);

(ii) secondary (i.e., mechanical treatment to obtain products of downgrade properties);

(iii) tertiary (i.e., recovery of chemicals); and

(iv) quaternary (i.e., energy recovery).

Plastic packaging contains numerous polymers and additives such as pigments, inks, dyes, metals, paper and adhesives, which create difficulty in the recycling process (Nanda and Berruti, 2021c). Plastics with chemical recovery are technically feasible but expensive without subsidies. This is mainly because of the cheaper petrochemical feedstock as compared to the overall cost to sustain a biorefining (Hopewell et al., 2009). Moreover, each thermochemical process has its own opportunities, threats, challenges, and benefits. In pyrolysis, the liquid oil contains higher energy density as compared to coal and other fuels, but it requires upgrading to remove acids, alcohols, water, and other oxygen-containing compounds, thereby adding expenses (Nanda et al., 2016c). In addition, the waste plastics require proper cleaning to remove any contaminants (i.e., soil, sand, silt, and glasses), which can affect the thermochemical conversion process and impact the machinery through slag and clogging issues. Therefore, the cleaning, washing, and pre-drying of the feedstock before the thermochemical conversion make the process energy-intensive and costly. The slagging in boilers and furnaces can adversely affect the process flow, efficiency, and filter systems by causing blockage and corrosion (Arena et al., 2010).

On the other hand, the gases produced from the gasification of waste plastics could contain toxic by-products, which require further purification and treatment to meet the fuel and environmental standards (Miandad et al., 2019). The solid residues and liquid effluents obtained from the thermochemical processing of a highly heterogeneous MSW can retain heavy metals and halogens. Heavy metals are relatively toxic, whereas halogen compounds corrode the equipment and are a potential threat in the form of acid rain if released into the atmosphere (Arena et al., 2010). Alkali compounds are the major issue for gas turbines and increase agglomeration in fluidized-bed reactors (Arena et al., 2010).

The optimization of catalytic thermochemical processing of waste plastics is essential to generate high-quality products at relatively lower temperatures and reaction time. However, the recovery and reusability of the heterogeneous catalysts from the solid product stream could add the factors of cost and energy requirement to the process (Okolie et al., 2019). The market for the products of biorefinery should be created and/or expanded to attract further interest and supplement the capital costs. Detailed techno-economic analysis and life-cycle assessment of each process involved in the waste plastic refinery are needed to understand the economic and environmental impacts of the process (Nanda et al., 2015).

4.8 Conclusions

Recycling of plastic waste is a growing concern for municipalities in urban and rural areas worldwide due to the highly heterogeneous nature of plastics. Plastics contain a mixture of different polymers and other materials such as additives, dyes, and pigments for special applications. Plastics that cannot be recycled can be diverted for thermochemical conversion to produce clean fuels and chemicals. Thermochemical pathways such as pyrolysis, liquefaction, and gasification are promising routes for the conversion of waste plastic to fuels. The role of the catalyst is crucial in the thermochemical conversion of plastics. The catalyst dramatically reduces the energy requirements of the process. Additionally, it enhances the product quality and properties that are comparable with the commercial fuels.

The plastic-based refinery is an emerging concept that can meet environmental standards, reduce waste plastic volumes, and help in building sustainable waste management infrastructures and enterprises. These refineries also develop new opportunities, recover usable energy, and save landfilling cost and volume. To scale up these processes, government funding, public support, adaption of new technology, and product marketability are indispensable. Co-processing of plastics with other waste biomass is also a better option with or without the help of a catalyst to enhance the quality of the products.

Acknowledgments

The financial support from the Natural Sciences and Engineering Research Council of Canada (NSERC), Canada Research Chairs (CRC) program, and Agriculture and Agri-Food Canada (AAFC) is greatly acknowledged.

REFERENCES

Achilias, D. S., Roupakias, C., Megalokonomos, P., Lappas, A. A., and Antonakou, E. V. 2017. Chemical recycling of plastic wastes made from polyethylene (LDPE and HDPE) and polypropylene (PP). *Journal of Hazardous Materials* 149: 536–542.

Agilyx. www.agilyx.com/ (accessed 3 October 2018)

Ahmaruzzaman, M., and Sharma, D. K. 2007. Coprocessing of petroleum vacuum residue with plastics, coal, and biomass and its synergistic effects. *Energy & Fuels* 21: 891–897.

Almeida, D., and Marques, M. D. F. 2016. Thermal and catalytic pyrolysis of plastic waste. *Polimeros* 26: 44–51.

Arena, U., Zaccariello, L., and Mastellone, M. L. 2010. Fluidized bed gasification of waste-derived fuels. *Waste Management* 30: 1212–1219.

Canada Fibers. www.canadafibersltd.com (accessed 3 October 2018)

Chen, D., Yin, L., Wang, H., and He, P. 2014. Pyrolysis technologies for municipal solid waste: A review. *Waste Management* 34: 2466–2486.

Chen, W. T., Jin, K., and Wang, N. H. L. 2019. Use of supercritical water for the liquefaction of polypropylene into oil. *ACS Sustainable Chemistry & Engineering* 7: 3749–3758.

DOW®. The Dow Chemical Company. Dow launches game changing fully recyclable polyethylene packaging solution in India. www.dow.com/en-us/news/press-releases/dow-launches-game-changing-fully-recyclable-polyethylene-packaging-solution-in-india (accessed 3 October 2018)

Geocycle. 2020. Co-processing. www.geocycle.com/co-processing (accessed 4 December 2020)

GreenMantra® Technologies. Creating value from recycled plastics. https://greenmantra.com/ (accessed 3 October 2018)

He, M., Xiao, B., Hu, Z., Liu, S., Guo, X., and Luo, S. 2009. Syngas production from catalytic gasification of waste polyethylene: Influence of temperature on gas yield and composition. *International Journal of Hydrogen Energy* 34: 1342–1348.

Hopewell, J., Dvorak, R., and Kosior, E. 2009. Plastics recycling: Challenges and opportunities. *Philosophical Transactions of the Royal Society B* 364: 2115–2126.

Karagoz, S., Karayildirim, T., Ucar, S., Yuksel, M., and Yanik, J. 2003. Liquefaction of municipal waste plastics in VGO over acidic and non-acidic catalysts. *Fuel* 82: 415–423.

Lee, K. H. 2009. Thermal and catalytic degradation of pyrolytic oil from pyrolysis of municipal plastic wastes. *Journal of Analytical and Applied Pyrolysis* 85: 372–379.

Lerici, L. C., Renzini, M. S., and Pierella, L. B. 2015. Chemical catalyzed recycling of polymers: Catalytic conversion of PE, PP and PS into fuels and chemicals over H-Y. *Procedia Materials Science* 8: 297–303.

Lin, Y. H. 2000. Conversion of waste plastics to hydrocarbons by catalytic zeolited pyrolysis. *Journal of the Chinese Institute for Environmental Engineers* 10: 271–277.

Lopez, A., Marco, I. D., Caballero, B. M., Laresgoiti, M. F., Adrados, A., and Aranzabal, A. 2011. Catalytic pyrolysis of plastic wastes with two different types of catalysts: ZSM-5 zeolite and Red Mud. *Applied Catalysis B: Environmental* 104: 211–219.

Lopez, G., Artetxe, M., Amutio, M., Alvarez, J., Bilbao, J., and Olazar, M. 2018. Recent advances in the gasification of waste plastics. A critical overview. *Renewable and Sustainable Energy Reviews* 82: 576–596.

Mansur, D., Simanungkalit, S. P., Fitriady, M. A., and Safitri, D. 2018. Liquefaction of plastic for fuel production and application of volcanic ash as catalyst. *AIP Conference Proceedings 2024*, 020001.

Marcilla, A., Beltran, M. I., and Navarro, R. 2009. Thermal and catalytic pyrolysis of polyethylene over HZSM5 and HUSY zeolites in a batch reactor under dynamic conditions. *Applied Catalysis B: Environmental* 86: 78–86.

Mastral, F. J., Esperanza, E., Garcia, P., and Juste, M. 2002. Pyrolysis of high-density polyethylene in a fluidised bed reactor. Influence of the temperature and residence time. *Journal of Analytical and Applied Pyrolysis* 63: 1–15.

Melo, F., and Morlanes, N. 2008. Synthesis, characterization and catalytic behaviour of NiMgAl mixed oxides as catalysts for hydrogen production by naphtha steam reforming. *Catalysis Today* 133–135: 383–393.

Miandad, R., Barakat, M. A., Aburiazaiza, A. S., Rehan, M., and Nizami, A. S. 2016. Catalytic pyrolysis of plastic waste: A review. *Process Safety and Environmental Protection* 102: 822–838.

Miandad, R., Rehan, M., Barakat, M. A., Aburiazaiza, A. S., Khan, H., Ismail, I. M. I., Dhavamani, J., Gardy, J., Hassanpour, A., and Nizami, A. S. 2019. Catalytic pyrolysis of plastic waste: Moving toward pyrolysis based biorefineries. *Frontiers in Energy Research* 7: 27.

Mohanty, P., Nanda, S., Pant, K. K., Naik, S., Kozinski, J. A., and Dalai, A. K. 2013. Evaluation of the physiochemical development of biochars obtained from pyrolysis of wheat straw, timothy grass and pinewood: Effects of heating rate. *Journal of Analytical and Applied Pyrolysis* 104: 485–493.

Nair, N., Kher, R., and Patel, R. 2016. Catalytic conversion of plastic waste to fuel. *International Journal of Advanced Research in Engineering, Science and Management*. ISSN: 2394-1766.

Nanda, S., Azargohar, R., Dalai, A. K., and Kozinski, J. A. 2015. An assessment on the sustainability of lignocellulosic biomass for biorefining. *Renewable and Sustainable Energy Reviews* 50: 925–941.

Nanda, S., Azargohar, R., Kozinski, J. A., and Dalai, A. K. 2014a. Characteristic studies on the pyrolysis products from hydrolyzed Canadian lignocellulosic feedstocks. *Bioenergy Research* 7: 174–191.

Nanda, S., and Berruti, F. 2021a. A technical review of bioenergy and resource recovery from municipal solid waste. *Journal of Hazardous Materials* 403: 123970.

Nanda, S., and Berruti, F. 2021b. Municipal solid waste management and landfilling technologies: A review. *Environmental Chemistry Letters* 19: 1433–1456.

Nanda, S., and Berruti, F. 2021c. Thermochemical conversion of plastic waste to fuels: A review. *Environmental Chemistry Letters* 19: 123–148.

Nanda, S., Dalai, A. K., Berruti, F., and Kozinski, J. A. 2016a. Biochar as an exceptional bioresource for energy, agronomy, carbon sequestration, activated carbon and specialty materials. *Waste and Biomass Valorization* 7: 201–235.

Nanda, S., Isen, J., Dalai, A. K., and Kozinski, J. A. 2016b. Gasification of fruit wastes and agro-food residues in supercritical water. *Energy Conversion and Management* 110: 296–306.

Nanda, S., Kozinski, J. A., and Dalai, A. K. 2016c. Lignocellulosic biomass: A review of conversion technologies and fuel products. *Current Biochemical Engineering* 3: 24–36.

Nanda, S., Mohammad, J., Reddy, S. N., Kozinski, J. A., and Dalai, A. K. 2014b. Pathways of lignocellulosic biomass conversion to renewable fuels. *Biomass Conversion and Biorefinery* 4: 157–191.

Nanda, S., Rana, R., Zheng, Y., Kozinski, J. A., and Dalai, A. K. 2017. Insights on pathways for hydrogen generation from ethanol. *Sustainable Energy & Fuels* 1: 1232–1245.

Nanda, S., Reddy, S. N., Dalai, A. K., and Kozinski, J. A. 2016d. Subcritical and supercritical water gasification of lignocellulosic biomass impregnated with nickel nanocatalyst for hydrogen production. *International Journal of Hydrogen Energy* 41: 4907–4921.

Okolie, J. A., Nanda, S., Dalai, A. K., and Kozinski, J. A. 2020. Hydrothermal gasification of soybean straw and flax straw for hydrogen-rich syngas production: Experimental and thermodynamic modeling. *Energy Conversion and Management* 208: 112545.

Okolie, J. A., Rana, R., Nanda, S., Dalai, A. K., and Kozinski, J. A. 2019. Supercritical water gasification of biomass: A state-of-the-art review of process parameters, reaction mechanisms and catalysis. *Sustainable Energy & Fuels* 3: 578–598.

Pedersen, T. H., and Conti, F. 2017. Improving the circular economy via hydrothermal processing of high-density waste plastics. *Waste Management* 68: 24–31.

Peterson, A. A., Vogel, F., Lachance, R. P., Fröling, M., Antal, M. J., and Tester, J. W. 2008. Thermochemical biofuel production in hydrothermal media: A review of sub- and supercritical water technologies. *Energy & Environmental Science* 1: 32–65.

Pyrowave™ – Closing the Loop. Technology: Turning plastic waste back into feedstock used to make new plastic again. https://pyrowave.com/?page_id=14 (accessed 3 October 2018)

Rangarajan, P., Bhattacharyya, D., and Grulke, E. 1998. HDPE liquefaction: Random chain scission model. *Journal of Applied Polymer Science* 70: 1239–1251.

Reddy, S. N., Nanda, S., Dalai, A. K., and Kozinski, J. A. 2014. Supercritical water gasification of biomass for hydrogen production. *International Journal of Hydrogen Energy* 39: 6912–6926.

ReVital Polymers. www.revitalpolymers.com (accessed 3 October 2018)

SABIC. SABIC demonstrates commitment to sustainable development at WEF with iconic structure, icehouse™. www.sabic.com/en/news/10040-sabic-demonstrates-commitment-to-sustainable-development-at-wef-with-iconic-structure-icehouse (accessed 3 October 2018)

Sancho, J. A., Aznar, M. P., and Toledo, J. M. 2008. Catalytic air gasification of plastic waste (polypropylene) in fluidized bed. Part I: Use of in-gasifier bed additives. *Industrial & Engineering Chemistry Research* 47: 1005–1010.

Singh, R. K., and Ruj, B. 2016. Time and temperature depended fuel gas generation from pyrolysis of real world municipal plastic waste. *Fuel* 174: 164–171.

Statista. 2020. Production of plastics worldwide from 1950 to 2018 (in million metric tons). www.statista.com/statistics/282732/global-production-of-plastics-since-1950/. (accessed 31 August 2020)

Syamsiro, M., Saptoadi, H., Norsujianto, T., Noviasri, P., Cheng, S., Alimuddin, Z., and Yoshikawa, K. 2014. Fuel oil production from municipal plastic wastes in sequential pyrolysis and catalytic reforming reactors. *Energy Procedia* 47: 180–188.

The World Bank. 2020. Tackling increasing plastic waste. http://datatopics.worldbank.org/what-a-waste/tackling_increasing_plastic_waste.html (accessed 3 December 2020)

Wallace, D. 2020.7 Types of plastics: Their toxicity & what they're most commonly used for. https://infographicjournal.com/7-types-of-plastics-their-toxicity-what-theyre-most-commonly-used-for/ (accessed 3 December 2020)

Williams, P. T., and Slaney, E. 2007. Analysis of products from the pyrolysis and liquefaction of single plastics and waste plastic mixtures. *Resources, Conservation and Recycling* 51: 754–769.

Wong, S. L., Ngadi, N., Abdullah, T. A. T., and Inuwa, I. M. 2015. Current state and future prospects of plastic waste as source of fuel: A review. *Renewable and Sustainable Energy Reviews* 50: 1167–1180.

You, Y. S., Kim, J. H., and Seo, G. 2000. Liquid-phase catalytic degradation of polyethylene wax over MFI zeolites with different particle sizes. *Polymer Degradation and Stability* 70: 365–371.

Young, R. 2019. Canada's plastic problem: Sorting fact from fiction. https://oceana.ca/en/blog/canadas-plastic-problem-sorting-fact-fiction (accessed 4 December 2020)

Zhang, L., Bao, Z., Xia, S., Lu, Q., and Walters, K. B. 2018. Catalytic pyrolysis of biomass and polymer wastes. *Catalysts* 8: 659.

5

Torrefied Solids: A Material Border Lining Biomass and Biochar

Tumpa R. Sarker, Sonil Nanda, Ramin Azargohar, Venkatesh Meda, Ajay K. Dalai

CONTENTS

Abbreviations

Dry torrefaction: DT
Empty fruit bunches: EFB

Equilibrium moisture content: EMC
Fixed carbon content: FC
Fuel ratio: FR
Hardgrove grindability index: HGI
Higher heating value: HHV
Hydrothermal carbonization: HTC
International energy agency: IEA
Rice straws: RS
Sugarcane bagasse: SB
Volatile matter: VM
Wet torrefaction: WT

5.1 Introduction

Due to the rapid growth of the economy and industrialization, the energy demand is increasing exponentially worldwide and is expected to rise by 20–30% from 2015 to 2040 (Newell et al., 2019). This huge energy demand is mainly satisfied by using fossil fuel. According to the International Energy Agency (IEA), around 19.3% of total energy such as solar, wind, geothermal, hydrothermal, and biomass comes from renewable sources (Ren et al., 2017). Among all renewable energy sources, biomass is considered a carbon-neutral source as CO_2 emission during conversion is already a portion of the carbon cycle (Cao et al., 2015). Worldwide biomass production from mostly wild plants was 146 billion metric tons per year (REN21 2017). Lignocellulosic feedstocks could potentially supplement roughly 14% of the world's total energy production (Okolie et al., 2019).

The efficient and economical conversion of the biomass into high-value fuel can enhance global energy security as well as mitigate the global climate impact and reduce the emission of greenhouse gases. Biomass usually comes from the forest, agricultural residue, as well as municipal solid waste (Okolie et al., 2021; Nanda and Berruti, 2021). These biomasses can be used for energy production using different conversion processes such as chemical processes (e.g., transesterification and esterification), biological processes (e.g., fermentation and anaerobic digestion), and thermochemical processes (e.g., pyrolysis, gasification, and combustion) (Nanda et al., 2014; Parakh et al., 2020).

Recently, biomass is widely being used for the production of solid char, syngas as well as biodiesel. Biomass is usually characterized by low bulk density, which has negative impacts on long-distance transportation (Medic et al., 2012; Pimchuai et al., 2010; Rousset et al., 2011). It has a comparatively low energy density than coal (Axelsson et al., 2012). Biomass can absorb moisture encouraging microbial decaying, preventing its long-term storage (Pimchuai et al., 2010; Rousset et al., 2012; Chen and Kuo, 2011). The high moisture content of biomass increases the cost of drying, transportation, and thermochemical conversion (Haykiri-Acma et al., 2016). The fibrous nature of biomass increases the grinding energy requirement (Wannachepeera et al., 2011). The heterogeneous nature of biomass creates complications during the design and control of the process (Stelte et al., 2012). The poor spherical shapes of biomass cause flowability issues (Repellin et al., 2010; Deng et al., 2009). Additionally, raw biomass also generates smoke during combustion and gasification (Stelte et al., 2011; Pimchuai et al., 2010). All these properties of biomass restrict its direct application for combustion and gasification (Rousset et al., 2012; Phanphanich and Mani, 2011).

Thermal pretreatment of biomass, that is, torrefaction is treated as an efficient and simple method for upgrading the characteristics of biomass and transform biomass to a solid material like coal. Recently, torrefaction has gained attention both from industry and research centers. Torrefaction is a thermochemical treatment of biomass, which enhances the chemical, physical, and biological properties of biomass for direct use in combustion and gasification purposes. Torrefaction reduces the moisture content of biomass and atomic ratio, significantly improving the grindability, hydrophobicity, ignitability, combustion and gasification reactivity. Torrefied biomass is considered as an exceptional solid fuel with characteristics between conventional biomass and charcoal (Nachenius et al., 2014).

Torrefaction also reduces the time and effort needed for drying and grinding; adds benefits in storage and long-distance transportation of biomass and other logistics significantly; facilitates the prolonged global trade of biomass and energy security; reduces ash-related problems; facilitates other thermo-chemical conversion systems of biomass (e.g., pyrolysis, gasification, combustion) effectively, reduces climate impact from the whole supply chain; and accelerates the engineering expansion of refinement schemes for biomass raw materials (Kim et al., 2012). Torrefaction has the potential to be a suitable pathway for producing diverse valuable clean products through the conversion of biomass by replacing the current fossil-fueled plants (Via et al., 2013). Generally, torrefied biomass is widely used for combustion and gasification for power production and improves the efficiency of the processes gasification, pyrolysis, and co-firing (Bergman et al., 2005).

There are several articles on torrefaction on various biomasses, focusing on characteristics such as torrefaction operating temperature (Kanwal et al., 2019; Singh et al., 2020; Wang et al., 2019), residence time (Kanwal et al., 2019; Singh et al., 2020; Wang et al., 2019), grindability (Phanphanich and Mani, 2011; Arias et al., 2008; Bridgeman et al., 2010), combustion (Pimchuai et al., 2010; Rousset et al., 2011), and gasification (Kuo et al., 2014; Muslim et al., 2017; Raut et al., 2016). Little literature is reported on mass and energy balance (Yan et al., 2010), as well as economic analysis (Svanberg et al., 2013; Uslu et al., 2008; Pirraglia et al., 2013) of the torrefaction process plant. In this chapter, some other aspects of torrefaction have been presented including the basics of torrefaction, torrefaction classification, mechanism or reaction in torrefaction, the effects of different parameters, available torrefaction reactors, properties of biochar obtained from torrefaction, and in the end the economics study.

5.2   Origin of Torrefaction

Torrefaction, the thermochemical conversion of biomass operated at low temperature, is quite similar to that used for the production of charcoal piles, applied as a dropping agent in the primary metal ore reduction process. However, the patents of torrefaction were technically recognized early in the 1800s and were initially used for coffee production (Britain et al., 1990). A few more patents can also be found in the area of torrefaction during the years 1922–1925 (>3), 1930–1932 (>3), and 1939–1952 (>10) (Nordin et al., 2013). Torrefaction was first performed for syngas production in the 1930s in France (Nordin et al., 2013; Ciolkosz and Wallace, 2011). Furthermore, the torrefaction process was re-pioneered by Bourgois and Doat in the 1980s where it was conducted at two different temperatures on two tropical wood samples. During the early- and mid-twentieth centuries, studies on biomass torrefaction for the conversion of energy were performed sporadically. Nevertheless, valuable fundamental data as well as additional information on thermal treatment of lignocellulosic biomass originated during this time, data being particularly focusing on high-temperature drying, thermal degradation, pyrolysis, dry distillation, conservation of wood, wood cooking, and heat stabilization (Nordin et al., 2013).

The later investigations on torrefaction can be divided into two groups, that is, the initial pioneering work conducted by Armines and Bourgois in 1981–1989 and the recent wide-ranging work done by many engineers and scientists initialized by the work carried out in Dutch energy center ECN and Eindhoven University of Technology (Bourgeois and Doat, 1984; Prins et al., 2006b; Bergman et al., 2005). During the late 1800s, a plant was demonstrated based on the original French work where they used torrefaction to generate reducing agents for the metallurgic industry. Pechiney company first established a torrefaction unit and operated successfully for a few years and then dismantled for economic reasons (Uslu et al., 2008). Besides, it must be stated and documented that other technical works were also completed (Ibrahim et al., 2013; Prins et al., 2006a).

In recent times, research and development in the field of torrefaction have been wide ranging, and many researchers are working dedicatedly on torrefaction, which is reflected by the increasing numbers of journal articles and scientific reports dedicated to this topic. Furthermore, the technical reports as well as published conference proceedings followed the increasing trend. However, scientific reports on torrefaction are still low in number as compared to other thermochemical conversion techniques of biomass, and around 5% as compared to studies on pyrolysis of biomass (Wannachepeera et al., 2011).

5.3   Torrefaction

Torrefaction is a method of thermochemical conversion of biomass, which is also called pretreatment of biomass intending to enhance the quality of raw biomass compositions to be comparable to coal. Torrefaction is similar to mild/slow pyrolysis, roasting, high-temperature drying, and wood cooking. Usually, torrefaction treatment of biomass requires inert atmosphere or oxygen-deficient conditions rendered at temperatures between 200°C and 300°C. However, torrefaction is conducted in presence of CO_2 and oxidative environment, but the best results were found for inert atmosphere (Cherubini and Strømman, 2011). The residence time can vary between 0.5 h and 2 h, but the reaction time can be reduced at higher torrefaction temperature (280–300°C) by taking some process-controlled measurements such as primary moisture content of biomass and material distribution. For example, the Topell torbed torrefaction has been conducted at a temperature above 350°C for only 2 min (Laughlin and Erasmus, 2009).

The goal of torrefaction is to modify the structure of lignocellulosic biomass aiming to preserve solid mass yield while facilitating its energy content to be conserved with an incomplete elimination of volatile matters. The whole torrefaction process is endothermic, indicating energy input is required to kickstart the method and to maintain it (Basu, 2018b). During the torrefaction process, biomass is partly devolatilized and decomposed, which leads to mass reduction ultimately increasing the energy density of biomass. Torrefaction reduces oxygen and hydrogen contents of biomass as a result of which less water vapor and smoke are released during combustion (Pimchuai et al., 2010). Therefore, the final solid product is characterized by hydrophobic, more brittle, more homogeneous, higher carbon content; better ignitions; higher atomic ratios (C/O and C/H); microbial degradation resistance; and high energy density. Torrefied biomass typically contains 30% more energy compared to its original biomass. The whole torrefaction process involves heating, drying, torrefaction, and cooling steps as shown in Figure 5.1. (Bergman et al., 2005).

(i) **Heating:** This stage continues until drying is reached. In this stage, biomass starts to heat with increasing temperature, and evaporation of moisture initiates at the end of this stage.

(ii) **Drying:** Drying is referred to as the removal of water from the biomass and starts at 100°C and continues up to 200°C. In this phase, physically bonded water is released, and biomass is partially decomposed leading to a small mass loss.

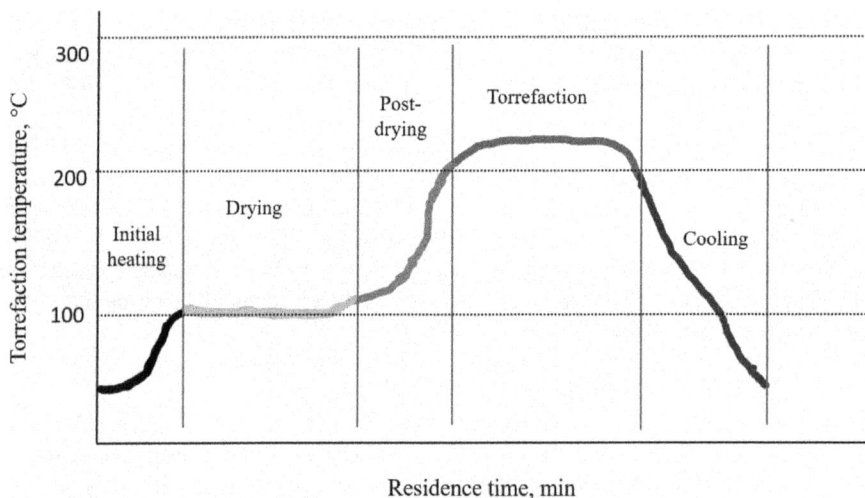

FIGURE 5.1   Different stages of torrefaction

(iii) **Torrefaction:** The main torrefaction process starts when the temperature becomes more than 200°C and continues up to a higher temperature. Decomposition, devolatilization, and carbonization of lignocellulosic biomass occur in this stage, which cause a significant mass loss and increase the energy density.

(iv) **Cooling:** The product of torrefaction is additionally cooled down below 200°C to room temperature.

5.4 Torrefaction Decomposition Mechanism

Lignocellulosic biomass mostly contains three basic building blocks, that is, cellulose, hemicellulose, and lignin. Many reactions occur during torrefaction, which causes thermal decompositions of biomass. Different reaction pathways occur in the torrefaction process, which are gathered in several reaction regimes as shown in Figure 5.2 (Bergman et al., 2005). Decomposition regimes for each polymer can be defined based on the temperature profile. Initial drying occurs at the very beginning of torrefaction. With an increase in temperature, depolymerization followed by devolatilization and carbonization occur in biomass structure. Usually, hemicellulose is a more responsive polymer compared to lignin while cellulose is the steadiest.

At lower torrefaction temperature, hemicellulose decomposes via devolatilization and carbonization while minor decomposition happens for lignin and cellulose. At higher torrefaction temperature, limited devolatilization and carbonization occur for lignin and cellulose whereas hemicellulose decomposes extensively and produces char-like material.

(i) **Depolymerization:** Depolymerization can be defined as the process of breaking the polymers (macromolecules) into component monomers (smaller molecules) e.g., its building blocks (Bergman et al., 2005). During torrefaction, lignocellulosic components of biomass degrade to their building blocks and break down the lignocellulosic structure at a higher temperature, which leads to the formation of volatile and non-volatile compounds.

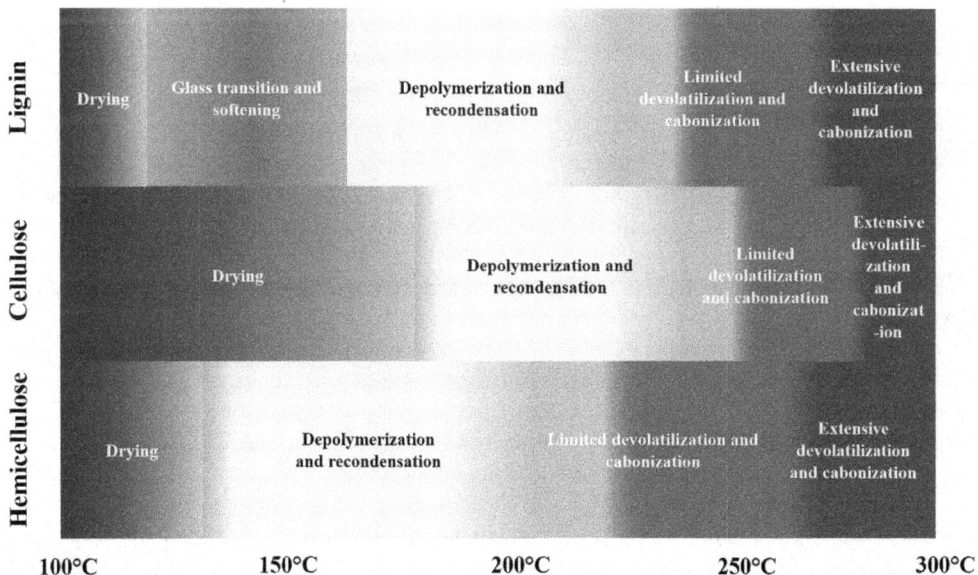

FIGURE 5.2 Torrefaction mechanism.

(ii) **Devolatilization:** The devolatilization is a fundamental process for all thermochemical processes (e.g., pyrolysis, combustion, and gasification), particularly for biomasses that contain large amounts of volatile matter. A large weight loss during torrefaction is mainly attributed to the devolatilization of biomass. Another name for this process is destructive drying process. Devolatilization is a process of removing volatile materials directly from an extruder machine during the melt extrusion of the polymer. In the torrefaction process, compounds having high molecular mass fragment into small compounds of low molecular mass, which evaporate from the biomass particles and are collected as torrefied volatiles (Sakthivadivel and Iniyan, 2017; Basu et al., 2014). Besides, some high-molecular-mass compounds remain in biomass reattached to the biochar at the end of devolatilization (Kim et al., 2009). All moisture and some light volatiles evaporate during this process, which causes mass loss. On average, 70% of the mass is retained as solid materials, which contain around 90% of its initial energy after torrefaction (Sakthivadivel and Iniyan, 2017; Bergman et al., 2005). The 30% mass converted into gases mainly occurs due to the thermal degradation of hemicellulose (Bergman et al., 2005)

(iii) **Carbonization:** Carbonization is known as a process of organic substance conversion into carbon or a carbon-containing residue through heating of biomass in an oxygen-free environment. The processes of torrefaction, carbonization as well as pyrolysis are almost the same, but their ultimate goal is different. Pyrolysis aims to maximize the liquid product whereas the objective of carbonization is to maximize the fixed carbon content in the solid product by minimizing the hydrocarbon content. The main purpose of torrefaction is to maximize the solid product by reducing the atomic ratio (O/C and H/C) (Mamvura and Danha, 2020). Carbonization generally occurs at more severe torrefaction operating conditions.

(iv) **Deoxygenation:** Deoxygenation is a chemical reaction involving the elimination of oxygen atoms from a molecule. Deoxygenation reaction during torrefaction increases the atomic ratio (O/C and H/C) of torrefied biomass by removing oxygen from the parent biomass and forming CO_2, CO, H_2O, and other compounds containing oxygen, carbon, and hydrogen (Cherubini and Strømman, 2011). The higher heating value (HHV) of torrefied biomass is also attributed to the removal of oxygen. The high oxygen content of biomass causes problems during blending with coal for combustion purposes as it has high polarity (Lange, 2007).

5.5 Types of Torrefaction

Torrefaction can be divided into three clusters based on the temperature such as light, mild, and severe torrefaction (Chen and Kuo, 2011). Light torrefaction takes place at temperatures ranging from 200°C to 235°C. In this case, almost all moisture and some light volatiles contained in biomass are evaporated. Hemicellulose is degraded to a certain degree in this situation while the cellulose and lignin remain unaffected or slightly affected (Rousset et al., 2011). The color of the biomass changes to brown in this type of torrefaction. Therefore, most of the mass is retained, and the energy of biomass increases slightly (Wen et al., 2014).

Mild torrefaction starts at temperatures of 235°C and continues up to 275°C. In this case, hemicellulose is decomposed considerably while cellulose is degraded to a certain extent and lignin is slightly disintegrated (Hill et al., 2013). Severe torrefaction occurs at higher temperature ranges from 275°C to 300°C. At this condition, hemicellulose is decomposed severely while cellulose is disintegrated largely. Lignin is the most thermally stable and hard to decompose. At high temperatures, lignin is consumed slightly. Since the key components of lignocellulosic biomass are hemicellulose and cellulose, they degrade significantly in severe torrefaction. Therefore, substantial mass loss occurs in this condition. At the same time, the energy of torrefied biomass increases to a certain extent resulting in intensified energy density after torrefaction (Chen et al., 2011a, 2011b, 2011c). The color of torrefied biomass becomes completely black, while, for the liquid phase collected during torrefaction, it turns from dark brown to black (Chen et al., 2015).

5.6   Mode of Torrefaction

There are different methods of conducting the torrefaction process based on the medium of heating as well as carrier gas such as thermal torrefaction or microwave torrefaction.

5.6.1   Thermal or Conventional Torrefaction

Based on the literature, the torrefaction of different biomasses has been carried out by the conventional method or electrical heating method. In this method, usually, biomass is placed in a metal reactor, which is being heated through a heating source. Heat passes to the biomass particles through the modes of conduction, convection, and radiation. The conventional heating process is slow and requires high energy to continue. Therefore, a higher heating rate can reduce the processing time of torrefaction because an incomplete reaction takes place (Wang et al., 2011). In this process, biomass attached to the reactor wall is heated faster as compared to the biomass in the middle, meaning that some portion of biomass becomes over torrefied whereas some are mildly torrefied, which is a result of a nonuniform heating. Thermal torrefaction is inefficient in terms of the heating method as it requires more energy and time to heat the reactor as compared to microwave heating (Jin et al., 2017).

5.6.2   Microwave Torrefaction

In the last few years, microwave torrefaction has become more popular as compared to conventional torrefaction. In this process, the only difference is the mode of heating method. Microwave pretreatment of lignocellulosic biomass was first introduced by Ooshima et al. (1984) and Azuma et al. (1984). Microwave is a part of the electromagnetic spectrum in which the wavelengths vary from 1 m to 1 mm with matching frequencies in the range of 300 MHz to 300 GHz. Microwave heating is a method of energy transfer instead of heat transfer as microwave heating takes place when electromagnetic energy is converted into heat within the irradiated material (Kostas et al., 2017). In microwave heating, heat energy is usually transferred from the center to the exterior of the material body, which is completely reverse of the conventional method (Figure 5.3). In the conventional thermal heating method, the outer layer is heated up significantly by conduction or convection (Kostas et al., 2017).

This distinctive inverted heating system offers many advantages over conventional heating – for example excellent efficiency of energy transfer for heating and decreased heating time to attain a certain temperature. At the same time, no additional time is required to heat up or cool down due to the instant heating effect (Kappe, 2004). The dielectric properties of a microwave can facilitate the penetration of these waves into the biomass and heat them uniformly and quickly. Microwaves usually consist of two perpendicular fields namely electric field and magnetic field. The electric field is mainly responsible for microwave heating, interacting with molecules via two identical mechanisms such as dipole rotation and

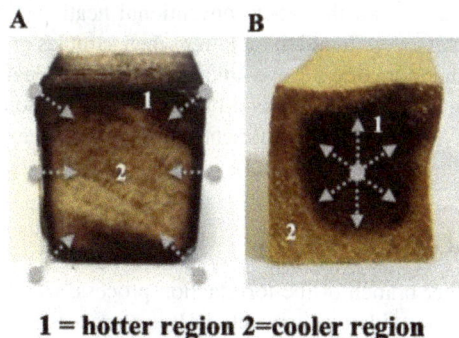

1 = hotter region 2=cooler region

FIGURE 5.3 Difference in temperature distribution and direction of heat transfer in (A) convention method and (B) microwave heating. Reproduced with permission from Kostas et al. (2017).

ions' migration (Huang et al., 2012). In the dipole polarization system, the microwave frequency makes the molecular dipoles rotate back and forth constantly with every oscillating electric field and heat is generated through both friction and random collision between rotating molecules. On the other hand, in ionic conduction, free ions move translationally through space attempting to align with the changing electric field. Similar to dipole rotation, the heat is generated through friction between moving ions. In this case, the temperature of the reaction mixture plays a vital role in energy transfer. In microwave heating, the rate of heat generation would be more efficient if there are more polar or ionic molecules.

In microwave torrefaction, the main concern is about the reactor and the materials of the reactor since all materials cannot evenly absorb the microwave irradiation. Materials can be graded into three different categories based on their interaction with microwaves as insulators/transparent in which material can absorb microwaves without any loss, for example, glass or ceramics; conductors/reflective where materials cannot absorb microwave (MW) therefore reflect, for example, metals; and dielectrics/absorptive where the microwave is absorbed and converted into heat such as polar component or water (Motasemi and Afzal, 2013). In general, dielectric heating is used for microwave torrefaction of biomass.

Many researchers have mentioned microwave torrefaction as energy efficient; however, the energy required to heat a certain amount of biomass is almost the same by either conventional or microwave heating (Chaturvedi and Verma, 2013; Li et al., 2016). Microwave pretreatment has been carried out on different biomasses such as oat hull (Abedi and Dalai, 2017), Douglas fir (Ren et al., 2017), rice husk and sugarcane residue (Wang et al., 2012), barley straw and wheat straw (Satpathy et al., 2014; Kashaninejad and Tabil, 2011). The microwave had been carried out on those biomasses in the range of 250–1000 W, while the heating time was 10–100 min.

Huang et al. (2012) conducted microwave torrefaction on rice straw and found that the heating value and energy density of rice straw were increased by 30% and 14%, respectively. To reach the expected solid and energy yields, the processing time and energy input of microwave torrefaction are kept lower than other conventional heating methods. Microwave torrefaction increased the hydrophobicity as well as the porosity of pellets (Abedi and Dalai, 2017; Azargohar et al., 2019). Ren et al. (2019) experienced the same results while conducting microwave torrefaction on Douglas fir sawdust.

Microwave pretreatment of biomass is also used in the presence of different chemicals to expedite the degradation of the lignocellulosic structure. Diverse organic compounds including formic acid, glycerol, ethanol, ethylene glycol, methanol, acetic acid, alkali, and formic acid can be used as solvents in organosolv pretreatment (Li et al., 2012a, 2012b). This alkali solution can be applied to biomass during microwave or before microwave pretreatment. The density of pellets increased, but heating value decreased due to this microwave alkali treatment, especially at 2% NaOH. Density is directly related to the moisture content of biomass and the applied pressure, and it continued to increase with the increase in pressure and moisture when the applied pressure ranged from 7 MPa to 4.1 MPa and moisture content ranged from 9% to 15% (wet basis) for canola, oat, and wheat straw (Guo et al., 2016).

Agu et al. (2017) used microwave pretreatment with NaOH and $Ca(OH)_2$ on canola straw and oat hull at 713 W for 6, 12, and 18 min. It is indicated in previous research that microwave treatment or combined microwave–alkali treatment enhances biomass decomposition by hastening the reactions throughout the torrefaction process compared with the other conventional heating. Cellulose, hemicellulose, and lignin contents of treated biomass were found to be lower when samples are subjected to alkali-assisted microwave pretreatment. Furthermore, microwave–alkali pretreatment degrades and depolymerizes lignin into small phenolic components, which act as a binder in pelletization. The pellets obtained from biomass subjected to microwave-chemical pretreatment showed higher tensile strength and relaxed density (Agu et al., 2017; Kashaninejad and Tabil, 2011).

5.6.3 Oxidative Torrefaction

Oxidative torrefaction is another branch of the torrefaction process, which is somehow different from the conventional torrefaction/non-oxidative torrefaction. It is a process in which biomass is torrefied in an oxidative environment at the same temperature and residence time range similar to non-oxidative torrefaction. The operating cost of the oxidative torrefaction process is low compared to inert gas torrefaction. There are three main advantages of oxidative torrefaction as compared to conventional torrefaction.

First, there is no need to use nitrogen or process for gas separation as air is used as the carrier gas; second, in oxidative torrefaction, some oxidative reactions also occur with devolatilization as well as thermal degradation of biomass, and oxidative reactions are generally exothermic, which means that heat is generated from torrefaction, which ultimately reduces the heating for endothermic reactions. In the end, the reaction rate of oxidative torrefaction is quicker than thermal torrefaction, which helps to reduce the overall torrefaction duration resulting reduction in cost.

In oxidative torrefaction at a higher temperature, the decomposition of biomass increases with a rise in oxygen concentration. Therefore, a shorter torrefaction duration is required to reach a target mass loss. However, in the case of lower torrefaction temperature, there is no effect of oxygen concentration on mass yield (Rousset et al., 2012). The performance of oxidative torrefaction concerning mass yield and characteristics of torrefied biomass depends on the nature of the raw feedstock. Oxidative torrefaction is more suitable for woody biomass than the fibrous one (non-woody biomass) (Álvarez et al., 2017; Chen et al., 2014a, 2014b).

Brachi et al. (2019) found a comparatively worse performance of oxidative torrefaction concerning mass and energy yield than the conventional one. On the contrary, physical properties (e.g., energy density, density, durability, and hardness) of torrefied pellet under oxidative environment were better than those from the inert atmosphere. Oxidative torrefaction also increases the hydrophobicity and the surface area of biochar to a certain extent while reducing the atomic ratio significantly (Álvarez et al., 2017; Zhang et al., 2019). Oxidative torrefaction of sawdust in a fluidized bed reactor made pellets strong and dense with high hydrophobicity (Wang et al., 2013). Superficial air velocity also plays an important role in the performance of torrefaction as the mass yield decreases with an increase in air velocity, indicating that surface oxidation enhanced the torrefaction reaction of biomass (Chen et al., 2013). The HHV of torrefied biomass increases by 1.07–1.24 times while the HHV of liquid is lower than that for non-oxidative torrefaction. Chen (2015) states that the behavior of enhancement factor for oxidative torrefaction is inverse to non-oxidative torrefaction, and with a rise in oxidative torrefaction severity both the enhancement factor and solid yield decrease due to the oxidative reactions in the biomass.

5.6.4 Wet Torrefaction

Another branch of torrefaction is wet torrefaction (WT) in which biomass is upgraded in compressed hot water or subcritical water under a small temperature range from 180 to 260°C while the pressure varies between 4 and 20 MPa for 5–240 min as residence time (Lynam et al., 2011; Chen et al., 2012). The other name of wet torrefaction is hydrothermal pretreatment (Murakami et al., 2012; Prawisudha et al., 2012), hydrothermal carbonization (HTC) (Liu and Balasubramanian, 2014; Liu et al., 2013), and hot compressed water pretreatment (Goh et al., 2012). The main products of wet torrefaction are a solid known as hydrochar and a combination of gas and liquid (Li et al., 2015). The main product is hydrochar, which accounts up for 89.1% of the energy and 88.3% of the mass in the raw biomass (Bach et al., 2013).

The gaseous product obtained from WT mainly contains CO_2, CO, H_2, and CH_4, among which around 90–95% is CO_2 due to the decarboxylation process, while the liquid product may contain various components such as sugars, furfurans, organic acids, phenolic acids as well as furfurals (Hoekman et al., 2013; Hoekman et al., 2011). The mechanism involved in the degradation of biomass in WT is different from dry torrefaction. In this case, a hydrolysis process is mainly responsible for the degradation of lignocellulosic structure owing to the existence of concentrated water. Therefore, the activation energies for the depolymerization reactions of the biomass polymers decline (Libra et al., 2011).

In WT, hemicellulose and cellulose trigger fast hydrolysis; however, the consequent change of hydrolysate is quicker than hydrolysis in WT (Reza et al., 2012). Additionally, the formation of furfural can also contribute to the yield of hydrochar by dissolving or polymerizing for coke (Reza et al., 2012). In comparison with DT, WT is less poisonous and more environmentally friendly because biomass, water, and inert gas are the only input streams (Kostas et al., 2017). The yield (mass and energy) of torrefaction is higher in WT than DT and requires a lower temperature and short residence time to obtain the same yield which is due to the higher reactivity of hydrothermal media than inert media (Bach et al., 2013). An alternative important advantage of WT is the percentage of ash presence in hydrochar, which is reduced to a certain extent compared to raw biomass (Aguado et al., 2020). This implies the potentiality of dissolving and washing out some inorganic components from biomass fuels (Bach et al., 2013).

Hydrochar produced from WT is characterized by HHV, high carbon content, lower atomic ratio, high hydrophobicity, grindability, and pellet-ability than those from DT (Bach and Skreiberg, 2016; Hoekman et al., 2013; Kambo and Dutta, 2015). WT became popular as it can convert all wet biomass with high moisture content (typically 50%) and waste material to solid fuel (hydrochar) without pre-drying, which makes the pretreatment economically alluring. This also implies the suitability of WT for a wide variety of biomass feedstocks (including aquatic-based biomass) to produce a high-quality fuel. The efficiency of WT can be enhanced by adding acidic media to the process (Lynam et al., 2011). Though torrefaction operating conditions (i.e., reaction time and temperature) can affect the properties of final products, the effect of temperature dominates over residence time in the form of mass yield, energy yield as well as energy density (Mäkelä et al., 2015; Bach et al., 2013).

Hydrochar has combustion characteristics similar to coal, which are even better than lignite (Zhang et al., 2016) and can directly be used in a coal-fired power plant for cofiring or combustion (He et al., 2018a, 2018b; Bach and Skreiberg, 2016). Though it has some advantages over DT, it also has some problems. WT system is more complicated than DT and requires high initial and maintenance costs. Besides, feedstocks containing impurities can create problems such as precipitating, depositing, and clogging the reactor resulting in high maintenance costs (Bach and Skreiberg, 2016). Bach et al. (2015) conducted WT on spruce and birch woods at three distinct temperatures and residence times in the range of 175–225°C and 10–60 min, respectively. The authors found that WT reduced the activation energy of hemicellulose but increased the energies for cellulose and lignin and also mentioned that WT can convert different biomasses into more homogeneous solid products.

Gong et al. (2019) conducted WT on rice straws (RS), empty fruit bunches (EFB), and sugarcane baggage (SB) to expand the biomass fuel properties and experienced that HHVs boosted by 4.7% in the RS, 4.4% in the EFB, and 5.3% in the SB. They also found that WT not only removed ash but also removed most of the chloride and potassium in the biomass particularly for EFB, and the removal efficiencies were the highest (98.6% and 99.3%, respectively).

Zhang et al. (2019) compared both DT and WT in terms of chemical characteristics and thermal nature of corn stalk digest and found that both torrefaction techniques increased the fuel properties of biomass. WT presented lower organics and alkali metals, whereas DT reserved more carbon and ash. Additionally, the results revealed the presence of organic materials of wet torrefied samples and displayed richer surface functional groups, comparatively complete lignin structure, and higher crystallinity than DT samples. In the case of thermal behavior, they found more rapid and concentrated burning for WT samples, whereas the DT samples showed more stable combustion. Higher bio-oil yield, higher H_2 generation, and higher sugar content were found for WT compared to DT (Wang et al., 2018a, 2018b). WT was also conducted with microwave on different biomasses by different researchers (Bach et al., 2017; Yu et al., 2020), and they found the process effective.

5.7 Products of Torrefaction

The key product of torrefaction is solid material named torrefied biomass, some condensable liquid, and non-condensable gases (Huang et al., 2012). However, the objective of torrefaction is to maximize the solid yield while facilitating its energy content to be improved and with incomplete elimination of volatile matters. Torrefaction reduces the hydrogen and oxygen contents of biomass resulting in less water vapor and smoke release during combustion. Torrefaction degrades hemicellulose that causes the formation of several condensable and non-condensable gases with high oxygen content. The yield of those product counts on the torrefaction operating conditions and its parameters. With increasing torrefaction severity, the solid product yield decreases while the gas and liquid yield increases. Figure 5.4 shows the different product compounds of torrefaction pretreatment of biomass.

This is known by various names such as torrefied biomass, torrefied material as well as torrefied char. It is a dry, partially carbonized solid which is left after the driven-off volatile compound (Bridgeman et al., 2010). The solid part comprises a complex structure of initial sugar structures and reaction products.

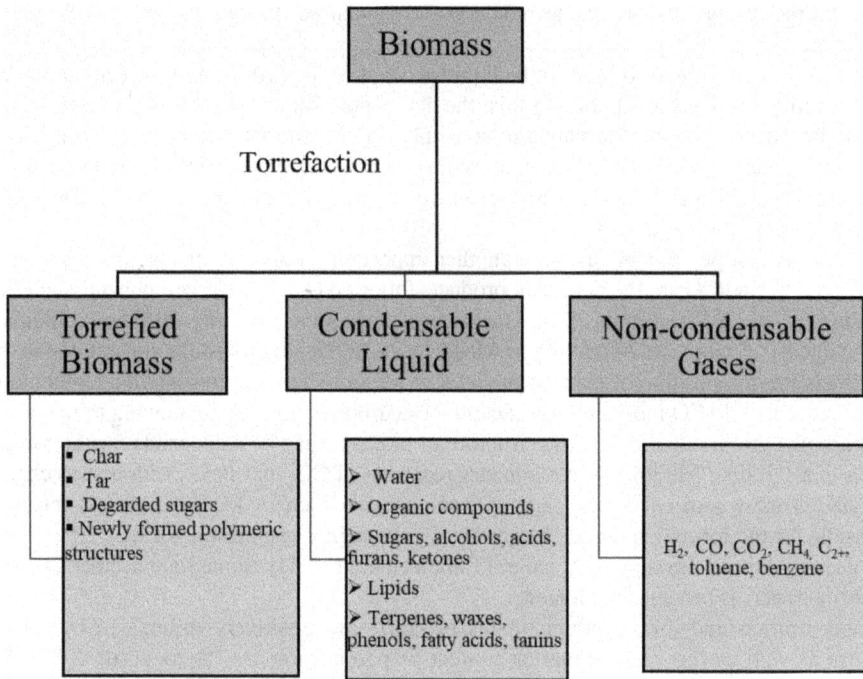

FIGURE 5.4 Products of torrefaction.

The solid products are huge altered sugar structures. Some newly formed polymeric structures are with possibly a certain grade of aromatic rings. It also contains unreacted lignin. The solid product is also characterized by carbon-rich char structures, ash composition, lower atomic ratio, higher fuel property, grindability, and hydrophobicity.

Apart from the solid product, a small amount of liquid–vapor is also produced during torrefaction, which is condensed later and is converted to the liquid product. The key component of the liquid product of torrefaction is water followed by acid, alcohols, ketones, methane, furfural, etc. (Stelt et al., 2011; Chang et al., 2012; Medic et al., 2012). The condensable liquids can be subdivided into three parts – that is, water product, organic compounds, and lipid compounds (Bergman et al., 2005). Water is formed in liquid during torrefaction because of evaporation of water containing the biomass, dehydration, and thermal decomposition of biomass, especially hemicellulose. The water content in liquid products gradually decreases with a rise in the degree of torrefaction, therefore tar yield increases (Chen et al., 2018). The organic compounds consist of sugar; polysugars; furans; aldehydes; ketones such as methanol, acetic acid, lactic acid, formic acid, furfural and hydroxyl acetate, which are mostly produced during devolatilization and carbonization (Tumuluru et al., 2011).

During torrefaction, a variety of acids, aldehydes, alcohols, and other organic compounds are formed due to decarboxylation of hemicellulose, cleavage of glycosidic bond, damage of C–C bond as well as the fermentation of C–O bond (Chen et al., 2018). The formation of those organic compounds increases with a rise in torrefaction temperature because of a higher degree of disintegration of biomass. The development of acetic acid and methanol is attributed to acetyl groups in the hemicellulose structure and methoxyl groups in hemicellulose and lignin, respectively (Medic et al., 2012). The HHV of these liquid products was minimal due to high water content and was found to be 8.7 MJ/kg at 300°C torrefaction temperature. Due to having a very poor HHV, these liquid products cannot be utilized as suitable fuel in a boiler or engine. But liquid products obtained from torrefaction can be used for washing raw materials to reduce the metallic species in cotton stalk as well as to improve the products of pyrolysis (Chen et al., 2018). In the end, the liquid compounds consist of waxes, terpenes, tannins, benzene, fatty acids, and phenols (e.g., toluene). Those lipid products are not produced from reaction but formed through the evaporation of inert compounds during torrefaction and condensed at room temperature (Bergman et al., 2005).

A small amount of tar can also form in the liquid product during torrefaction, which mainly contains carbon followed by oxygen. In addition to this, the decomposition of lignin at higher torrefaction temperature is also attributed to the formation of some phenol and aromatic compounds, which leads to a growth in the carbon content of tar. Chen et al. (2018) observed that the HHV of tar obtained from rice husk torrefaction was 21.1–23.9 MJ/kg, while torrefaction temperature ranged from 210°C to 300°C, thus indicating the potentiality of tar as fuel.

Non-condensable or permanent gases are another important product of torrefaction. Among the permanent gases, CO_2 and CO are the dominant products followed by other gas components such as H_2 and CH_4 (Medic et al., 2012; Wang et al., 2011). The gas product has a low heating value and contains around 10% of the energy of biomass. According to Medic et al. (2012), the moisture content of raw biomass slightly affects the composition of permanent gases. With an increase in torrefaction temperature, CO_2 slightly decreased while CO increased significantly. Decarboxylation reaction owing to the presence of unstable carboxyl group in hemicellulose structure of biomass may be responsible for the formation of CO_2 (Prins et al., 2006a, 2006b), while secondary reactions of CO_2 and steam with porous char and the decarboxylation of low molecular mass carbonyl compounds, which are formed during torrefaction, may be responsible for the formation of CO. Moreover, CO is mainly responsible for the HHV of non-condensable gases produced through torrefaction (Tumuluru et al. 2011). Some light aromatic components are also formed such as benzene and toluene.

The concentration and total amount of gas products are positively influenced by torrefaction temperature as well as the volatile matter content of parent biomass (Deng et al. 2009). Besides, low methane was detected at high torrefaction conditions. At lower torrefaction temperatures such as 200–240°C, the amount of gaseous product yield is very low but with an increase in torrefaction temperature, the gaseous product yield heightened as the hemicellulose and cellulose disintegrated much at higher temperatures (Chen et al., 2018). Additionally, a kinetic study on the production of permanent gases during torrefaction demonstrated that the gases are usually produced via similar autonomous first-order reactions (Prins et al., 2006b). Prins et al. (2006a) found higher non-condensable volatiles in deciduous wood (willow) compared to coniferous wood indicating the higher degree of devolatilization for deciduous, xylan-containing wood compared to coniferous wood. The alternative use of torrefied gas is in the production of heat that can be utilized for drying wet biomass.

5.8   Effects of Torrefaction Parameters

Different parameters play a vital role in the yields of torrefaction products as well as their properties. The main torrefaction parameters are temperature, residence time, moisture content of raw biomass, the particle size of biomass, and heating rate during torrefaction.

5.8.1   Effects of Temperature

Biomass degradation reaction depends on plant components and their compositions. Thermal degradation of lignocellulosic structures occurs when biomass is exposed to an elevated temperature, which ultimately causes mass loss. The various components of biomass perform individually and interact differently depending on torrefaction temperature. Torrefaction operating temperature plays a dominant role in the quality and yield of the torrefied product. With the increase in the torrefaction temperature, the yield of the solid product falls; on the contrary, the yield for volatile compounds increases (Wang et al., 2011; Medic et al., 2012). This mass loss is assigned to thermal cracking and devolatilization of cellulose, hemicellulose, and lignin during torrefaction (Crawford et al., 2016), whereas the contribution of hemicellulose is the highest (Chen and Kuo, 2010).

The most thermally stable compound is lignin, which usually decomposes over a wide range of temperatures. High reactivity, decarbonization and more intense devolatilization of hemicellulose, and decomposition of cellulose and lignin at high temperature cause huge mass loss during torrefaction (Medic et al. 2012). The biomass degradation rate is also affected by the composition of hemicellulose. For example, hemicellulose of deciduous wood containing a higher amount of xylan is more sensitive and reactive in temperature and causes high mass loss than coniferous wood containing glucomannan-based hemicellulose (Basu, 2018a). Mass yield of rice husk decreases from 90 wt.% to 74 wt.% while torrefaction temperature increases from 230°C to 270°C.

Bergman et al. (2005) studied the impact of different torrefaction parameters (e.g., temperature, reaction time, particle size) on the mass yield and grindability of torrefied biomass via the multifactor method and found the effect of temperature to be superior to others. Numerous studies have been conducted to understand the effects of different torrefaction parameters (Hill et al., 2013; Yue et al., 2017; Chen et al., 2017a, 2017b; Prapakarn et al., 2018; Bai et al., 2018; Dyjakon et al., 2019). These studies concluded that the influence of torrefaction temperature was highest on yield and properties of biomasses followed by reaction time and heating rate, whereas particle size has the minimum influence. Manatura (2020) linked the effect of temperature and residence time on the mass and energy yield of sugarcane baggage and brought to light that a superior role had been played by temperature rather than time.

Although mass and energy yield decline with torrefaction temperature, the energy density of torrefied biomass upsurges with the degree of torrefaction because a decrease in energy is inferior to a decline in mass. Singh et al. (2020) mentioned that the optimum torrefaction temperature of woody biomass was 250°C with 40 min residence time concerning energy yield and HHV of torrefied biomass. Furthermore, for woody biomass, the optimum torrefaction condition was 250°C and 60 min considering grindability (Wang et al., 2011).

Tian et al. (2020) conducted torrefaction on corncob in a tubular furnace at torrefaction temperatures (210, 240, 270, and 300°C) for 30 min and reported that the most suitable torrefaction temperature was 240°C concerning mass and energy yield. Kai et al. (2019) also mentioned that torrefaction temperature considerably affected the characteristic of rice straw. Additionally, the recommended temperature for upgrading spent coffee residue by torrefaction was 260°C (Barbanera and Muguerza, 2020). Other researchers also obtained similar results. Therefore, the temperature was found to be the dominating parameter among all torrefaction process parameters, and the optimum operating temperature was approximately found at 250°C based on mass and energy yield, energy density and grindability.

5.8.2 Effects of Residence Time

Besides temperature, the role of residence time is also vital for the degradation of hemicellulose. Torrefaction requires longer residence time compared to other thermochemical conversation processes such as combustion and pyrolysis (Nhuchhen et al., 2014). A long residence time allows biomass material to reside within the reactor for an elongated time. However, the net influence of residence time is not as important as temperature. The yield of torrefied products and their characteristics were affected at longer torrefaction time. Mass loss of solid product is usually accelerated with an increase in residence time resulting in lower mass as well as energy yield (Tian et al., 2020; Kai et al., 2019; Singh et al., 2020). This happens due to growth in the magnitude of devolatilization (Prins et al., 2006b).

Solid and energy yield of rice straw after torrefaction decreased by 5.4% and 3%, respectively with a rise in residence time from 10 min to 70 min (Kai et al., 2019). Longer duration time (60 min) at higher-severity torrefaction condition is usually better than shorter torrefaction duration (30 min) for biomass torrefaction process because longer reaction duration maintains the higher energy yield (Chen et al., 2017a, 2017b; Manatura, 2020). Previous studies show that torrefaction above 1 h requires more energy, but biomass degradation is slow.

Similar results are also obtained by Cardona et al. (2019), who showed that residence time had a mild impact in respect to mass yield and energy gain. Volatile (condensable and non-condensable) products significantly contribute to mass yield at higher residence time as the yield of volatiles product upsurges with residence time. The release rate of CO and CO_2 was maximum at residence time for 8–12 min at 300°C for rice straw and then decreased gradually (Chen et al., 2018). Bates and Ghoniem (2012) had also drawn the same conclusion and mentioned that the quantity of methanol and lactic acid also increased up to 10 min and then remain unchanged.

In the case of the elemental composition of torrefied biomass, the carbon content surged with the rise in residence time while oxygen followed the reverse trend. The carbon content of sugarcane baggage increased from 37.1 wt.% to 42.6 wt.% at torrefaction temperature 250°C when residence time boosted up from 15 min to 60 min while oxygen content decreased from 57.2 wt.% to 52 wt.% (Kanwal et al., 2019). Though the carbon content rises with an increase in reaction time, the absolute carbon content always declines under expansion in the reaction of carbon dioxide and the steam with the porous char (Prins et al., 2006b).

5.8.3 Effects of Particle Size

The particle size of biomass affects the torrefaction yield. The amount of heat required to torrefy or devolatilize the biomass usually depends on the properties of biomass, its size, and shape. The mode of heat transfer (convection and conduction) and the rate of heat transfer from the reactor to biomass or within biomass are also affected by these parameters (Nhuchhen et al., 2014). The lower the particle size, the higher is the contact surface area and heat transfer efficiency during the process of torrefaction. There is a possibility of a nonuniform distribution of heat within the biomass for larger particle sizes due to the heterogeneous and anisotropic properties of the biomass (Nhuchhen et al., 2014). Additionally, secondary reactions that can take place during torrefaction may also be affected by particle size and influence the product yield.

Weight loss of solid product after torrefaction of pine sample decreased from 12.2 wt.% to 9.2 wt.% when particle size increased from 0.23 mm to 0.81 mm because of higher heat transfer rate and lower diffusion resistance of volatiles in small particles compared to a large one (Peng et al., 2012). Bergman et al. (2005) carried out a torrefaction experiment with three different particle sizes while other variables remained constant and found no significant effects up to 5 cm. High percentages of mass loss were observed in smaller particles compared to that of a larger particle at the same torrefaction conditions (Medic et al., 2012; Kokko et al., 2012). Besides, biomass particle size has a great influence on heat transfer efficiency for microwave-assisted torrefaction. Wang et al. (2012) found that mass reduction ratios of rice straw was 65 wt.%, 69 wt.%, and 72 wt.% at different particle sizes at 50/100 mesh, 100/200 mesh, and more than 200 mesh, respectively, which indicates that mass loss increases with decrease in particle size as high reaction temperature and heating rate can be obtained at smaller particle size.

5.8.4 Effects of Heating Rates

The heating rate also affects the yield of torrefaction significantly. The yield of condensable liquid increases with a rise in torrefaction heating rate. This is because the higher heating rate reduces the side reactions as well as the effects of heat and mass transfer in the particles (Kumar et al., 2017). Furthermore, the morphological structure of torrefied biomass is also affected by the heating rate.

5.8.5 Effects of Moisture Content

The efficiency of torrefaction depends on the initial moisture content of fresh biomass. If the moisture content of raw biomass is high, then almost all moisture evaporates at the first stage of torrefaction, and thereby mass yield decreases. The high moisture content of biomass can create a high-temperature profile within the materials, which can affect the torrefaction performance resulting in a shorter residence time.

5.8.6 Effects of Reactors

Different types of reactors are used for torrefaction. Heat is usually transferred to the reactors by two main mechanisms of direct and indirect heating. The working media, movement of biomass, and the mechanism of heat transfer are vital distinctive features of the reactors. There are several types of reactors used for torrefaction of biomass such as fixed bed reactor, fluidized bed, microwave, rotary drum, and screw-type (Chen et al., 2013; Basu, 2018b; Nachenius et al., 2013). All reactors have several advantages and disadvantages. The schematic diagram of different torrefaction reactors is shown in Figure 5.5.

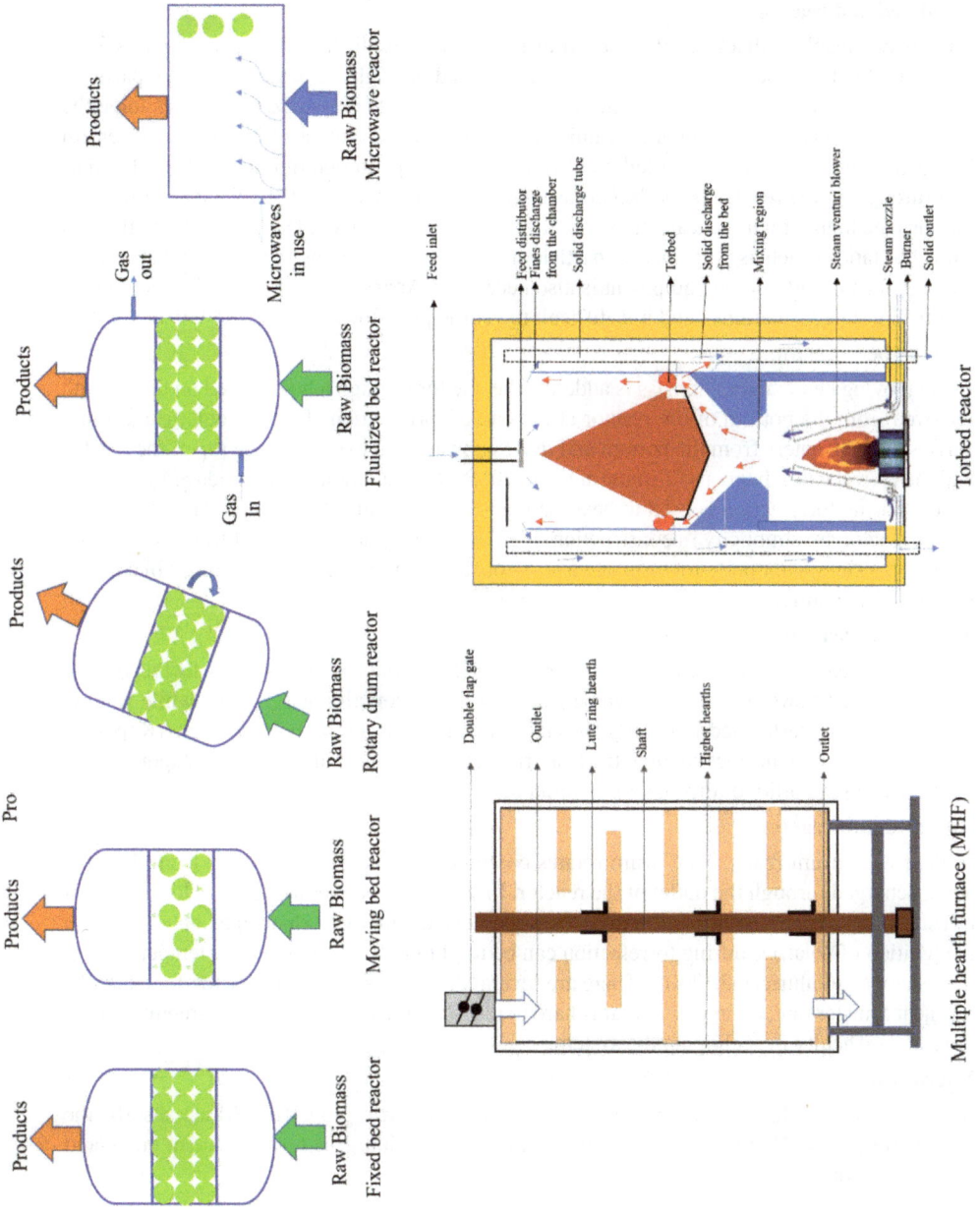

FIGURE 5.5 Different types of torrefaction reactors.

Reference: Mamvura and Danha (2020) and Koppejan et al. (2012)

(i) Fixed bed reactor

It is the simplest reactor among all for torrefaction, and the operation is easy for any type of biomass. In this reactor, the heating bed is fixed. The biomass is fed into the reactor to dry and torrefy in the furnace. At the end of torrefaction, the reactor is cooled down, and torrefied biomass is collected (Ribeiro et al., 2018). The main disadvantages of this reactor are lower heat transfer rate and the possibility of having a hotspot.

(ii) Fluidized bed reactor

It is easily scalable with an excellent heat transfer rate. In a fluidized bed reactor, biomass feedstocks are fed from the bottom of the reactor and moved through the heated inorganic particles by blowing hot inert gas from the bottom of the reactor. These pressurized hot gases force the ground or pulverized raw biomass to fluidize causing the movement of biomass to a certain height throughout the reactor called bed height, which helps to obtain uniform heat distribution throughout the reactor bed (Nachenius et al., 2013). In this case, the velocity of this inert gas must be higher than the least fluidization velocity (Mamvura and Danha, 2020). It also has some limitations such as very smaller particle size is required to blow the gas with high velocity, and, for this, additional equipment is also necessary. Apart from this, there is a possibility of attrition (fine formations), and it is difficult to get the plug flow.

(iii) Moving bed reactor

In the moving bed reactor, biomass is added to the reactor through a hole placed on the top and removed from the bottom of the reactor at the end of torrefaction. The gaseous medium that carries the heat enters from the bottom and moves upward and exits from the top of the reactor and thus torrefaction happens (Tumuluru et al., 2011). Recirculation of the produced gases and vapors during torrefaction is used to heat the reactor (Nachenius et al., 2013). The advantages of this reactor are simplicity in construction, high bed density, and good heat transfer capacity. However, a significant pressure drop can also be observed, and there are difficulties in controlling the temperature.

(iv) Screw-type reactor

Screw-type reactor is a continuous reactor in which biomass is transported to the reactor through auger screws. The reactor can be placed both horizontally or vertically while heating of the reactor is performed indirectly through a medium inside the hollow screw (Koppejan et al., 2012). Due to indirect heating, the heat transfer rate is lower, and several hot spots can be formed. The scalability of this reactor is limited.

(v) Rotary drum reactor

In the rotary drum reactor, the drum rotates continuously, and biomass is fed from the inlet and discharged through the outlet of the reactor. In this case, the pressure drop is low, and heat transfer occurs both directly and indirectly. Superheated vapor or exhaust gas obtained from the ignition of volatiles during torrefaction can be used to achieve the required temperature of the reactor (Tumuluru et al., 2010). There are several disadvantages of this reactor, for example, low heat transfer rate, temperature that is hard to measure and regulate, a lesser amount of plug flow, and difficulty in scaling up the system.

(vi) Microwave reactor

In microwave torrefaction, water in the biomass has dielectric properties, which absorbs the microwave power. Therefore, heat is generated for torrefaction. It is feasible due to quick and faster operation.

(vii) Torbed reactor

In this reactor, intense heat transfer occurs, which enables torrefaction at a short residence time. A medium carrying heat is driven from the lowest end of the torbed at high velocity (50–80 m/s), which forces the biomass particles to move both horizontally and vertically inside the reactor because toroidal swirls are formed and heat the biomass particles rapidly on the outer walls of the reactor. However, the particle size of biomass is important in this process (Koppejan et al., 2012).

(viii) Multiple hearth furnace (MHF) or Herreshoff oven

This is a continuous reactor and an established technology for numerous applications. A solo phase in the torrefaction process takes place in every individual layer of the reactor. Over the layers, the temperature gradually increases from 220°C to 300°C. Biomass usually arrives from the topmost part of the reactor on a horizontal plate, which is driven to the inside mechanically. Furthermore, the biomass particles fall on a second plate through a hole. The same procedure is repeated over all the multiple layers to get even mixing and steady heating. In the upper layer of the reactor, the drying of biomass takes place while torrefaction takes place in the lower layer of the reactor. Temperature and residence time can easily be controlled, and fine particles can be added. The major limitations of this type of reactor are the requirement of more space for installation, low heat transfer rate, and shaft scaling.

5.9 Properties of Torrefied Biomass or Biochar

The physicochemical properties of biomass after torrefaction somehow differ from their parent biomass due to thermal degradation of the lignocellulosic structure. Torrefaction significantly improves the physical, chemical, and thermal properties of biomass, which are promising for further processing of the torrefied product. Furthermore, the color of the torrefied product changes from light brown to black depending on the torrefaction process conditions.

5.9.1 Proximate and Ultimate Analyses

The complex polymer of lignocellulosic biomass is converted to monomer and therefore from monomer to different condensable and non-condensable products through the process of torrefaction. This transformation modifies the proximate and ultimate components of torrefied products. The proximate and ultimate analyses of raw and torrefied biomass are presented in Table 5.1.

The proximate analysis consists of the measurement of volatile matter (VM), moisture, and ash while the fixed carbon (FC) can be determined by the differences. Torrefaction drives away almost all the moisture as well as the light volatile matter from the biomass by dehydration, decomposition, and devolatilization reactions of biomass. Therefore, volatile matter content declines while fixed carbon content rises with torrefaction severity. The decline in volatile matter content as well as in other chemical transformations of remaining compounds creates the more brittle coal-like carbonaceous solid torrefied product. Ash content of raw biomass has a comparative effect on the final torrefied product, and it increases slightly with an increase in the degree of torrefaction (Pahla et al., 2020; Manatura, 2020; Acharya and Dutta, 2016). Raw biomass usually has high volatile matter ranging from 70 to 88 wt.% and a low amount of fixed carbon around 10–21 wt.% (Chen et al., 2015).

In the case of torrefied biomass, VM lies in between 40–85 wt.%, while FC is in the range of 13–45 wt.% (Chen et al., 2015). The lignin content of lignocellulosic biomass is increased comparatively due to the decomposition of hemicellulose and incomplete depolymerization of cellulose and lignin throughout torrefaction (Zhang et al., 2015), which triggered the FC positively and decreased the VM with a rise in torrefaction temperature. Torrefied biomass contains 1–12% more ash than raw biomass. Torrefaction changes the properties of biomass to be similar to coal to replace coal for heat and power generation. Fuel ratio (FR) is another important parameter for torrefied biomass, which is defined as the ratio of FC to VM, usually used to evaluate the materials in the combustion process. The FR of raw biomass lies in the range of 0.23–0.25 while the FR was found to be in the range 0.98–1.56 for a torrefied wheat straw at severe torrefaction condition (Bai et al., 2018).

The ultimate analysis of biomass gives an idea of biomass composition/elements. The chemical compounds of lignocellulosic biomass are carbon, hydrogen, nitrogen, oxygen, and sulfur. The amount of oxygen present in biomass can be calculated by the differences. The major sources of heat from combustion are carbon and hydrogen in biomass. Although oxygen is beneficial for fuel burning, high oxygen content decreases the heating value of biomass. Coal usually has carbon 60–85 wt.% while oxygen ranges from 5–20 wt.% (Prins et al., 2007). On the contrary, the carbon content of biomass is low (50 wt.%), and oxygen content is high (45 wt.%). The calorific value of biomass is lower in comparison with coal.

Tian et al. (2020) found that carbon content increased by 54 wt% while oxygen content decreased by 67 wt.% in corncobs when torrefied at 300°C for 30 min. Kanwal et al. (2019) also found 54 wt.% more carbon and 25 wt.% less oxygen in sugarcane baggage after torrefying at 300°C for 60 min. The properties of biochar are not influenced significantly due to residence time. However, according to Table 5.1,

TABLE 5.1

Proximate and Ultimate Analyses of Biomass before and after Torrefaction

Reference	Temperature (°C)	Time (min)	Volatile Matter (wt%)	Ash (wt%)	Fixed Carbon (wt%)	Carbon (wt.%)	Hydrogen (wt%)	Oxygen (wt%)	Nitrogen (wt%)	Higher Heating Value (MJ/kg)
Wang et al. (2018b)	Corn stalk		87.2	2.3	2.3	44.5	6.3	46.4	0.3	17.7
	200		84.4	2.8	12.8	47.7	5.2	44.1	0.2	18.7
	230		78.9	3.3	17.9	50.7	4.8	41.0	0.3	19.8
	260		66.4	4.3	29.3	54.6	4.5	36.3	0.3	21.3
	290		42.9	6.3	50.8	61.2	3.8	28.3	0.4	23.9
Zhang et al. (2019)	Corn stalk digest	30	63.2	9.9	16.7	37.7	4.9	35.5	2.0	15.3
	220		59.2	22.3	18.3	40.9	4.1	31.0	1.6	16.4
	240		50.3	26.5	22.7	42.0	3.8	26.2	1.7	17.0
	260		45.3	28.8	25.4	42.7	3.2	23.6	1.8	17.4
	280		31.7	34.7	32.6	44.6	2.8	16.2	1.8	17.8
Tian et al. (2020)	Corncobs		85.7	4.9	9.4	44.3	6.4	44.4	0.4	18.7
	210	30	85.7	5.8	13.2	47.8	6.3	40.2	0.4	20.1
	240		81.3	7.2	11.5	54.2	6.0	32.7	0.5	22.8
	270		57.2	10.8	32.0	65.1	5.2	18.8	0.7	26.9
	300		37.8	12.2	50.0	68.4	5.0	14.4	0.8	28.1
Singh et al. (2020)	Woody biomass		81.8	0.7	11.4	43.8	7.9	47.9	0.4	19.3
	220	40	58.6	1.2	36.7	50.3	6.8	42.3	0.6	20.8
	250		46.1	1.4	50.7	58.6	5.2	35.5	0.7	22.5
	280		36.8	1.9	60.4	64.8	4.9	29.8	0.6	24.8
	Coal		39.6	6.6	51.1	76.5	5.2	2.3	8.3	30.4
Kanwal et al. (2019)	Sugarcane bagasse					32.5	5.0	61.6	0.4	16.5
	200	15				32.8	5.0	61.3	0.4	19.1
		30				33.1	5.0	61.0	0.4	19.3
		45				33.9	5.0	60.2	0.4	19.9
		60				34.5	5.0	59.7	0.4	20.4
	225	15				34.8	5.0	59.3	0.4	19.2
		30				36.2	5.0	58.0	0.4	19.3
		45				36.9	4.9	57.4	0.4	20.4
		60				38.0	4.8	56.3	0.4	21.7
	250	15				37.1	4.9	57.2	0.4	19.3
		30				39.2	4.7	55.3	0.3	19.7
		45				40.5	4.7	54.0	0.3	20.6
		60				42.6	4.7	52.0	0.3	22.1
	275	15				39.6	4.6	55.0	0.3	19.3
		30				42.1	4.3	52.8	0.3	20.6
		45				43.2	4.2	51.9	0.3	21.3
		60				45.6	4.1	49.6	0.3	22.6
	300	15				40.6	3.9	57.8	0.3	20.8
		30				48.3	3.8	47.3	0.3	21.2
		45				49.6	3.6	46.3	0.3	23.0
		60				50.3	3.4	45.8	0.2	24.0

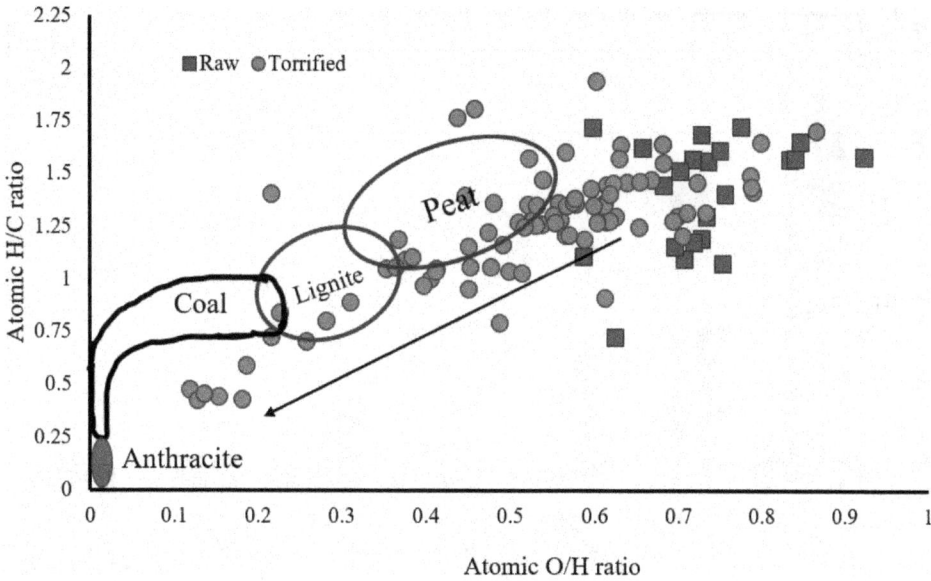

FIGURE 5.6 Van Krevelen diagram for raw and torrefied biomass samples (data adapted from Table 5.1).

there is a gradual growth in carbon and hydrogen content and a decrease in oxygen content with an increase in both torrefaction temperature as well as residence time. The rise in carbon content and decline in oxygen content in torrefied biochar is mainly due to the release of CO, CO_2, and water vapor during torrefaction (Park et al., 2013). Similarly, a declining trend in the hydrogen content was observed in torrefied biomass mostly at higher temperatures due to the release of some hydrocarbons at higher torrefaction temperatures (Ru et al., 2015).

The atomic ratio of biomasses is illustrated in Figure 5.6. A linear relationship was observed between H/C and O/C atomic ratios. The O/C and H/C ratio of raw biomass and torrefied biomass range from 0.4 to 0.8 and 1.2 to 2 as well as 0.1 to 0.84 and 0.4 to 1.7, respectively. From Figure 5.7, it can be concluded that the atomic ratio decreases with increasing torrefaction severity. Torrefaction at medium-to-high temperatures can produce coal-like solid fuel (Pahla et al., 2020).

5.9.2 Higher Heating Value

The HHV of biomass before and after torrefaction is represented in Table 5.1. Raw biomass usually has a lower heating value due to high C/O content. The relative decrease in oxygen and enhancement of carbon content in torrefied biomass result in a significant increase in the HHV of torrefied biomass. HHV of the torrefied product increases due to having more C–C and C–H bonds with the aromatic molecules, which can release more energy than raw biomass containing more C–O and H–O bonds. When fuel is burned, carbon plays an endothermic role whereas an exothermic role is played by hydrogen and oxygen, thus controlling C/O and H/O improves the calorific value of biomass (Senneca, 2017). HHV of raw biomass for the presented biomass falls in the range of 15.25–19.31 MJ/kg while HHV of torrefied biomass ranges between 17 and 29 MJ/kg, which is close to coal as it has HHV of 30.4 MJ/kg (Singh et al., 2020). It is reported that residence time has less influence on energy loss compared to temperature.

5.9.3 Grindability

Grindability is defined as the amount of energy needed for the grinding process of any materials. Biomass needs to be ground to uniform small particle size before utilization. The grinding energy requirement for raw biomass is usually high owing to its highly tenacious and fibrous nature. During torrefaction, the

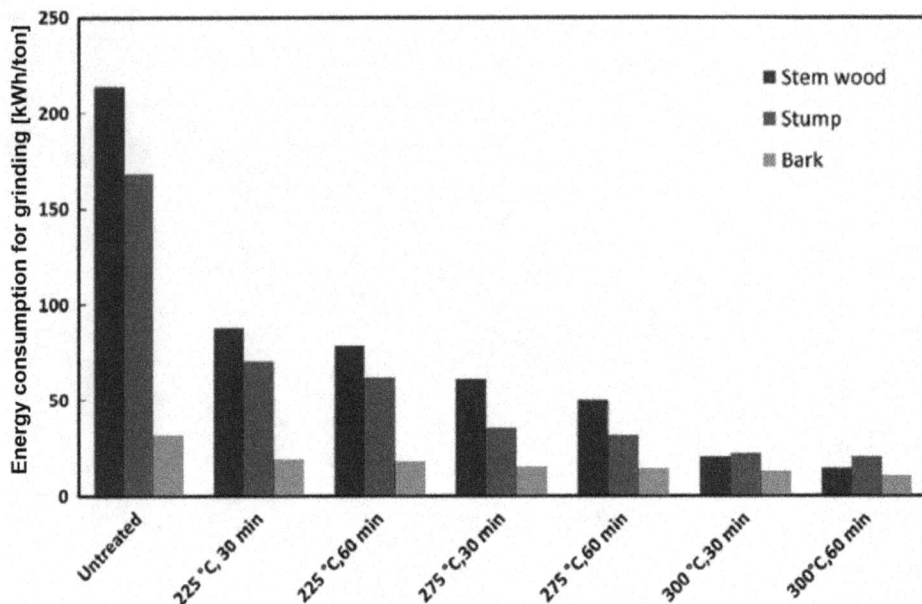

FIGURE 5.7　Grinding energy consumption of biomass. Reproduced with permission from Wang et al. (2018a).

tenacious nature of biomass is dropped due to the breakdown of the hemicellulose matrix and depolymerization of the cellulose, which also decrease the fiber length. Besides, the removal of hemicellulose during torrefaction makes the cell wall weaken. Thus, the fiber particle becomes shorter, spherical, and brittle compared to parent biomass.

Biomass tends to shrink during the torrefaction, which makes the biomass flaky, lightweight, and fragile with low mechanical strength. This change in microstructure improves grindability of torrefied biomass. Several factors can influence the grindability of biomass such as initial particle size, material properties, feeding rate, and grinding machine (Mani et al., 2004). Apart from this, the moisture content of biomass also plays a vital role in grinding as high moisture content increases the shear strength resulting in high grinding energy (Mani et al., 2004).

In general, grindability is expressed as Hardgrove grindability index (HGI), which represents the difficulty in grinding the solid biomass into powdery form (Wu et al., 2012). The higher the HGI, the easier to crush the solid into power (Wu et al., 2012). The HGI increases with a rise in torrefaction severity resulting in a decrease in the specific grinding energy requirements (Wu et al., 2012; Manouchehrinejad and Mani, 2018). Grinding energy requirements of biomass at different torrefied conditions are shown in Figure 5.7.

It is mentioned that particle size changes to smaller dimensions after torrefaction helps to increase the sphericity and surface area, and these properties of torrefied biomass help to increase the combustion and gasification efficiency. Time has less impact on energy consumption compared to torrefaction temperature. The grinding energy of torrefied biomass was reduced by 80–90% compared to parent biomass (Bergman et al., 2005; Ciolkosz and Wallace, 2011). From Figure 5.7, it is observed that the grinding energy of medium torrefied biomass decreases by around 50% (Wang et al., 2018b). Torrefaction noticeably changes the physicochemical properties of biomass and therefore affects the grinding characteristics.

5.9.4　Hydrophobicity and Fungal Durability

Hydrophobicity of torrefied biomass is another important parameter to be considered as it improves microbial degradation. The primary cell wall composition of lignocellulosic biomass is accountable for its hygroscopic nature (Andersson and Tillman, 1989). The free and bound moisture inside the biomass causes absorption of water in humid conditions, which further causes microbial degradation.

Degradation of hemicellulose and decrease in the atomic ratio ultimately restricts to form water via hydrogen bonds (Phanphanich and Mani, 2011; Kanwal et al., 2019). Equilibrium moisture content measurement (EMC) or immersion test is generally used for measuring hydrophobicity. In the case of the immersion test, the hydrophobicity of biomass is determined by measuring the total moisture absorption while the samples are immersed in water for a certain period. On the contrary, EMC is measured by introducing material in a controlled environment (at 90% relative humidity and 30°C) for 48–72 h. EMC and moisture uptake rate of torrefied biomass were found much lower than their parent biomass (Yan et al., 2009). Additionally, the EMC of torrefied biomass decreases with an increase in torrefaction severity due to the high reduction of the –OH group at a higher temperature (Chen et al., 2014a, 2014b). On the other hand, the moisture uptake rate also reflects the hydrophobicity of biomass. The moisture uptake rate of raw sugarcane was found to be 11.8 wt.% and reduced to 1.01 wt.% when torrefied at 300°C for 60 min (Kanwal et al., 2019).

A similar trend was found for densified material as moisture uptake rate was reduced by 54% due to torrefaction (Azargohar et al., 2019). Torrefaction depolymerized the long chains of polysaccharides and thus created a short polymeric structure of biomass, especially from the hemicellulose's fractions (Bergman et al., 2005), which were coupled with limited devolatilization and carbonization of the lignin and cellulose. During torrefaction, some unsaturated biomass was also produced, which had more nonpolar properties due to chemical rearrangement reaction resulting in reduced affinity to moisture. Besides, the tar present in the liquid condensed on the top surface of biomass and blocked the pores in the torrefied product and reduced its capacity of moisture absorption through capillary action (Felfli et al., 2005; Ohm et al., 2015).

The reasons behind the hydrophobicity of torrefied biomass are (Ciolkosz and Wallace, 2011):

(i) The formation of more nonpolar molecules as a result of the breakdown of hemicellulose,

(ii) The elimination of –OH functional group during torrefaction, which restricts hydrogen bond formation with water,

(iii) Collapse of hemicellulose structures unfastens the cellulose and lignin allowing the water molecule to be released, and

(iv) Decomposition and deconstruction of hemicellulose make the cellulose and lignin brittle leading to hydrophobic nature.

This hydrophobic nature of torrefied biomass facilitates the utilization and allows transportation over long distances and storage for a long period in open space without deterioration, decomposition, decay, or microbial attack (Tumuluru et al., 2011). The moisture content of biomass during storage is important for introducing chemical and microbial reactions. High moisture content along with high-temperature storage may create extreme off-gassing and self-heating from bio-based fuels (Tumuluru et al., 2011).

5.9.5 Particle Size Distribution, Sphericity, and Particle Surface Area

To know the flowability and ignition behavior during combustion or co-firing, sphericity, particle size, and its distribution are important (Tumuluru et al., 2011). The particle size distribution of different biomasses in various size ranges at various torrefaction conditions is shown in Figure 5.8 (Wang et al., 2018a). According to this study, a large number of particles fall in a higher size range in the case of untreated biomass. With an increase in torrefaction severity, particles on a larger sieve size decrease while increasing the number of particles with a smaller size. Residence time has less influence on the particle size of torrefied biomass compared to coal. The smaller particle as well as more uniform particle size distribution can be obtained at higher torrefaction conditions having lower heat resistance in the additional thermochemical process (Singh et al., 2020). Besides, the cumulative particle size distribution curve shifts in the direction of smaller size because of acute fragility of torrefied biomass, which is comparable to coal (Repellin et al., 2010; Phanphanich and Mani, 2011). Similar results have been observed by other researchers (Kanwal et al., 2019; Singh et al., 2020).

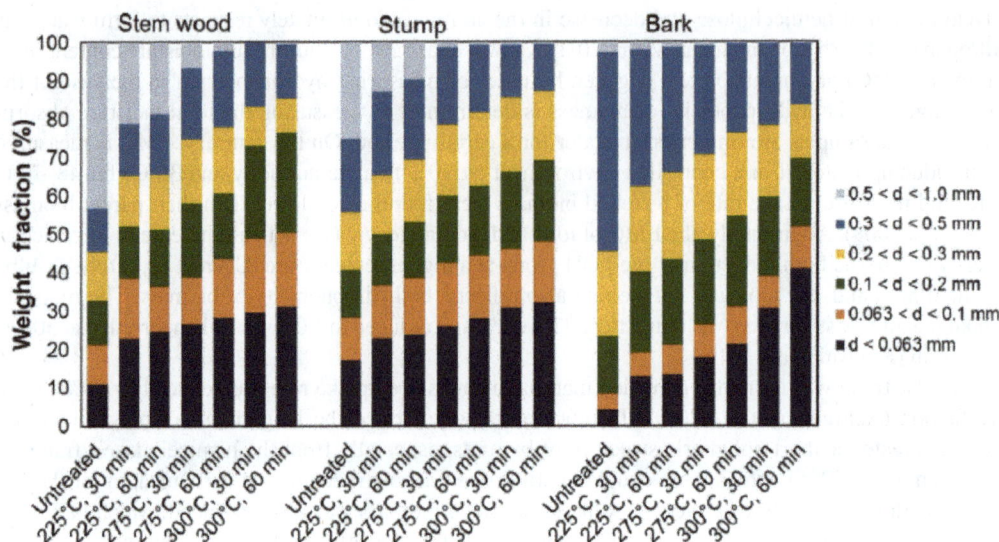

FIGURE 5.8   Distribution of biomass particle size at different torrefaction conditions. Reproduced with permission from Wang et al. (2018a).

Sphericity and surface area are also influenced by torrefaction conditions. Sphericity of pine chips increased from 0.5% to 0.6% at severe torrefaction conditions (300°C) (Phanphanich and Mani, 2011). Torrefied biomass usually has a higher particle surface area compared to untreated biomass, which indicates the efficient application of torrefied biomass for combustion and co-firing.

5.9.6   Density, Porosity, and Morphology

Density (i.e., bulk and particle) is a vital physical parameter of biomass for constructing a structure for handling, storage, and transport of biomass. Bulk density of biomass depends on various parameters such as particle size, shape, moisture content, and surface morphology. The density of biomass decreases with an increase in torrefaction temperature (Singh et al., 2020) due to the removal of gases and volatile materials through devolatilization. During torrefaction, the devolatilization process also causes the shrinkage in the physical dimension (size and shape) of biomass resulting in a reduced volume (Azargohar et al., 2019).

Morphology as well as the color of biomass also change after torrefaction. The tubular structure is noticed at the highest torrefaction condition due to degradation of lignin. This change of microstructure improves the grindability of biomass. Apart from this, the total volume of pore space in torrefied biomass is greater than that for the parent biomass owing to the release of gaseous and volatile products (Chen et al., 2011a, 2011b, 2011c). The combustion and gasification are also affected positively by the high porosity of torrefied biomass (Niu et al., 2019a, 2019b; Li et al., 2018).

5.9.7   Molecular Composition and Changes

Lignocellulosic biomass comprises three basic compounds cellulose $(C_6H_{10}O_5)_m$, hemicellulose $(C_5H_8O_4)_m$, and lignin $((C_9H_{10})_3(OCH_3)_{0.9-1.7})_m$ (m = degree of polymerizations). Weight percentages of different components of lignocellulosic biomass typically include 40–60% cellulose, 15–30% hemicellulose, and 10–25% lignin (Loppinet-Serani et al., 2010). Cellulose is a long-chain polysaccharide composed of β-1,4 linked D-glucose subunits, and its polymer of high molecular weight usually provides strength to the material by holding the bundles of fibers together. Hemicellulose is a combination of different polymerized monosaccharides such as glucose, mannose, xylose, arabinose, galactose, and glucuronic acid (Chen et al., 2015), which usually acts as a connector between the cellulose and lignin compound. Lignin has a very complex three-dimensional structure constructed of phenolic monomer units joined by ester bonds, which behave like adhesive binding cellulose and hemicellulose together. Additionally, it also delivers rigidity to the structure along with resistance and impermeability to the plant as well as support.

All these chemical compositions are affected by torrefaction, and the degree depends on torrefaction severity in different temperature ranges. Among all, hemicellulose is the most temperature-sensitive compound and easily degrades at a lower temperature around 220–315°C while cellulose starts to decompose at 315–400°C. A gradual decomposition of lignin appears at a temperature ranging from 160°C to 900°C. The decomposition peaks of cellulose and hemicellulose overlap in some biomasses due to the similarity in structure, while sometimes they are visibly recognized for some biomass samples (Chen and Kuo, 2010). Besides, the thermal behavior of cellulose is endothermic while exothermic reactions were found for hemicellulose and lignin. The decomposition of chemical compounds happens due to (Nordin et al., 2013; Nunes et al., 2018):

- The separation of acetyl groups from hemicellulose that encourages depolymerization of the polysaccharides of wood;
- Dehydration reactions for the obliteration of –OH groups;
- Demethoxylation of the lignin which produces methanol, acetic acid, and furfural; and
- Mitigation of the hydrogen bonds in lignin structure and alteration of the networks of the aromatic rings of lignin.

Hemicellulose is the most fragile polymer because of being amorphous. Torrefaction of hemicellulose promotes the development of acetic acid by de-acetalization. Depolymerization reactions promote the generation of pentoses (e.g., xylose and arabinose) and hexoses (e.g., mannose, glucose, and galactose). Cellulose is a polymer of a longer chain. The degree of crystallinity increases first with temperature. In addition, the hydroxyl groups present in cellulose and hemicellulose degrade.

5.10 Conclusions

Though biomass has huge potential in the energy sector, it is not widely used due to its unfavorable properties – for example, high moisture content, hygroscopic, and high volume. Torrefaction is an attractive way of thermochemical pretreatment, which expands the inherent properties of biomass – for example, high bulk density, HHV, hydrophobicity, low atomic ratio, and high grindability. Torrefaction also permits the use of torrefied biomass in combustion, co-firing with or without coal in the coal-fired power plants as well as gasification for the production of syngas. In case of improvement of properties of biomass, the temperature of torrefaction plays the highest influence, followed by residence time, moisture content, and in the end the particle size.

Acknowledgments

The authors would like to thank the Natural Sciences and Engineering Research Council (NSERC) of Canada, Canada Research Chairs (CRC) Program, Agriculture and Agri-Food Canada (AAFC), BioFuelNet Canada, and the College of Engineering's Dean's Scholarship program at the University of Saskatchewan for funding this research.

REFERENCES

Abedi, A., and Dalai, A. K. 2017. Study on the quality of oat hull fuel pellets using bio-additives. *Biomass and Bioenergy* 106: 166–175.

Acharya, B., and Dutta, A. 2016. Fuel property enhancement of lignocellulosic and nonlignocellulosic biomass through torrefaction. *Biomass Conversion and Biorefinery* 6: 139–149.

Agu, O., Tabil, L., and Dumonceaux, T. 2017. Microwave-assisted alkali pre-treatment, densification and enzymatic saccharification of canola straw and oat hull. *Bioengineering* 4: 1–32.

Aguado, R., Cuevas, M., Villarejo, L. P., Cartas, M. L. M., and Sánchez, S. 2020. Upgrading almond-tree pruning as a biofuel via wet torrefaction. *Renewable Energy* 145: 2091–2100.

Álvarez, A., Gutiérrez, I., Pizarro, C., Lavín, A. G., and Bueno, J. L. 2017. Comparison between oxidative and non-oxidative torrefaction pretreatment as alternatives to enhance properties of biomass. *WIT Transactions on Ecology and the Environment* 224: 247–255.

Andersson, M., and Tillman, A. M. 1989. Acetylation of jute: Effects on strength, rot resistance, and hydrophobicity. *Journal of Applied Polymer Science* 37: 3437–3447.

Arias, B., Pevida, C., Fermoso, J., Plaza, M. G., Rubiera, F., and Pis, J. J. 2008. Influence of torrefaction on the grindability and reactivity of woody biomass. *Fuel Processing Technology* 89: 169–175.

Axelsson, L., Franzén, M., Ostwald, M., Berndes, G., Lakshmi, G., and Ravindranath, N. H. 2012. Perspective: Jatropha cultivation in Southern India: Assessing farmers' experiences. *Biofuels, Bioproducts and Biorefining* 6: 246–256.

Azargohar, R., Soleimani, M., Nosran, S., Bond, T., Karunakaran, C., Dalai, A. K., and Tabil, L. G. 2019. Thermo-physical characterization of torrefied fuel pellet from co-pelletization of canola hulls and meal. *Industrial Crops and Products* 128: 424–435.

Azuma, J. I., Tanaka, F., and Koshijima, T. 1984. Enhancement of enzymatic susceptibility of lignocellulosic wastes by microwave irradiation. *Journal of Fermentation Technology* 62: 377–384.

Bach, Q. V, Chen, W. H., Lin, S. C., and Sheen, H. K. 2017. Wet Torrefaction of microalga *Chlorella vulgaris* ESP-31 with microwave-assisted heating. *Energy Conversion and Management* 141: 163–170.

Bach, Q. V., and Skreiberg, O. 2016. Upgrading biomass fuels via wet torrefaction: A review and comparison with dry torrefaction. *Renewable and Sustainable Energy Reviews* 54: 665–677.

Bach, Q. V., Tran, K. Q., Khalil, R. A., Skreiberg, Ø., and Seisenbaeva, G. 2013. Comparative assessment of wet torrefaction. *Energy and Fuels* 27: 6743–6753.

Bach, Q. V., Tran, K. Q., Skreiberg, Ø., and Trinh, T. T. 2015. Effects of wet torrefaction on pyrolysis of woody biomass fuels. *Energy* 88: 443–456.

Bai, X., Wang, G., Sun, Y., Yu, Y., Liu, J., Wang, D., and Wang, Z. 2018. Effects of combined pretreatment with rod-milled and torrefaction on physicochemical and fuel characteristics of wheat straw. *Bioresource Technology* 267: 38–45.

Barbanera, M., and Muguerza, I. F. 2020. Effect of the temperature on the spent coffee grounds torrefaction process in a continuous pilot-scale reactor. *Fuel* 262: 116493.

Basu, P. 2018a. Biomass characteristics. In *Biomass Gasification, Pyrolysis and Torrefaction: Practical Design and Theory.* Elsevier, pp. 49–87.

Basu, P. 2018b. Torrefaction. In *Biomass Gasification, Pyrolysis and Torrefaction: Practical Design and Theory.* Elsevier, pp. 93–154.

Basu, P., Sadhukhan, A. K., Gupta, P., Rao, S., Dhungana, A., and Acharya, B. 2014. An experimental and theoretical investigation on torrefaction of a large wet wood particle. *Bioresource Technology* 159: 215–222.

Bates, R. B., and Ghoniem, A. F. 2012. Biomass torrefaction: Modeling of volatile and solid product evolution kinetics. *Bioresource Technology* 124: 460–469.

Bergman, P. C. A, Boersma, A. R., R. Zwart, W. R., and Kiel, J. H. A. 2005. *Torrefaction for Biomass Co-firing in Existing Coal-fired Power Stations.* Energy Research Centre of the Netherlands ECN ECNC05013.

Bourgeois, J. P., and Doat, J. 1984. Torrefied wood from temperate and tropical species. advantages and prospects. *Bioenergy 84. Proceedings of conference, Goteborg, Sweden. Volume III. Biomass conversion,* June 15–21, pp. 153–159.

Brachi, P., Chirone, R., Miccio, M., and Ruoppolo, G. 2019. Fluidized bed torrefaction of biomass pellets: A comparison between oxidative and inert atmosphere. *Powder Technology* 357: 97–107.

Bridgeman, T. G., Jones, J. M., Williams, A., and Waldron, D. J. 2010. An investigation of the grindability of two torrefied energy crops. *Fuel* 89: 3911–3918.

Britain, V., and Offrion, F. O. 1990. Improvements in the process of and apparatus for rationally and continuously treating or torrefying coffee. *Patent GB190001714.*

Cao, L., Yuan, X., Li, H., Li, C., Xiao, Z., Jiang, L., Huang, B., et al. 2015. Complementary effects of torrefaction and co-pelletization: Energy consumption and characteristics of pellets. *Bioresource Technology* 185: 254–262.

Cardona, S., Gallego, L. J., Valencia, V., Martínez, E., and Rios, L. A. 2019. Torrefaction of eucalyptus-tree residues: A new method for energy and mass balances of the process with the best torrefaction conditions. *Sustainable Energy Technologies and Assessments* 31: 17–24.

Chang, S., Zhao, Z., Zheng, A., He, F., Huang, Z., and Li, H. 2012. Characterization of products from torrefaction of sprucewood and bagasse in an auger reactor. *Energy & Fuels* 26: 7009–7017.

Chaturvedi, V., and Verma, P. 2013. An Overview of Key Pretreatment processes employed for bioconversion of lignocellulosic biomass into biofuels and value added products. *Biotech* 3: 415–431.

Chen, D., Gao, A., Ma, Z., Fei, D., Chang, Y., and Shen, C. 2018. In-depth study of rice husk torrefaction: Characterization of solid, liquid and gaseous products, oxygen migration and energy yield. *Bioresource Technology* 253: 148–153.

Chen, D., Mei, J., Li, H., Li, Y., Lu, M., Ma, T., and Ma, Z. 2017a. Combined pretreatment with torrefaction and washing using torrefaction liquid products to yield upgraded biomass and pyrolysis products. *Bioresource Technology* 228: 62–68.

Chen, Q., Zhou, J. S., Liu, B. J., Mei, Q. F., and Luo, Z. Y. 2011. Influence of torrefaction pretreatment on biomass gasification technology. *Chinese Science Bulletin* 56: 1449–1456.

Chen, W. H. 2015. Torrefaction. In: *Pretreatment of Biomass: Processes and Technologies.* Elsevier Inc., pp. 173–192.

Chen, W. H., Hsu, H. J., Kumar, G., Budzianowski, W. M., and Ong, H. C. 2017b. Predictions of biochar production and torrefaction performance from sugarcane bagasse using interpolation and regression analysis. *Bioresource Technology* 246: 12–19.

Chen, W. H., Hsu, H. C., Lu, K. M., Lee, W. J., and Lin, T. 2011a. Thermal pretreatment of wood (Lauan) block by torrefaction and its influence on the properties of the biomass. *Energy* 36: 3012–3021.

Chen, W. H., and Kuo, P. C. 2010. A study on torrefaction of various biomass materials and its impact on lignocellulosic structure simulated by a thermogravimetry. *Energy* 35: 2580–2586.

Chen, W. H., and Kuo, P. C. 2011b. Torrefaction and co-torrefaction characterization of hemicellulose, cellulose and lignin as well as torrefaction of some basic constituents in biomass. *Energy* 36: 803–811.

Chen, W. H., Lu, K. M., Lee, W. J., Liu, S. H., and Lin, T. C. 2014a. Non-oxidative and oxidative torrefaction characterization and SEM observations of fibrous and ligneous biomass. *Applied Energy* 114: 104–113.

Chen, W. H., Lu, K. M., Liu, Tsai, C. M., Lee, W. J., and Lin, T. C. 2013. Biomass torrefaction characteristics in inert and oxidative atmospheres at various superficial velocities. *Bioresource Technology* 146: 152–160.

Chen, W. H., Peng, J., and Bi, X. T. 2015. A state-of-the-art review of biomass torrefaction, densificaation and applications. *Renewable and Sustainable Energy Reviews* 44: 847–866.

Chen, W. H., Tu, Y. J., and Sheen, H. K. 2011c. Disruption of sugarcane bagasse lignocellulosic structure by means of dilute sulfuric acid pretreatment with microwave-assisted heating. *Applied Energy* 88: 2726–2734.

Chen, W. H., Ye, S. C., and Sheen, H. K. 2012. Hydrothermal carbonization of sugarcane bagasse via wet torrefaction in association with microwave heating. *Bioresource Technology* 118: 195–203.

Chen, Y., Liu, B., Yang, H., Yang, Q., and Chen, H. 2014b. Evolution of functional groups and pore structure during cotton and corn stalks torrefaction and its correlation with hydrophobicity. *Fuel* 137: 41–49.

Cherubini, F., and Strømman, A. H. 2011. Principles of biorefining. *Biofuels*: 3–24.

Ciolkosz, D., and Wallace, R. 2011. A review of torrefaction for bioenergy feedstock production. *Biofuels, Bioproducts and Biorefining* 5: 317–329.

Crawford, N. C., Nagle, N., Sievers, D. A., and Stickel, J. J. 2016. Biomass and bioenergy the effects of physical and chemical preprocessing on the flowability of corn stover. *Biomass and Bioenergy* 85: 126–134.

Deng, J., Wang, G. J., Kuang, J. h., Zhang, Y. L., and Luo, Y. H. 2009. Pretreatment of agricultural residues for co-gasification via torrefaction. *Journal of Analytical and Applied Pyrolysis* 86: 331–337.

Dyjakon, A., Noszczyk, T., and Smêdzik, M. 2019. The influence of torrefaction temperature on hydrophobic properties of waste biomass from food processing. *Energies* 12.

Felfli, F. F., Luengo, C. A., Suárez, J. A., Beatón, P. A., Suárez, J. A., and Beatón, P. A. 2005. Wood briquette torrefaction. *Energy for Sustainable Development* 9: 19–22.

Goh, C. S., Tan, H. T., and Lee, K. T. 2012. Pretreatment of oil palm frond using hot compressed water: An evaluation of compositional changes and pulp digestibility using severity factors. *Bioresource Technology* 110: 662–669.

Gong, S. H., Im, H. S., Um, M., Lee, H. W., and Lee, J. W. 2019. Enhancement of waste biomass fuel properties by sequential leaching and wet torrefaction. *Fuel* 239: 693–700.

Guo, L., Wang, D., Tabil, L. G., and Wang, G. 2016. Compression and relaxation properties of selected biomass for briquetting. *Biosystems Engineering* 148: 101–110.

Haykiri-Acma, H., Yaman, S., and Kucukbayrak, S. 2016. Combustion characteristics of torrefied biomass materials to generate power. *4th IEEE International Conference on Smart Energy Grid Engineering, SEGE 2016*, pp. 226–230.

He, C., Tang, C., Li, C., Yuan, J., Tran, K. Q., Bach, Q. V., Qiu, R., and Yang, Y. 2018a. Wet torrefaction of biomass for high quality solid fuel production: A review. *Renewable and Sustainable Energy Reviews* 71: 259–271

He, Q., Guo, Q., Ding, L., Gong, Y., Wei, J., and Yu, G. 2018b. Co-pyrolysis behavior and char structure evolution of raw/torrefied rice straw and coal blends. *Energy and Fuels* 32: 12469–12476.

Hill, S. J., Grigsby, W. J., and Hall, P. W. 2013. Chemical and cellulose crystallite changes in *Pinus* radiata during torrefaction. *Biomass and Bioenergy* 56: 92–98.

Hoekman, S. K., Broch, A., and Robbins, C. 2011. Hydrothermal carbonization (HTC) of lignocellulosic biomass. *Energy and Fuels* 25: 1802–1810.

Hoekman, S. K., Broch, A., Robbins, C., Zielinska, B., and Felix, L. 2013. Hydrothermal carbonization (HTC) of selected woody and herbaceous biomass feedstocks. *Biomass Conversion and Biorefinery* 3: 113–126.

Huang, Y. F., Chen, W. R., Chiueh, P. T., Kuan, W. H., and Lo, S. L. 2012. Microwave torrefaction of rice straw and Pennisetum. *Bioresource Technology* 123: 1–7.

Ibrahim, R. H. H., Darvell, L. I., Jones, J. M., and Williams, A. 2013. Physicochemical characterisation of torrefied biomass. *Journal of Analytical and Applied Pyrolysis* 103: 21–30.

Jin, S., Guo, C., Lu, Y., Zhang, R., Wang, Z., and Jin, M. 2017. Comparison of microwave and conventional heating methods in carbonization of polyacrylonitrile-based stabilized fibers at different temperature measured by an in-situ process temperature control ring. *Polymer Degradation and Stability* 140: 32–41.

Kai, X., Meng, Y., Yang, T., Li, B., and Xing, W. 2019. Effect of torrefaction on rice straw physicochemical characteristics and particulate matter emission behavior during combustion. *Bioresource Technology* 278: 1–8.

Kambo, H. S., and Dutta, A. 2015. Comparative evaluation of torrefaction and hydrothermal carbonization of lignocellulosic biomass for the production of solid biofuel. *Energy Conversion and Management* 105: 746–755.

Kanwal, S., Chaudhry, N., Munir, S., and Sana, H. 2019. Effect of torrefaction conditions on the physicochemical characterization of agricultural waste (Sugarcane bagasse). *Waste Management* 88: 280–290.

Kappe, C. Oliver. 2004. Controlled microwave heating in modern organic synthesis. *Angewandte Chemie International Edition* 43: 6250–6284.

Kashaninejad, M., and Tabil, L. G. 2011. Effect of microwave-chemical pre-treatment on compression characteristics of biomass grinds. *Biosystems Engineering* 108: 36–45.

Kim, Y. H., Jun, K. W., Joo, H., Han, C., and Song, I. K. 2009. A simulation study on gas-to-liquid (natural gas to Fischer-Tropsch synthetic fuel) process optimization. *Chemical Engineering Journal* 155: 427–432.

Kim, Y. H., Lee, S. M., Lee, H. W., and Lee, J. W. 2012. Physical and chemical characteristics of products from the torrefaction of yellow poplar (*Liriodendron tulipifera*). *Bioresource Technology* 116: 120–215.

Kokko, L., Tolvanen, H., Hämäläinen, K., and Raiko, R. 2012. Comparing the energy required for fine grinding torrefied and fast heat treated pine. *Biomass and Bioenergy* 42: 219–223.

Koppejan, J., Sokhansanj, S., Melin, S., and Madrali, S. 2012. Status overview of torrefaction technologies. *IEA Bioenergy Task* 32.

Kostas, E. T., Beneroso, D., and Robinson, J. P. 2017. The application of microwave heating in bioenergy: A review on the microwave pre-treatment and upgrading technologies for biomass. *Renewable and Sustainable Energy Reviews* 77: 12–27.

Kumar, L., Koukoulas, A. A., Mani, S., and Satyavolu, J. 2017. Integrating torrefaction in the wood pellet industry: A critical review. *Energy and Fuels* 31: 37–54.

Kuo, P. C., Wu, W., and Chen, W. H. 2014. Gasification performances of raw and torrefied biomass in a downdraft fixed bed gasifier using thermodynamic analysis. *Fuel* 117: 1231–1241.

Lange, J. P. 2007. Lignocellulose conversion: An introduction to chemistry, process and economics. *Biofuels, Bioproducts and Biorefining* 1: 39–48.

Laughlin, B., and Erasmus, N. 2009. Energy products from wood waste using torbed reactor technology. *TAPPI Engineering, Pulping & Environmental Conference*, Memphis, TN.

Li, H., Liu, X., Legros, R., Bi, X. T., Lim, C. J., and Sokhansanj, S. 2012a. Pelletization of torrefied sawdust and properties of torrefied pellets. *Applied Energy* 93: 680–685.

Li, H., Qu, Y., Yang, Y., Chang, S., and Xu, J. 2016. Microwave irradiation – a green and efficient way to pretreat biomass. *Bioresource Technology* 199: 34–41.

Li, M. F., Shen, Y., Sun, J. K., Bian, J., Chen, C. Z., and Sun, R. C. 2015. Wet torrefaction of bamboo in hydrochloric acid solution by microwave heating. *Sustainable Chemistry and Engineering* 3: 2022–2029.

Li, M. F., Sun, S. N., Xu, F., and Sun, R. C. 2012b. Organosolv fractionation of lignocelluloses for fuels, chemicals and materials: A biorefinery processing perspective. In: *Biomass Conversion: The Interface of Biotechnology, Chemistry and Materials Science*, (Eds.) C. Baskar, S. Baskar, and R. S. Dhillon. Heidelberg: Springer-Verlag Berlin, pp. 341-379.

Li, T., Niu, Y., Wang, L., Shaddix, C., and Løvås, T. 2018. High temperature gasification of high heating-rate chars using a flat-flame reactor. *Applied Energy* 227: 100–107.

Libra, J. A., Ro, K. S., Kammann, C., Funke, A., Berge, N. D., Neubauer, Y., Titirici, M. M., Fuhner, C., Bens, O., Kern, J., and Emmerich, K. H. 2011. Hydrothermal carbonization of biomass residuals: A comparative review of the chemistry, processes and applications of wet and dry pyrolysis. *Biofuels* 2: 71–106.

Liu, Z., and Balasubramanian, R. 2014. Upgrading of waste biomass by hydrothermal carbonization (HTC) and low temperature pyrolysis (LTP): A comparative evaluation. *Applied Energy* 114: 857–864.

Liu, Z., Quek, A., Hoekman, S. K., and Balasubramanian, R. 2013. Production of solid biochar fuel from waste biomass by hydrothermal carbonization. *Fuel* 103: 943–949.

Loppinet-Serani, A., Aymonier, C., and Cansell, F. 2010. Supercritical water for environmental technologies. *Journal of Chemical Technology & Biotechnology* 85: 583–589.

Lynam, J. G., Coronella, C. J., Yan, W., Reza, M. T., and Vasquez, V. R. 2011. Acetic acid and lithium chloride effects on hydrothermal carbonization of lignocellulosic biomass. *Bioresource Technology* 102: 6192–6199.

Mäkelä, M., Benavente, V., and Fullana, A. 2015. Hydrothermal carbonization of lignocellulosic biomass: Effect of process conditions on hydrochar properties. *Applied Energy* 155: 576–584.

Mamvura, T. A., and Danha, G. 2020. Biomass torrefaction as an emerging technology to aid in energy production. *Heliyon* 6: e03531.

Manatura, K. 2020. Inert torrefaction of sugarcane bagasse to improve its fuel properties. *Case Studies in Thermal Engineering* 19: 100623.

Mani, S., Tabil, L. G., and Sokhansanj, S. 2004. Grinding performance and physical properties of wheat and barley straws, corn stover and switchgrass. *Biomass and Bioenergy* 27: 339–352.

Manouchehrinejad, M., and Mani, S. 2018. Torrefaction after pelletization (TAP): Analysis of torrefied pellet quality and co-products. *Biomass and Bioenergy* 118: 93–104.

Medic, D., Darr, M., Shah, A., Potter, B., and Zimmerman, J. 2012. Effects of torrefaction process parameters on biomass feedstock upgrading. *Fuel* 91: 147–154.

Motasemi, F., and Afzal, M. T. 2013. A Review on the microwave-assisted pyrolysis technique. *Renewable and Sustainable Energy Reviews* 28: 317–330.

Murakami, K., Kasai, K., Kato, T., and Sugawara, K. 2012. Conversion of rice straw into valuable products by hydrothermal treatment and steam gasification. *Fuel* 93: 37–43.

Muslim, M. B., Saleh, S., and Samad, N. A. F. A. 2017. Torrefied biomass gasification: A simulation study by using empty fruit bunch. *MATEC Web of Conferences* 131.

Nachenius, R. W., Ronsse, F., Venderbosch, R. H., and Prins, W. 2013. Biomass pyrolysis. *Advances in Chemical Engineering* 42: 75–139.

Nachenius, R. W., Wardt, T. V. D., Ronsse, F., and Prins, W. 2014. Torrefaction of biomass in a continuous rotating screw reactor. *22nd European Biomass Conference and Exhibition, Germany,* ETA-Florence Renewable Energies, Germany, pp. 1018–1024.

Nanda, S., and Berruti, F. 2021. A technical review of bioenergy and resource recovery from municipal solid waste. *Journal of Hazardous Materials* 403: 123970.

Nanda, S., Mohammad, J., Reddy, S. N., Kozinski, J. A., and Dalai, A. K. 2014. Pathways of lignocellulosic biomass conversion to renewable fuels. *Biomass Conversion and Biorefinery* 4: 157–191.

Newell, R. G., Raimi, D., and Aldana, G. 2019. Global Energy Outlook 2019: The Next Generation of Energy. *Resources for the Future*. https://www.rff.org/publications/reports/global-energy-outlook-2019/ (accessed 1 December 2020)

Nhuchhen, D., Basu, P., and Acharya, B. 2014. A comprehensive review on biomass torrefaction. *International Journal of Renewable Energy & Biofuels*: 1–56.

Niu, Y., Liu, S., Shaddix, C. R., and Hui, S. 2019a. An intrinsic kinetics model to predict complex ash effects (ash film, dilution, and vaporization) on pulverized coal char burnout in air (O_2/N_2) and oxy-fuel (O_2/CO_2) atmospheres. *Proceedings of the Combustion Institute* 37: 2781–2790.

Niu, Y., Lv, Y., Lei, Y., Liu, S., Liang, Y., Wang, D., Hui, E., and Hui, S. 2019b. Biomass torrefaction: Properties, applications, challenges, and economy. *Renewable and Sustainable Energy Reviews* 115: 1–18.

Nordin, A., Pommer, L., Nordwaeger, M., and Olofsson, I. 2013. *Biomass Conversion through Torrefaction. Technologies for Converting Biomass to Useful Energy.* Boca Raton: CRC Press, pp. 217–244.

Nunes, L. J. R., Matias, J. C. D. O., and Catalão, J. P. D. S. 2018. Biomass torrefaction process. *Torrefaction of Biomass for Energy Applications:* 89–124.

Ohm, T. I., Chae, J. S., Kim, J. K., and Oh, S. C. 2015. Study on the characteristics of biomass for co-combustion in coal power plant. *Journal of Material Cycles and Waste Management* 17: 249–257.

Okolie, J. A., Nanda, S., Dalai, A. K., and Kozinski, J. A. 2021. Chemistry and specialty industrial applications of lignocellulosic biomass. *Waste and Biomass Valorization* 12, 2145–2169.

Okolie, J. A., Rana, R., Nanda, S., Dalai, A. K., and Kozinski, J. A. 2019. Supercritical water gasification of biomass: A state-of-the-art review of process parameters, reaction mechanisms and catalysis. *Sustainable Energy and Fuels* 3: 578–598.

Ooshima, H., K. A., Harano, Y., and Yamamoto, T. 1984. Microwave treatment of cellulosic materials for their enzymatic hydrolysis. *Biotechnology Letters* 6: 289–294.

Pahla, G., Ntuli, F., and Muzenda, E. 2020. Torrefaction of landfill food waste for possible application in biomass. *Waste Management* 71: 512–520.

Parakh, P. D., Nanda, S., and Kozinski, J. A. 2020. Eco-friendly transformation of waste biomass to biofuels. *Current Biochemical Engineering* 6: 120–134.

Park, J., Meng, J., Lim, K. H., Rojas, O. J., and Park, S. 2013. Transformation of lignocellulosic biomass during torrefaction. *Journal of Analytical and Applied Pyrolysis* 100: 199–206.

Peng, J. H., Bi, H. T., Sokhansanj, S., and Lim, J. C. 2012. A study of particle size effect on biomass torrefaction and densification. *Energy and Fuels* 26: 3826–3839.

Phanphanich, M., and Mani, S. 2011. Impact of torrefaction on the grindability and fuel characteristics of forest biomass. *Bioresource Technology* 102: 1246–1253.

Pimchuai, A., Dutta, A., and Basu, P. 2010. Torrefaction of agriculture residue to enhance combustible properties. *Energy and Fuels* 24: 4638–4645.

Pirraglia, A., Gonzalez, R., Saloni, D., and Denig, J. 2013. Technical and economic assessment for the production of torrefied ligno-cellulosic biomass pellets in the US. *Energy Conversion and Management* 66: 153–164.

Prapakarn, N., Prapakarn, S., Liplap, P., and Arjharn, W. 2018. Effects of torrefaction temperature and residence time on agricultural residue after pelletization process: Corncobs/cornhusks, rice straw, and sugarcane trash. *Suranaree Journal of Science and Technology* 25: 373–382.

Prawisudha, P., Namioka, T., and Yoshikawa, K. 2012. Coal alternative fuel production from municipal solid wastes employing hydrothermal treatment. *Applied Energy* 90: 298–304.

Prins, M. J., Ptasinski, K. J., and Janssen, F. J. J. G. 2006a. More efficient biomass gasification via torrefaction. *Energy* 131: 3458–3470.

Prins, M. J., Ptasinski, K. J., and Janssen, F. J. J. G. 2006b. Torrefaction of wood. Part 1. Weight loss kinetics. *Journal of Analytical and Applied Pyrolysis* 77: 28–34.

Prins, M. J., Ptasinski, K. J., and Janssen, F. J. J. G. 2007. From coal to biomass gasification: Comparison of thermodynamic efficiency. *Energy* 32: 1248–1259.

Raut, M. K., Basu, P., and Acharya, B. 2016. The effect of torrefaction pre-treatment on the gasification of biomass. *International Journal of Renewable Energy & Biofuels*:1–14.

Ren, S., Lei, H., Zhang, Y., Wang, L., Bu, Q., Wei, Y., and Ruan, R. 2019. Furfural production from microwave catalytic torrefaction of Douglas fir sawdust. *Journal of Analytical and Applied Pyrolysis* 138: 188–195.

Ren, X., Sun, R., Meng, X., Vorobiev, N., Schiemann, M., and Levendis, Y. A. 2017. Carbon, sulfur and nitrogen oxide emissions from combustion of pulverized raw and torrefied biomass. *Fuel* 188: 310–323.

REN21.2017. REN21Renewables global status report, REN21 Secretariat, Paris. http://www.Ren21.Net/Wp-Content/Uploads/2017/06/17–8399_GSR_2017_Full_Report_0621_Opt.Pdf(2017) (accessed 21 June 2018).

Repellin, V., Govin, A., Rolland, M., and Guyonnet, R. 2010. Energy requirement for fine grinding of torrefied wood. *Biomass and Bioenergy* 34: 923–930.

Reza, M. T., Lynam, J. G., Vasquez, V. R., and Coronella, C. J. 2012. Pelletization of biochar from hydrothermally carbonized wood. *Environmental Progress & Sustainable Energy* 31: 225–234.

Ribeiro, J., Godina, R., Matias, J., and Nunes, L. 2018. Future perspectives of biomass torrefaction: Review of the current state-of-the-art and research development. *Sustainability* 10.

Rousset, P., Aguiar, C., Labbé, N., and Commandré, J. M. 2011. Enhancing the combustible properties of bamboo by torrefaction. *Bioresource Technology* 102: 8225–8231.

Rousset, P., MacEdo, L., Commandré, J. M., and Moreira, A. 2012. Biomass torrefaction under different oxygen concentrations and its effect on the composition of the solid by-product. *Journal of Analytical and Applied Pyrolysis* 96: 86–91.

Ru, B., Wang, S., Dai, G., and Zhang, L. 2015. Effect of torrefaction on biomass physicochemical characteristics and the resulting pyrolysis behavior. *Energy and Fuels* 29: 5865–5874.

Sakthivadivel, D., and Iniyan, S. 2017. Combustion characteristics of biomass fuels in a fixed bed microgasifier cook stove. *Journal of Mechanical Science and Technology* 31: 995–1002.

Satpathy, S. K., Tabil, L. G., Meda, V., Naik, S. N., and Prasad, R. 2014. Torrefaction of wheat and barley straw after microwave heating. *Fuel* 124: 269–278.

Senneca, O. 2017. Oxidation of carbon: What we know and what we still need to know. *Energy* 120: 62–74.

Singh, S., Chakraborty, J. P., and Mondal, M. K. 2020. Torrefaction of woody biomass (*Acacia nilotica*): Investigation of fuel and flow properties to study its suitability as a good quality solid fuel. *Renewable Energy* 153: 711–724.

Stelt, M. J. C. V. D., Gerhauser, H., Kiel, J. H. A., and Ptasinski, K. J. 2011. Biomass upgrading by torrefaction for the production of biofuels: A review. *Biomass and Bioenergy* 35: 3748–3762.

Stelte, W. 2012. *Guideline: Densification of Torrefied Biomass.* Resultat Kontrakt (RK) Report Danish Technological Institute, Taastrup.

Stelte, W., Clemons, C., Holm, J. K., Sanadi, A. R., Ahrenfeldt, J., Shang, L., and Henriksen, U. B. 2011. Pelletizing properties of torrefied spruce. *Biomass and Bioenergy* 35: 4690–4698.

Stelte, W., Dahl, J., Peter, N., Nielsen, K., and Hansen, H. O. 2012. Densification Concepts for Torrefied Biomass. *Torrefaction Workshop European Biomass Conference*, Milano.

Svanberg, M., Olofsson, I., Flodén, J., and Nordin, A. 2013. Analysing biomass torrefaction supply chain costs. *Bioresource Technology* 142: 287–296.

Tian, X., Dai, L., Wang, Y., Zeng, Z., Zhang, S., Jiang, L., Yang, X., Yue, L., Liu, Y., and Ruan, R. 2020. Influence of torrefaction pretreatment on corncobs: A study on fundamental characteristics, thermal behavior, and kinetic. *Bioresource Technology* 297: 122490.

Tumuluru, J. S., Sokhansanj, S., Hess, J. R., Wright, C. T., and Boardman, R. D. 2011. A Review on biomass torrefaction process and product properties for energy applications. *Industrial Biotechnology* 7: 384–401.

Tumuluru, J. S., Sokhansanj, S., Wright, C. T., and Boardman, R. D. 2010. *Biomass Torrefaction Process Review and Moving Bed Torrefaction System Model Development.* U.S Department of Energy Office of Biomass Program.

Uslu, A., Faaij, A. P. C., and Bergman, P. C. A. 2008. Pre-treatment technologies, and their effect on international bioenergy supply chain logistics. techno-economic evaluation of torrefaction, fast pyrolysis and pelletisation. *Energy* 33: 1206–1223.

Via, B. K., Adhikari, S., and Taylor, S. 2013. Modeling for proximate analysis and heating value of torrefied biomass with vibration spectroscopy. *Bioresource Technology* 133: 1–8.

Wang, C., Peng, J., Li, H., Bi, X. T., Legros, R., Lim, C. J., and Sokhansanj, S. 2013. Oxidative torrefaction of biomass residues and densification of torrefied sawdust to pellets. *Bioresource Technology* 127: 318–325.

Wang, G. J., Luo, Y. H., Deng, J., Kuang, J. H., and Zhang, Y. L. 2011. Pretreatment of biomass by torrefaction. *Chinese Science Bulletin* 56: 1442–1448.

Wang, L., Barta-rajnai, E., Skreiberg, Khalil, R., Czégény, Z., Jakab, E., Barta, Z., and Grønli, M. 2018a. Effect of torrefaction on physiochemical characteristics and grindability of stem wood, stump and bark. *Applied Energy* 227: 137–148.

Wang, M. J., Huang, Y. F., Chiueh, P. T., Kuan, W. H., and Lo, S. L. 2012. Microwave-induced torrefaction of rice husk and sugarcane residues. *Energy* 37: 117–184.

Wang, X., Wu, J., Chen, Y., Pattiya, A., Yang, H., and Chen, H. 2018b. Comparative study of wet and dry torrefaction of corn stalk and the effect on biomass pyrolysis polygeneration. *Bioresource Technology* 258: 88–97.

Wang, Z., Lim, C. J., and Grace, J. R. 2019. A comprehensive study of sawdust torrefaction in a dual-compartment slot-rectangular spouted bed reactor. *Energy* 189: 116306.

Wannachepeera, J., Fungtammasan, B., and Worasuwannarak, N. 2011. Effects of temperature and holding time during torrefaction on the pyrolysis behaviors of woody biomass. *Journal of Analytical and Applied Pyrolysis* 92: 99–105.

Wen, J. L., Sun, S. L., Yuan, T. Q., Xu, F., and Sun, R. C. 2014. Understanding the chemical and structural transformations of lignin macromolecule during torrefaction. *Applied Energy* 121: 1–9.

Wu, K. T., Tsai, C. J., Chen, C. S., and Chen, H. W. 2012. The characteristics of torrefied microalgae. *Applied Energy* 100: 52–57.

Yan, W., Acharjee, T. C., Coronella, C. J., and Vásquez V. R. 2009. Thermal pretreatment of lignocellulosic biomass. *Environmental Progress & Sustainable Energy* 28: 435–440.

Yan, W., Hastings, J. T., Acharjee, T. C., Coronella, C. J., and Vásquez, V. R. 2010. Mass and energy balances of wet torrefaction of lignocellulosic biomass. *Energy and Fuels* 24: 4738–4742.

Yu, K. L., Chen, W. H., Sheen, H. K., Chang, J. S., Lin, C. S., Ong, H. C., Show, P. L., Ng, E. P., and Ling, T. C. 2020. Production of microalgal biochar and reducing sugar using wet torrefaction with microwave-assisted heating and acid hydrolysis pretreatment. *Renewable Energy* 156: 349–360.

Yue, Y., Singh, H., Singh, B., and Mani, S. 2017. Torrefaction of sorghum biomass to improve fuel properties. *Bioresource Technology* 232: 372–379.

Zhang, D., Wang, F., Zhang, A., Yi, W., Li, Z., and Shen, X. 2019. Effect of pretreatment on chemical characteristic and thermal degradation behavior of corn stalk digestate: Comparison of dry and wet torrefaction. *Bioresource Technology* 275: 239–246.

Zhang, S., Chen, T., Li, W., Dong, Q., and Xiong, Y. 2016. Physicochemical properties and combustion behavior of duckweed during wet torrefaction. *Bioresource Technology* 218: 1157–1162.

Zhang, S., Dong, Q., Zhang, L., Xiong, Y., Liu, X., and Zhu, S. 2015. Effects of water washing and torrefaction pretreatments on rice husk pyrolysis by microwave heating. *Bioresource Technology* 193: 442–448.

6

Pelletization of Torrefied Biomass Using Binders

Jennifer Anno-Kusi, Tumpa R. Sarker, Sonil Nanda and Ajay K. Dalai

CONTENTS

6.1 Introduction

Due to the growing global population and the industrial revolution, the reserve of fossil fuels keeps on diminishing. Fuel from fossil reserves besides running out has severe impacts in terms of environmental pollution and climate change. To prevent the risk of extreme climate change, attention is being shifted to renewable energy sources (Liu et al., 2014). Renewable energy sources although a great substitute to fossil fuels are not being utilized to a high extent due to high processing cost and transportation, weather or seasonal implications, and poor technology (Stelte et al., 2012).

Bioenergy obtained from biomass, which is relatively inexpensive, available abundantly, and renewable is gaining ground as an alternate source of energy to fossil fuel (Azargohar et al., 2018). Biomass can be converted to solid, liquid, or gaseous biofuels using different thermochemical conversion technologies such as torrefaction, gasification, and combustion. (Patel et al., 2016). Aside from this, energy from biomass has a zero-net release of CO_2 into the environment when burnt and has relatively lower SO_x, NO_x, and ash emissions (Clark and Deswarte, 2008).

Lignocellulosic biomass has a high moisture content, low calorific value, low bulk, and energy density, and is hydrophilic, which makes it have a low energy efficiency (Mamvura and Danha, 2020) and restricts its usages for the production of biofuel on a large scale. There are various techniques available that can enhance these inherent properties of raw biomass and increase its energy content. Torrefaction followed by densification is a promising technique to produce high-quality solid fuel comparable to coal (Svanberg et al., 2013), which is gaining attention recently due to its simplicity and low operating cost (Chai and Saffron, 2016).

Torrefaction, the thermochemical treatment of biomass, helps to improve the physical and chemical composition of biomass for fuel applications (Chin et al., 2013). However, handling, storage, and transportation of torrefied biomass is difficult due to high dust formation, self-heating, and explosion risk

(Kumar et al., 2017). Bio-coal has a low volumetric energy density (Kumar et al., 2017). This reduces the cost of storage, handling, and transportation and enhances profitability. Aside from this, dust generated during the handling and transportation of bio-coal can prevent fire outbreaks or explosions. This is due to the combination of torrefaction and densification (Thrän et al., 2016).

Densification is the use of mechanical force to compact biomass into uniform sizes and shapes (Chen et al., 2015). Densification of biomass helps to achieve efficient storage, handling, and transportation of the solid fuel by improving the density by 4 to 10 times. It also makes biomass energy-dense and enhances mechanical strength (Lu et al., 2014). Methods of biomass densification include extrusion, bailing, briquetting, and pelletization (Stelte et al., 2012). Torrefied pellets can therefore be co-fired with coal in high proportions. The top five wood-exporting countries in the world are the United States (26% of the total world export), Canada (11.9% of the total world export), Latvia (8.2% of the total world export), Estonia (6% of the total world export), and Austria (6% of the total world export) (Natural Resources Canada, 2020). Canada exports $396.7 million of wood pellet per annum, thus making it the second-largest biomass pellet exporter globally (Natural Resources Canada, 2020).

There are several studies based on the torrefaction of lignocellulosic biomass. Densification of torrefied biomass has been conducted by various researchers. Densification of torrefied biomass requires more energy as it requires high pressure, high die temperature as well as a binder. However, there are several advantages too. This study extensively discusses the pros and cons of densification of torrefied biomass, process parameters of torrefaction as well as densification. Furthermore, the effects of different kinds of binders on the quality of fuel pellets are also investigated. In the end, a detailed techno-economic analysis (TEA) and lifecycle assessment (LCA) have been studied to examine the feasibility of this method and environmental impact.

6.2   Chemistry of Lignocellulosic Biomass

Biomass is defined as an organic material derived from plants or animals. Lignocellulosic biomass mainly consists of cellulose, hemicellulose, and lignin. Other minor constituents of biomass are organic extractives and inorganic minerals. Lignin, cellulose, and hemicellulose contents in biomass are about 10–25 wt.%, 35–55 wt.%, and 20–40 wt.%, respectively (Nanda et al., 2013; Chen et al., 2015). The mass composition of the components is provided in a range as they vary based on the origin of the biomass, type of biomass, and harvesting time (Ribeiro et al., 2018). Table 6.1 shows the compositions of different biomasses. These constituents have different structures and so have different decomposition characteristics. Hemicellulose requires the lowest temperature to decompose (i.e., 220–315°C) resulting in a more weight loss of the biomass. Cellulose decomposes between 200°C and 300°C, while lignin decomposes in a temperature range of 200–500°C (Lu et al., 2012; Nanda et al., 2017).

6.3   Torrefaction

Torrefaction is the thermal pretreatment of biomass in which it is slowly heated between 200 and 300°C in the absence of oxygen resulting in the modification of the physical and chemical properties of biomass (Tumuluru et al., 2011). Figure 6.1 shows the difference in properties of raw and torrefied biomass. It shows that torrefied biomass has improved heating value, moisture content, grindability, and energy density. The torrefied product obtained after torrefaction is called torrefied biomass, green coal, or bio-coal, and it serves as a source of fuel (Barskov et al., 2019). It has a higher calorific value, lower moisture content, lower oxygen content and is less fibrous and tenacious (Sarker et al., 2020, 2021). Torrefaction also results in the production of condensable and non-condensable gases like CO, CO_2, methanol, hydrogen, methane, acetic acid, and water. (Medic et al., 2010).

During torrefaction as temperature increases, intermolecular bonds between lignin get weaker. Lignin moves from a glassy state to an elastic state. The temperature at which this occurs is termed glass transition temperature (Stelte et al., 2012). Glass transition temperature of lignin is between 100 and 140°C (Peng et al., 2015). Torrefaction leads to the removal of the most hydroxyl groups that act as binding

TABLE 6.1

Fiber Content of Different Lignocellulosic Biomasses

Biomass	Cellulose (wt%)	Hemicellulose (wt%)	Lignin (wt%)
Bagasse	55	17	23
Banana waste	13	15	14
Barely straw	40	20	15
Coastal Bermuda grass	25	36	6.4
Corn cobs	45	35	15
Corn stalks	35	15	19
Corn stover	38	26	19
Cotton seed hairs	80–95	5–20	0
Grasses	25–40	35–50	10–30
Hardwood stems	40–55	24–40	18–25
Leaves	15–20	80–85	0
Logging residue	38	13	26
Oat hull	30	31	7
Oat straw	41	16	11
Paper	85–99	0	0–15
Pine	49	15	26
Rice husk	36	15	20
Rice straw	32	24	18
Sawdust	55	14	21
Softwood stems	45–50	25–30	25–35
Sorghum straw	33	18	15
Sponge gourd fibers	67	17	16
Sugarcane baggage	42	25	20
Switchgrass	45	31	12
Wheat straw	45	28	9
Wood chips	44	23	29

References: Adapa et al. (2011); Ciolkosz and Wallace (2011); Abedi and Dalai (2017); Kumar and Sharma (2017).

sites for water. Hence, bio-coal absorbs less water than non-torrefied biomass (Stelte et al., 2012). With increasing torrefaction temperature, carbon content increases, and there is a decrease in hydrogen and oxygen content (Emadi et al., 2017). Thus, the final solid product will have lower oxygen–carbon and hydrogen–carbon ratios which ensure that less smoke and water vapor are produced during combustion or gasification (Tumuluru et al., 2010). Torrefaction also reduces the mass of the biomass while conserving its energy yield (Ribeiro et al., 2018).

During torrefaction, pre-drying occurs at 50–150°C to remove unbound water. At temperatures above 280°C, CO, CO_2, H_2, and other hydrocarbons such as CH_4, toluene, and benzene are formed. Condensable vapors like alcohol, acetic acids, aldehydes, water, phenols, and ketones are also formed. The biochar yield, which is the ratio of the mass of the solid product to the mass of the raw biomass, is averaged at 70 wt.% (Peng et al., 2015). The remaining 30% of the biomass is converted to volatile organic compounds and non-condensable components. However, the biomass loses only 10% of its initial energy. Therefore, energy densification is achieved as a result of torrefaction (Barskov et al., 2019). Thus, the energy density is approximately 1.3 times higher on a mass basis (Adhikari et al., 2019).

6.4 Densification

Densification of biomass is usually conducted to get a uniform size with high density to facilitate handling, long-distance transportation as well as storage. The products from densification can be pellets,

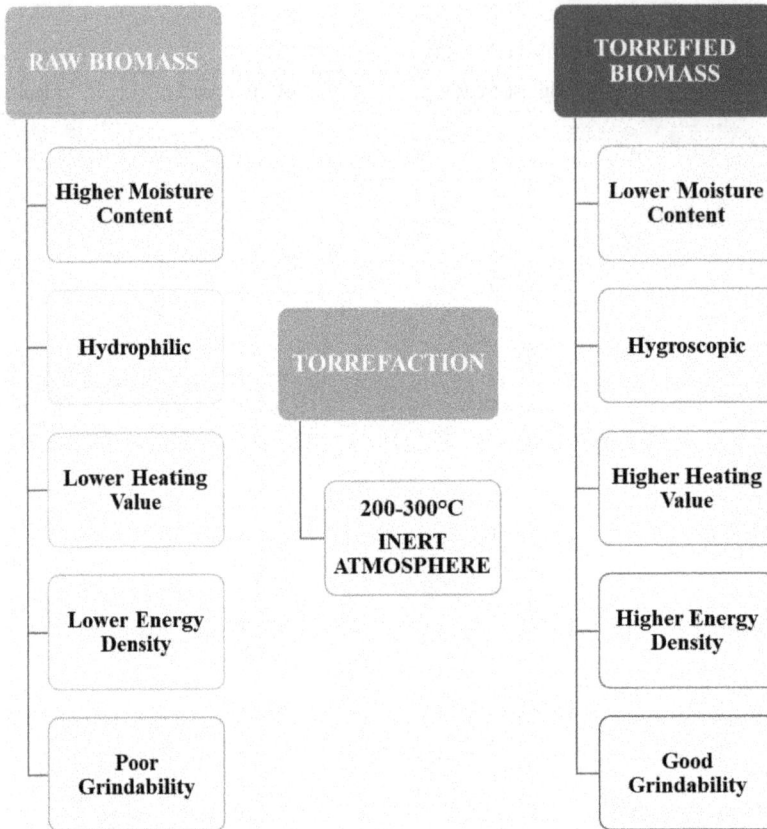

FIGURE 6.1   Property variation of torrefied biomass as compared to raw biomass (adapted from Chen et al., 2015).

briquettes, etc. A typical pelletizer consists of a flat plate or ring-shaped die with cylindrical press channels and two rollers through which the torrefied biomass flows. The biomass is squeezed between the die and rollers and moves into the openings of the press channel. As the roller passes over the channel, the biomass is further compressed inside the channel, and a pellet is produced. Pellets are cylindrically shaped and have a diameter of 6–25 mm and a length of 3–50 mm (Stelte et al., 2012). According to Peng et al. (2015), the bonding mechanism can be put into five main categories such as:

(i)   Force of attraction between particles
(ii) Forces of adhesion and cohesion between particles
(iii) Interfacial forces and capillary pressure to move liquids onto the surface
(iv) Mechanical interlocking to form close bonds
(v)   Formation of solid bridges

As the particles continue to be compressed, particles on adjacent sides combine by interlocking and solid bridge forces. Due to the angular structure of the particles, gaps are created between them, and this makes the solid bridge forces weaker. Stronger solid bridges can be formed by the use of appropriate binders, which results in pellets with higher strength (Dai et al., 2019). Pellets have a large surface area, which makes them efficient for thermochemical conversions (Lu et al., 2014). The homogeneous shape and structure of pellets are also beneficial when they are used in boiler systems (Stelte et al., 2012).

The formation of pellets from torrefied biomass is considered more energy-intensive than that of raw biomass (Shang et al., 2014). This is because raw biomass has a large amount of lignin, which is a natural binder and supports densification. However, lignin undergoes a structural transformation during torrefaction, and its glass transition temperature increases. The binding property of lignin deteriorates, which causes a decrease in pellet strength. In addition, extreme temperature increases the number of pores created, which affects the density of the pellets (Dai et al., 2019).

Glass transition temperature has a very high impact on pelletization. Pelletization of the high glass-transitioned biomass requires a higher die temperature of over 170°C to soften the lignin (Dai et al., 2019). This can cause the risk of dust explosion, make operation more difficult, and the production cost higher. Therefore, the pelletization of torrefied biomass requires an external binder to bind the biomass particles together. There are different binders available that have been used by various researchers for pelletization. In addition, other factors can affect the quality of densified material such as pressure, temperature, moisture content, and pelletizing equipment. The effects of different parameters on pellet quality are discussed as follows:

(i) **Moisture content:** Moisture has an important role to play during pelletization. It acts as a binder and also activates internal binders and external binders added to the bio-coal (Azargohar et al., 2018). The energy required for compacting the bio-coal into pellets also decreases as moisture content increases due to the lubrication effect (Stelte et al., 2012). The optimum moisture content for the pelletization of wood species ranges between 5–10 wt.% and 10–20 wt.% for grasses. When the moisture content goes beyond the optimum level, it affects the mechanical properties of the pellets. This is because the moisture gets trapped between the particles, and the binding force is affected.

(ii) **Temperature:** During pelletization, the movement between the biochar and the press channels generates heat. This generated heat influences the ease of the flow of binders, resulting in better binding of pellets. When binders are present at the right temperature, they can deform and bind well. As a result, more mechanically durable pellets are produced (Sarker et al., 2021). Low temperatures cannot make the binders melt, and extreme temperatures decompose the binder as well as certain components of the biomass. An increase in temperature helps to reduce friction in the press channel, which lowers the energy and pressure requirement for the pelletization process. The density of the pellets increases as the temperature is increased until it gets to a temperature of 90°C. An increase in temperature beyond 90°C usually does not increase the pellet density (Stelte et al., 2012).

(iii) **Pelletizing pressure:** The pressure applied during pelletization influences the compactness of the pellets (Sarker et al., 2021). High pressure removes water from the bio-coal and increases the temperature and friction in the die. High pressure results in pellets of greater density and mechanical strength. Building up this pressure requires the use of energy. There is a maximum pressure above which any additional energy input is converted into excess heat (Stelte et al., 2012).

(iv) **Particle size:** Particle size is one parameter that influences pellet density a lot. Small particle size increases the friction in the press channel due to the increase in surface area, contact between the particles, and the walls of the press channel. However, smaller particle size results in pellets of higher density due to the easier flow of binders through particles. Larger particle sizes result in the production of low durable pellets due to limited interaction among particles. Particle sizes of a wider variation are optimal for the best quality of pellet. The amount of fines (particles with a diameter less than 0.5 mm) should not be more than 10% to 20% unless a binder is added (Stelte et al., 2012).

(v) **Dimensions of the press channel:** The diameter of the press channel should range between 6 and 25 mm. The ratio of the length to the diameter of the press channel is known as the aspect ratio or compression ratio. This ratio influences the pressure build-up in the press channel. There is an exponential relation between the aspect ratio and pressure built up. When the length of the press channel is increased, the mechanical properties of the pellets also improve. The optimum length depends on the type of raw material, temperature, moisture content, and size of particles of the char (Stelte et al., 2012).

TABLE 6.2

Effects of Moisture and Temperature on Pellet Quality

Raw Material	Operating Conditions	Pelletizer	Operations	Reference
Canola meal	Temperature: 60–90°C	Lab-scale single-pelleting unit	An increase in temperature produced pellets of higher density	Tilay et al. (2015)
European beech	Temperature: 20 and 180°C	Single pellet press unit	When the temperature is increased, there is a decrease in pressure required for pelletization.	Stelte et al. (2011)
Oak	Moisture: 1–16 wt.%	Punch-and-die process	5–12 wt.% moisture is the optimum for producing quality pellets	Li and Liu (2000)
Scots pine	Moisture: 8–14 wt.%	Pilot mill	When the moisture content is increased there is a decrease in pellet density and energy used	Samuelsson et al. (2012)
Wheat straw	Temperature: 20 and 180°C	Single pellet press unit	An increase in temperature decreased the pressure requirements for pelletization	Stelte et al. (2011)
Wood sawdust	Moisture: 11–41 wt.%	Universal testing machine	Moisture content does not affect the heating value of pellets	Poddar et al. (2014)

TABLE 6.3

Effects of Pressure and Particle Size on Pellet Quality

Raw Material	Operating Conditions	Pelletizer	Operations	Reference
Agricultural and wood biomass blends	Particle size: 150–300, 300–425, and 425–600 µm	In-house built single unit pelletizer	With a decrease in particle size, there was an increase in density and mechanical strength of pellets	Harun and Afzal (2016)
Corn stover briquette	Pressure: 5–15 MPa	Hydraulic press	Density and durability of pellets increased with increasing pressure for moisture content ≤ 10 wt.%	Mani et al. (2006)
Oil palm shell	Particle size: 160–570 µm	In-house built cylindrical pelletizer unit	Increasing particle size resulted in a decrease in mechanical durability and density of pellets	Arzola et al. (2012)
Rice straw	Particle size: 1.8–15 mm	Self-designed single-pelleting unit	Effect of particle size was greater for milled material in comparison to chopped material	Wang et al. (2018)
Wheat straw	Compaction pressure: 50–550 MPa	Single pellet press unit	For pressures ≤ 250 MPa, the pellet density increased. When pressure > 250 MPa, increase in density was minimal	Stelte et al. (2011)
Wood sawdust	Pressure: 6–20 kN	Universal testing machine	Pellet density increased with increasing pressure, but the effect was minimal at higher pressures. The pressure did not affect the heating value of pellets	Poddar et al. (2014)

Some notable research works carried out on the effects of different parameters affecting pelletization are summarized in Tables 6.2 and 6.3. Although moisture content does not affect the heating value of the pellets, it influences the density of pellets as well as the energy required for pelletization. An increase in moisture content results in decreased pellet density as well as the energy required. High temperature and pressure also result in the production of denser pellets while an increase in particle size results in decreased density and mechanical strength of pellets.

6.5 Effects of Binders

A binder is a liquid or solid that forms a bridge, matrix, film, or chemical reaction resulting in strong interparticle bonds (Lu et al., 2014). Binders play an important role in densification by increasing the adhesion forces between particles. Some components of biomass act as natural binders. These include lignin, starch, soluble carbohydrates, water, fat, and protein (Azargohar et al., 2018). The presence of 1–3 wt.% water in the torrefied biomass acts as a binder and also activates external binders added.

Torrefaction affects the binding properties of the natural binders present in biomass (Ghiasi et al., 2014). Lignin, cellulose, and hemicellulose decompose thermally during the torrefaction to produce organic acids and charcoal (Li et al., 2012). Therefore, binders can be added to improve the densification process as well as the durability, strength, and moisture resistance of torrefied pellets. Binders are classified as either organic or inorganic. Examples of binders include glycerol, molasses, raw biomass, NaOH, $Ca(OH)_2$, lignosulfonate, bentonite, and polyvinyl alcohol. Pellet strength increases with an increase in the amount of binder added and vice versa. However, the fraction of binders added can range from 0.5 to 20 wt.% or more (Peng et al., 2015). Effects of different binders on pellet quality are presented as follows:

(i) Starch:

Wu (2013) explored the use of starch as a binder for the pelletization of torrefied sawdust obtained from southern pine. The use of starch as a binder was also investigated by Mallory (2013). Starch binder produced harder pellets because it increased the forces of adhesion between the biochar particles. The ash content of the torrefied pellets, however, increased with the addition of lignin binder.

(ii) Sawdust:

Peng et al. (2015) analyzed and compared the pelletization of biochar from pine sawdust using starch, lignin, and raw pine sawdust as binders. The biochar was formed at 330°C with a residence time of 30 min. Their studies showed that the pellet density decreased for all binders because they were unable to fill the pores of the biochar. This can be associated with a high degree of torrefaction. For practical applications, the sawdust should measure less than 1 mm in diameter to help curb this problem to some extent.

From their experiments, the density of all pellets formed with binders decreased compared to the density of the control wood pellet. The energy densities of the pellets made using starch and lignin binders were higher than that of the control wood pellets. While the energy density of the pellets formed using sawdust binders was slightly lower because raw sawdust has a lower heating value. The moisture content of all pellets formed using binders was lower compared to the raw wood pellet. Energy consumption required for all pellets with binders was higher but reduced with increasing binder content. For an optimal biochar yield of 71% (280°C and 52 min), Meyer hardness of the torrefied pellets increased from 8 N/mm^2 (no binder added) to 15 N/mm^2 when 20% of sawdust binder was added.

(iii) Waste plastics:

The use of plastic wastes, especially linear low-density polyethylene, as a binder for pelletization of torrefied biomass was investigated by Emadi et al. (2017). Linear low-density polyethylene has excellent binding and energy characteristics (Massaro et al., 2014). It can be obtained from municipal solid waste or waste plastics from agriculture. Utilizing this waste as a binder will reduce not only environmental pollution but also greenhouse gas emissions. The feedstocks for the experiment were wheat and barley straw. Although they have a lower quality compared to wood, they are adequately available and acceptable to produce bioenergy (Satpathy et al., 2014). The wheat and barley straws were torrefied at 250°C for 15 min. The biochar was mixed with 10% linear low-density polyethylene and stored for 2 days to ensure uniformity of the mixture before pelletizing.

The addition of the linear low-density polyethylene binder increased the density of the pellet and decreased its ash content. The tensile strength of the wheat straw pellets increased by

280% while that of the barley straw pellets increased by 253%. This is due to the high tensile strength of linear low-density polyethylene (17.8 MPa) as well as its ability to create strong interlocking of the torrefied biomass particles. The higher heating value (HHV) of the pellets also increased due to the HHV of linear low-density polyethylene (42 MJ/kg). The addition of low-density polyethylene helped achieve a higher density of pellets. It increased the density of torrefied wheat straw pellets from 1106.4 kg/m^3 to 1126.8 kg/m^3 while the density of barley straw pellets increased from 1096 kg/m^3 to 1115 kg/m^3.

(iv) Proteinaceous binder:

Proteinaceous feedstock can be developed into wood adhesive or binders. Adhikari et al. (2019) developed a sustainable binder from proteinaceous waste retrieved from specified risk materials (SRMs) from the rendering industry. The SRM was hydrolyzed to obtain peptide (Mekonnen et al., 2013). Seventy-eight percent peptide was mixed with 22% of a polyamidoamine epichlorohydrin (PAE) resin to form the binder.

Adhikari et al. (2019) carried out a study using fir sawdust as the feed for torrefaction. Three percent of the binder was mixed with the biochar for pelletization. The results of the studies were that the bulk density, pellet density, and gross heating value of pellets formed using the proteinaceous binder were higher than that of the control pellets formed with no binder. The hydroxyl, amine, amide, and carboxyl functional groups present in the peptide enhanced polar interconnection and hydrogen bonds between particles of the biochar and the binder. However, high chlorine content was observed in the pellets produced using a proteinaceous binder. Chlorine content for the proteinaceous pellets was 1200 µg/g as against 8 µg/g of pellets formed with no binder. This high chlorine content can result in corrosion of equipment due to the release of HCl gas during the combustion of the pellets. The peptide has approximately 1.77% inorganic chloride which accounts for such results. The chlorine level in the pellets can be reduced by pre-washing the proteinaceous material with Milli-Q water and passing it through thermal hydrolysis. This results in a peptide with a reduced chlorine content of 0.513% (Adhikari et al., 2019).

(v) Bentonite, glycerol, lignosulfonate, and wood residue as binders:

Lu et al. (2014) studied the effects of the binders such as glycerol, bentonite, wood residue, and lignosulfonate on the pelletization of wheat straw. Results from the studies showed that the higher heating value increased significantly with the addition of glycerol binder. The tensile strength, which reflects the hardness and resistance of the pellets to breakage, increased with all binders used. The specific energy consumption required for pelletization was also significantly lower for pellets produced using binders compared to that without any binder. Among all the binders, glycerol reduced the specific energy consumption by the lowest level. The ash content of the pellets did not increase when the wood residue binder was used. Lignosulfonate and bentonite resulted in higher density pellets because they were able to fill in the gaps between the wheat straw particles. Crude glycerol has a higher heating value of 27.1 MJ/kg due to the presence of biodiesel and fatty acids. Hence, it increased the higher heating value of the pellets, significantly. Bentonite on the other hand has no heating value.

(vi) Biomass gasification residue:

Dai et al. (2019) investigated the use of lignin, starch, and polyvinyl alcohol (PVA) binders combined with an additive. This was to reduce the amount of organic binder used while still achieving pellets of high quality. The additive, biomass gasification residue (BGR), was obtained as a by-product from a wood pellet gasification plant. Organic binders are known to cause tar formation, so the addition of this additive is aimed at reducing the amount of organic binder used. The feedstock for the pelletization was sawdust. The pellets had the highest strength when 12.9% each of lignin and polyvinyl alcohol was used, while they had the highest strength when 9.1% of starch was used.

Binders with concentration of 9.1% were used to test for the hydrophobicity and tar production of the pellets. After the pellets were exposed for 24 h under 75% relative humidity, the polyvinyl alcohol

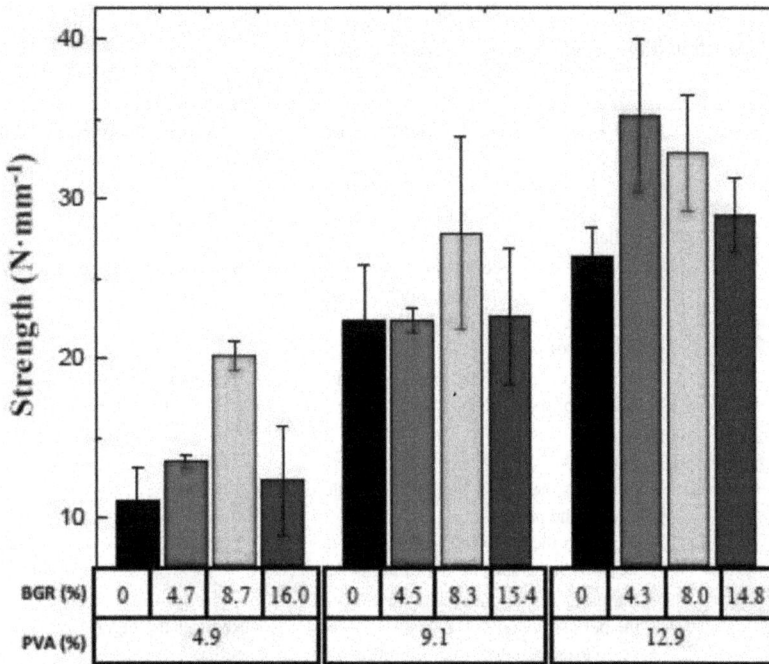

FIGURE 6.2 Influence of biomass gasification residue on the strength of polyvinyl alcohol torrefied pellets. Reproduced with permission from Dai et al. (2019).

binder showed the highest hydrophobicity. Tar generation was lower for the lignin binder due to its higher thermal stability. Pellets formed from polyvinyl alcohol had the highest tensile strength of 26.3 N/mm, followed by starch pellets with 14.9 N/mm. Pellets formed using lignin binder had the least tensile strength of just 3.58 N/mm.

Adding additives is a way of improving the strength of the pellets. Biomass gasification residue works by enhancing the contact between the char particles and strengthening the solid bridge force. Varying quantities of biomass gasification residue were added to pellets formed from polyvinyl alcohol binder at concentrations of 4.9%, 9.1%, and 12.9%. Figure 6.2 shows the change in strengths with the addition of polyvinyl alcohol binder and biomass gasification residue additive at different concentrations. The strength of pellets increased in all cases. For the 12.9% polyvinyl alcohol concentration, the addition of 4.3% of biomass gasification residue resulted in the highest strength of pellets of 35.27 N/mm. The strengths and weaknesses of all binders already discussed are summarized in Tables 6.4 and 6.5.

6.6 Techno-Economic Analysis of Torrefaction and Densification of Biomass

TEA is an evaluation of the technical and economic viability and sustainability of a process, product, or service. Before commercializing technologies, their economic and technical feasibility must be well understood and analyzed (Shah et al., 2016). The economic analysis includes capital cost, operating cost, cost of production, revenue, and return on investment. The technical analysis includes maintenance requirements, transportation, installation, service life, market viability, skill requirement, and intrinsic risks. Thus, TEA is vital in preventing wastage of efforts and investments by evaluating alternative options and deciding on resource allocation. Although combined torrefaction and pelletization is a promising technology, there is little knowledge about its techno-economic performance. A techno-economic assessment is necessary for further development and commercialization. Few studies have been done, and their findings show that torrefaction and pelletization can be commercialized.

TABLE 6.4

Comparison of the Strengths and Weaknesses of Various Binders

Binders	Merits	Demerits
Bentonite	• Decreases the energy consumption for pelletization • Increases the tensile strength, density, and bursting temperature resistance of pellets	• Increases the ash content of the pellets
Glycerol	• Decreases the specific energy consumption for pelletization • Increases the HHV, density, tensile strength of pellets	• Increases the ash content of the pellets
Lignin	• Increases the strength, hardness, and density of the pellets • Ensures lower energy consumption during pelletization	• Increases tar production • Reduces the HHV of the pellets
Low-density polyethylene	• Increases the tensile strength and failure load of the pellets • Increases the HHV of the pellets • Decreases the ash content of the pellets	-
Sawdust	• Increases the hardness and durability of the pellets • Decreases the energy consumption during pelletization and therefore reduce the overall cost	• Causes tar formation • Reduces the higher heating value of the pellets • Unable to fill nanometer pores of biochar; hence produces pellets of lower density
Starch	• Increases the strength, density, and hardness of the pellets • Decreases the energy consumption during pelletization	• Increases tar production • Reduces the HHV of the pellets

TABLE 6.5

Comparison of the Strengths and Weaknesses of Various Binders

Binders	Merits	Demerits
Biomass gasification residue	• Enhances pellet strength • Helps to reduce the use of organic binders • Reduces cost and tar generation	-
Lignosulfonate	• Decreases the energy consumption for pelletization • Increases the tensile strength and density of pellets	• Increases the ash content of the pellets
Polyvinyl alcohol	• Increases pellet strength • Increases the hydrophobicity of pellets	• Increases tar production
Proteinaceous binder	• Increases the durability, bulk density, heating value of pellets	• Produces pellets of high chlorine content
Wood residue	• Decreases the energy consumption for pelletization • Decreases the ash content of pellets • Increases the density, tensile strength of the pellets	• Causes tar formation • Has a low heating value

Xu et al. (2014) performed an economic analysis of torrefaction based on a capacity-factored estimation. The production plant in this study had a capacity of 100,000 tons/year and operated 6 months each year with a plant lifespan of 20 years. The total capital cost calculated was $6.6/ton, and the total operation cost was $9.7/ton of feedstock. The operation cost was calculated based on the assumption that the biomass had a moisture content of 20%. The direct capital cost of the plant was approximated to be $8 million. The overall cost of the torrefaction process was estimated to be $16.3/ton of feedstock or $23/ton of product. They also performed studies on standalone torrefaction and pelletization plants as well as combined torrefaction and pelletization plant. The results showed that the capital investment required for the torrefaction, pelletization, and combined torrefaction and pelletization plant was $8 million, $7 million, and $10 million, respectively. The operation costs were also estimated as $9.66/ton, $10.36/ton, and $13.27/ton for the torrefaction, pelletization, and combined torrefaction and pelletization plant, respectively. The transportation cost for the combined torrefaction and pelletization plant was just $0.28/GJ, which was 26% cheaper than a conventional pelletization plant and 72% cheaper than a torrefaction plant. The superior fuel properties of the torrefied pellets far outweigh the additional costs needed to produce them (Agar, 2017). Therefore, the combined torrefaction and pelletization plant is preferred.

Batidzirai et al. (2013) focused their study on compact moving bed reactor and identified thermal efficiency, mass yield, mass balance, and energy balance using woody biomass and straw as raw materials. The results of their study showed that for woody biomass, the calorific efficiency of torrefaction was 94%, and the mass efficiency on an ash-free basis was 48%. The energy efficiency for straw was 96% and the mass efficiency 65%. The cost of production of the torrefied pellets from woody biomass was estimated using the factorial approach to be between the US $3.3/GJ and the US $4.8/GJ. Over an extended period, the cost of production is anticipated to fall to between the US $2.1/GJ and the US $5.1/GJ. When production cost falls to that level, the price of torrefied pellets will be competitive to that of raw pellets.

The capital investment for a 60,000 ton/year plant is estimated to be $6.2–8.9 million. Of this, about 39% is the installation cost, and around 31% is the equipment cost. If the cost of biomass is not included, the production cost for torrefied biomass is about $47–$66/ton (Uslu et al., 2008). The production cost of torrefied pellets obtained from woody biomass was estimated by Batidzirai et al. (2013) to be US $3.3–4.8/GJ$_{LHV}$. In the long term, however, the production cost can reduce significantly to over 50% due to either scale-ups or advancement in technology. Thus, in the future, the production cost of torrefied pellets from woody biomass is estimated to be between the US $2.1/GJ$_{LHV}$ and the US $4.1/GJ$_{LHV}$.

(i) Capital cost:

It involves all costs associated with purchasing equipment as well as building infrastructures. Shah et al. (2012) used ballpark estimates and suggested that for plants that process 1500–2000 tons/day of waste wood or 750–1000 tons/day of dry chips, the capital cost is between $10 million and $16 million based on the technology used.

(ii) Operating cost:

These are costs that come with operating the plant. They include energy, taxes, insurance, labor, maintenance, and repair costs. For the labor cost, Shah et al. (2012) assumed that the plant had five people as its staff, and the rate of pay was $20/h. The cost of repair, maintenance, insurance, and taxes was taken to be 2% of the total capital cost. Drying accounts for the highest cost of energy. It accounts for $3.11/MWh$_{LHV}$ (Svanberg et al., 2013). The process becomes cheap when the feedstock requires no previous drying, or the fuel used for drying is cheap. Another way of reducing energy cost is when the wet feedstock is combusted directly to fuel after the torrefaction process (Agar, 2017). Torrefaction accounts for $1.56/MWh$_{LHV}$ and pelletizing $1.49/MWh$_{LHV}$ (Svanberg et al., 2013).

(iii) Transportation and distribution cost:

Transportation and distribution costs depend on the volume of the final product. The cost of the torrefied pellets is lower compared to conventional pellets. For a plant with a capacity of 200,000 tons/year, road transportation of biomass to the torrefaction plant costs $4.60/MWh$_{LHV}$. Also, for a plant with a capacity of 25,000 tons/year, the distribution cost is $4.32/MWh$_{LHV}$. However, when the plant capacity is increased to 200,000 tons/year the distribution cost reduces by 16.7% to $3.6/MWh$_{LHV}$ (Svanberg et al., 2013).

(iv) Other factors affecting the economics of the process:

TABLE 6.6

Cost Comparison of the Production of Raw and Torrefied Pellets

Description	Torrefied Pellets	Conventional Pellets
Production capacity (ton/year)	64,000	80,000
Production volume (m³/year)	91,129	123,077
Production cost ($)	7,283,555	8,394,771
Utility cost (M$)	1.48	1.05
Logistics and transportation ($/ton)	3.07	5.04
Price of pellets ($/ton)	158.97	137.39
Sales revenue ($)	10,173,762	10,991,764
Depreciation period (year)	10	10
Depreciation ($)	667,092	464,550
Taxes ($)	777,997	746,294
Profit before tax ($)	2,223,022	2,132,447
Net earnings ($)	1,444,972	1,386,091
Cashflow ($)	2,112,101	1,850,703

Note: Reproduced with permission from Agar (2017).

Factors that can affect the production cost include the degree or severity of torrefaction in terms of temperature and residence time and moisture content (Shah et al., 2012). An increase in these factors increases the cost of production. The cost of biomass as well as the amount of binder used and its cost also affect production cost (Pirraglia et al., 2013).

The annual costs involved in producing torrefied pellets and conventional pellets are compared and presented in Table 6.6. The specific costs required to produce torrefied pellets are greater than conventional pellets. While the payback period for the torrefied pellets plant was 3.2 years that of conventional pellets was 2.5 years. The return on investment for torrefied pellets was 21.7% and that of conventional pellets was 29.8%. These show that conventional wood pellet production is more attractive in terms of return on investment. However, for logistics cost, that of torrefied pellets was cheaper. The advantage torrefied pellets have over conventional pellets is their extraordinary fuel properties as well as the higher thermal efficiency of the process (Agar, 2017).

6.7   Lifecycle Assessment of Biomass Torrefaction and Densification

Lifecycle Assessment (LCA) is a tool that is used to point out the possible environmental impacts of the end products of a process from the acquisition of raw materials to production, usage, and final disposal (Shah et al., 2012). It shows different categories of impacts throughout the product's lifecycle (Patel et al., 2016). LCA provides details on emissions and resources required by the system (Adams et al., 2015). It can also be used to assess the mass and energy inputs and outputs of the yield. When using the ISO 14040 series for LCA analysis, four steps are used (Patel et al., 2016).

  (i)   The goal, scope, and boundary definition
 (ii)   Lifecycle inventory analysis
(iii)   Lifecycle impact assessment
 (iv)   Interpretation of the result

Features like system boundaries, functional units, feedstocks, and environmental effects influence an LCA analysis (Shah et al., 2012). For the torrefaction and densification of biomass, the general system boundary includes three phases.

(i) Planting, harvesting, and transporting of biomass

(ii) Pretreatment, production, and upgrading of primary products, as well as distribution to end-users

(iii) Recycling and razing of the plant.

Figure 6.3 shows the different phases as well as inputs and outputs required for each phase. For the first phase, the land is used and there are changes in the land due to the application of pesticides and fertilizers as well as the removal of weed from the land. This usage of land has both direct and indirect environmental effects. Cultivation affects the carbon stock present in the land. Frequent application of fertilizers like nitrogen, phosphorous, and potassium unto the land also affects soil quality due to the leaching of nitrates. It is recommended that instead of using mineral fertilizers, organic fertilizers or waste from biomass should be used.

The transportation of the biomass to the plant site is also found in Phase 1. It is generally assumed that the plant is close to the plantation site. This will help reduce environmental impacts resulting from transportation (Patel et al., 2016). In Phase 2 pretreatment, torrefaction, pelletization, delivery, and usage of the product are considered. Pretreatment involves drying, grinding, and crushing. Energy consumption differs based on the moisture content of the biomass. Impact assessment on the environment varies based on operating conditions such as heating rate, temperature, pressure, and type of reactor used. Phase 3 involves demolishing and recycling the plant after its plant life. Most works from literature focus on Phases 1 and 2 as pelletization of torrefied biomass is a new technology yet to be commercialized. There is inadequate information on its maintenance and end life.

Adams et al. (2015) modeled a torrefied pellets plant in Norway that had its torrefied pellets delivered to a power plant based in the UK. The feed for the process was Scots pine because it has a high calorific value. The plant had an annual capacity of 60,000 tons, and operated 7,200 hours in a year. The torrefaction occurred at 250°C for 30 min. It was also assumed that the gases produced during the torrefaction process were recirculated thus reducing fuel required for production. The functional unit was 1 ton or 1 MJ of torrefied pellets delivered.

The goal of the LCA was to analyze the environmental impacts of the torrefied pellets and identify the components that contribute most to the impact score. The software used for the analysis was SimaPro 7.3, and the background data was obtained from the Ecoinvent 2.0 database. The usage phase was not included so the model ended at the delivery of the torrefied pellets to the power plant in the UK for co-firing. LCA identifies the materials and energy used as well as emissions into the environment. The emissions include particulate matter, wastewater discharge, and air emissions. The use of water as well as that of wastewater discharged was not included in this analysis because there were inadequate data on them. Some of the environmental impacts considered in this study were particulates, land use, climate change, and energy use. To produce 60,000 tons of torrefied pellets fuel per annum, 95 km^2 of land was required. Land being a limited resource implies that the implementation of torrefaction could be restricted; however, its positive impact on the ecosystem makes its application justified. Findings from the study are presented in Table 6.7. It shows that torrefied pellets result in less greenhouse gas emissions as well as lower consumption of fossil fuel. The results were sensitive to the drying requirements of the feedstock as well as the amount of gas that could be recirculated as fuel for the process.

The energy here represents the amount of energy required to produce 1 ton of torrefied biomass. With increasing moisture content, the energy required also increased. Some of this energy was used for the conversion of biomass into volatiles. This occurred at an average of 60% combustion efficiency. The higher the moisture content in the biomass the lower the conversion efficiency.

The energy required for pelletization depends greatly on the type of biomass used as well as process conditions such as moisture content and feed size associated with the process. This study was done on aspen wood, sawdust, Douglas fir, and municipal solid waste. Averagely, the electrical energy required for pelletization was between 16 and 49 kWh/ton. The energy required for pelletization has a linear relationship with the density of the pellets. The majority of the pelletization energy is used to cause the bio-coal to flow within the press channels (Stelte et al., 2012).

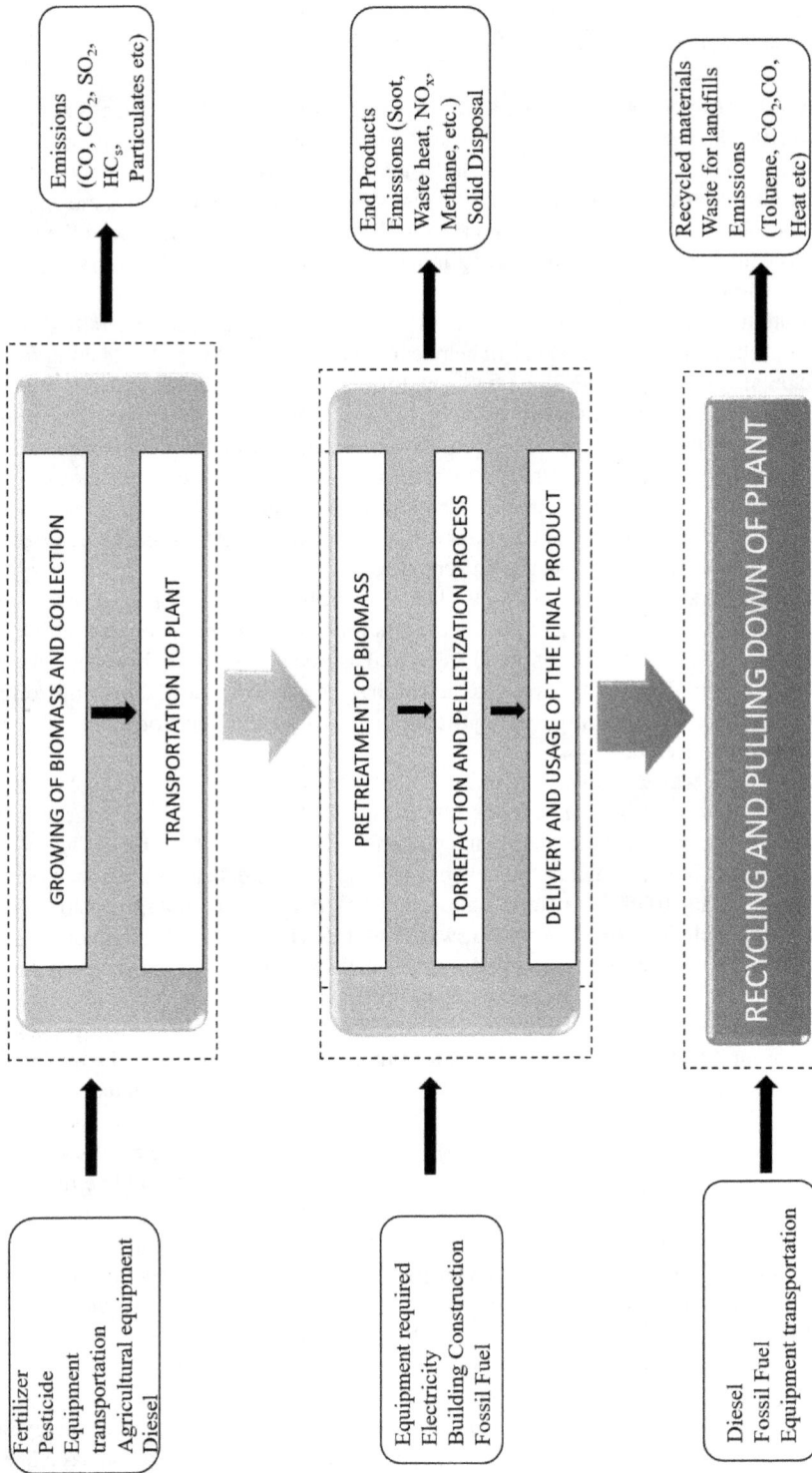

FIGURE.6.3 General system boundary for lifecycle assessment (adapted from Patel et al., 2016).

TABLE 6.7

Summarized LCA Results for the Production of Torrefied Pellets

Torrefied Pellet Production	Climate Change	Particulate Matter	Fossil Depletion
Lifecycle stage	g CO_2 equivalent	g PM_{10} equivalent	G oil equivalent
Growth of feedstock	1.2999	0.0485	0.3952
Felling tree – pine logs	0.0114	0.0000	0.0043
Felling trees forest residues	0.0029	0.0043	0.0011
Wood chips – pine logs	1.3506	0.0037	0.4542
Wood chips – forest residue	0.5232	0.0015	0.1785
Screening wood chips	0.5919	0.0015	0.1976
Biomass storage	0.0262	0.0001	0.0059
Reactor electricity – torrefied biomass	0.7368	0.0007	0.2274
Reactor electricity – torgas	0.4421	0.004	0.1364
Process heat of combustor	3.0929	0.0009	0.6700
Cooling	0.0194	0.0000	0.0060
Grinding	2.1557	0.0021	0.6652
Pelletization	4.2103	0.0042	1.2992
Plant infrastructure	0.0404	0.0001	0.0116
Delivery to power plant	2.9990	0.0068	1.1119

Note: Reproduced with permission from Adams et al. (2015).

6.8 Critical Discussion

The temperature at which torrefaction is carried out is vital to the yield of biochar. For a fixed time, the biochar yield decreases as the reaction temperature increases. Increasing temperature also implies that biochar of higher carbon content and heating value is produced. Biochar obtained under very high temperatures also absorbs water easily. The mass yield, energy yield, and volatile content of the biochar decrease with an increase in temperature. The optimal torrefaction temperature is between 250°C and 300°C with a yield of at least 60%.

Raw biomasses like wood residue and sawdust have their lignin intact and serve as a cheap source of the binder. Softwood has a higher lignin composition compared to hardwood. While all binders build up the ash content of pellets, wood as a binder reduces the ash content of pellets. However, these raw sources have a lower heating value, so they reduce the heating value of the pellets. Organic binders also have the disadvantage of tar generation during the gasification process. Solid binders are unable to fill the nanometer pores of the biochar which leads to lower-dense pellets.

From the investigations carried out by Dai et al. (2019), pellets obtained from the polyvinyl alcohol binder had higher hydrophobicity and strength than pellets from the commercially used lignin and starch binders. Higher tar composition, however, was observed for polyvinyl alcohol pellets, and this was addressed by introducing biomass gasification residue additive. The addition of this additive meant a smaller amount of the organic binder could be used, which decreased tar production. Starch and lignin are high-priced binders. Starch also forms a part of the food supply chain, so its usage is not encouraged. Although starch forms very strong bonds, it reduces the heating value of pellets due to its low heating value.

NaOH as a binder works by softening the lignin present in the torrefied biomass. Its usage, however, causes high hydrophilicity of the pellets, which is not a desired property of fuel. Compared to other binders, glycerol has HHV, so it increases the HHV of the pellets to a greater extent. The high tensile strength of the pellets is expected because torrefaction causes depolymerization of different molecules. This causes a change in the physicochemical properties of the biomass making it harder and brittle.

It is observed that the amount of binder used in the experiments was on a higher scale. In some cases, 30% of binder must be used before good pellet properties can be observed which is not preferred. On the other hand, the proteinaceous binder required just a 3% additional amount of binder. Lignosulfonate and bentonite binders used were also just 2%.

From the TEA, one factor that affects the production cost severely is the moisture content of the feedstock. This has a direct influence on the extent of drying, torrefaction temperature and residence time required, and consequently fuel or energy used for production. Also, when the cost involved in the production of torrefied pellets is compared to that of conventional pellets, the production of conventional pellets is cheaper. However, comparing the cost involved in the production of torrefied pellets with stand-alone torrefaction and standalone pelletization for enhanced pellet fuel quality shows that combined torrefaction and pelletization plant is more profitable.

From the lifecycle assessment, the production and application of fertilizer contribute greatly to global warming. Therefore, species of crops that require minimal agrochemicals would have to be developed. The amount of greenhouse gases emitted during the entire lifecycle of the torrefaction and pelletization plant is lower. As such it is necessary to optimize the process conditions to encourage minimal usage of fossil fuel during torrefaction and pelletization as that serves as the major contributing factor to greenhouse gases emitted.

Some challenges that the torrefaction and pelletization technology is facing that need to be addressed before it can be fully utilized commercially include (Batidzirai et al., 2013):

(i) Less understanding of the torrefaction process – For instance, the effects of the reaction conditions on the torrefaction process need to be further investigated. When residence time is decreased, it cuts down on operational cost, and the biochar yield is higher. However, before a lower residence time can be used, the biomass fed into the reactor should be of smaller particle sizes which require additional use of energy. Therefore, optimum conditions of torrefaction should be developed using trial and error.

(ii) Torrefaction reactors need to be expanded to meet commercial use.

(iii) Optimum torrefaction process control is difficult to achieve. As a result, consistent products are not obtained.

(iv) Apart from woody biomass, passing a wide variety of heterogeneous biomass through torrefaction tends to create problems. Inconsistency in the size of the biomass results in uneven torrefaction. Agricultural residues also have long fibers, are low in density, and ignite easily. These shortcomings come in the way of the torrefaction technology to have a large resource base.

(v) There is also the issue of ensuring safety by preventing dust explosions, self-ignition, and spontaneous combustion of the biochar.

The following are some of the applications and opportunities of torrefaction and pelletization technologies:

(i) They are used for co-firing and co-gasification in coal-fired power plants (Thrän et al., 2016). Torrefied pellets have combustion characteristics like coal. They can, therefore, be co-fired directly with pulverized coal at electric power plants. Eighty percent of torrefied pellets and 20% of coal can be mixed for this purpose, unlike non-torrefied biomass where only 5–10% is allowed (Mamvura and Danha, 2020).

(ii) Torrefied pellets are used in standalone combustion units. Combustion is the most common method of releasing energy from fuel. Power stations can combust torrefied pellets for electricity production. Thus, using torrefied pellets alone is not economical, but it is a great solution for eliminating greenhouse gases. More research is required to make the torrefaction and pelletization process as economical as possible so stand-alone combustion of pellets can be done profitably.

(iii) Torrefied pellets can be used in gasifiers to produce syngas. This ensures higher gasification efficiency as the quantity of syngas produced increases with increasing temperature. There is also less char generation from the torrefied pellets due to higher heating value and low volatile content. The use of torrefied pellets during gasification also reduces the formation of agglomerates. Chen et al. (2015) studied bamboo gasification experimentally using raw and torrefied bamboo and compared it with that of bituminous coal. While the syngas production from raw bamboo was low, syngas production from the torrefied bamboo was high and comparable to coal. The gasification efficiency of the torrefied bamboo was 88% higher than that of raw bamboo.

(iv) Torrefied pellets serve as solid fuel in industrial furnaces and metallurgical processes. They are also used for residential heating. During the production of iron, coke and pulverized coal are fed into a blast furnace. Averagely 300–350 kg of coke and 150–200 kg of coal are usually consumed per ton of hot iron. These fossil fuels contribute greatly to CO_2 emissions which lead to global warming. The iron-making process accounts for 5% of the overall anthropogenic CO_2 emissions. The release of CO_2 can be reduced by using torrefied pellets in place of pulverized coal. This would also help to reduce SO_2 emissions (Basu et al., 2013).

(v) Pellets can be turned into high-quality biofuels through fast pyrolysis. When bio-oils are produced from torrefied pellets instead of raw biomass, they have an improved oxygen-to-carbon ratio (Zheng et al., 2013). The bio-oil from the pellets also has lesser water content and highly concentrated lignin content (Meng et al., 2012).

(vi) Torrefied pellets can be used as activated carbon to treat water. They also act as adsorbents to remove organic and inorganic substances from liquids and gases. When drilling oil or gas wells, torrefied pellets can be used for absorbing contaminated water.

6.9   Conclusions

Torrefaction serves as an efficient method of improving biomass for use in energy production. Densification of bio-coal obtained from torrefaction helps to improve the mechanical properties of the fuel as well as reduces handling storage and transportation costs of the final product. Some components of biomass like lignin, water, and protein are natural binders, but they lose most of their binding capabilities after being exposed to high temperatures during torrefaction. External binders can, therefore, be employed. When binders are added to torrefied biomass before pelletization, the energy required for pelletizing is reduced compared to using no binders. Thus, adding binders helps to reduce power consumption. Apart from this advantage, binders also enhance the mechanical strength, hydrophobicity, density, and fuel property of the pellets.

Combined torrefaction and densification are more economical compared to standalone torrefaction and densification. Considering that torrefied pellets have properties comparable to coal, they have a huge market potential. Advancements in technology as well as incorporating binders can help reduce the production cost of torrefied pellets and make them competitive. The analysis of the environmental impacts of the torrefaction and pelletization of biomass showed that there were reduced emissions of greenhouse gases although relatively larger land is required for the cultivation of feedstock. Greenhouse gas emissions can be further reduced if more research is carried out so that the process conditions are optimized and the use of fossil fuel during torrefaction and pelletization reduced.

Acknowledgments

The authors would like to thank the Natural Sciences and Engineering Research Council (NSERC) of Canada, Canada Research Chairs (CRC) Program, and Agriculture and Agri-Food Canada (AAFC) and BioFuelNet Canada for funding this research.

REFERENCES

Abedi, A., and Dalai, A. K. 2017. Study on the quality of oat hull fuel pellets using bio-additives. *Biomass and Bioenergy* 106: 166–175.

Adams, P. W. R., Shirley, J. E. J., and McManus, M. C. 2015. Comparative cradle-to-gate lifecycle assessment of wood pellet production with torrefaction. *Applied Energy* 138: 367–380.

Adapa, P., Tabil, L., and Schoenau, G. 2011. Grinding performance and physical properties of non-treated and steam exploded barley, canola, oat and wheat straw. *Biomass and Bioenergy* 35: 549–561.

Adhikari, B. B., Chae, M., Zhu, C., Khan, A., Harfield, D., Choi, P., and Bressler, D. C. 2019. Pelletization of torrefied wood using a proteinaceous binder developed from hydrolyzed specified risk materials. *Processes* 7: 229.

Agar, D. A. 2017. A comparative economic analysis of torrefied pellet production based on state-of-the-art pellets. *Biomass and Bioenergy* 97: 155–161.

Arzola, N., Gómez, A., and Rincón, S. 2012. The effects of moisture content, particle size and binding agent content on oil palm shell pellet quality parameters. *Ingeniería e Investigación* 32: 24–29.

Azargohar, R., Nanda, S., and Dalai, A. K. 2018. Densification of agricultural wastes and forest residues: A review on influential parameters and treatments. In: *Recent Advancements in Biofuels and Bioenergy Utilization,* (Eds.) P. K. Sarangi, S. Nanda, and P. K. Mohanty. Singapore: Springer Nature, pp. 27–51.

Barskov, S., Zappi, M., Buchireddy, P., Dufreche, S., Guillory, J., Gang, D., Hernandez, R., Bajpai, R., Baudier, J., Cooper, R., and Sharp, R. 2019. Torrefaction of biomass: A review of production methods for bio-coal from cultured and waste lignocellulosic feedstocks. *Renewable Energy* 142: 624–642.

Basu, P., Rao, S., and Dhungana, A. (2013). An investigation into the effect of biomass particle size on its torrefaction. *The Canadian Journal of Chemical Engineering* 91: 466–474.

Batidzirai, B., Mignot, A. P. R., Schakel, W. B., Junginger, H. M., and Faaij, A. P. C. 2013. Biomass torrefaction technology: Techno-economic status and future prospects. *Energy* 62: 196–214.

Chai, L., and Saffron, C. M. 2016. Comparing pelletization and torrefaction depots: Optimization of depot capacity and biomass moisture to determine the minimum production cost. *Applied Energy* 163: 387–395.

Chen, W. H., Peng, J., and Bi, X. T. 2015. A state-of-the-art review of biomass torrefaction, densification and applications. *Renewable and Sustainable Energy Reviews* 44: 847–866.

Chin, K. L., P. S. H'ng, Go, W. Z., Wong, W. Z., Lim, T. W., Maminski, M., and Luqman, A. C. 2013. Optimization of torrefaction conditions for high energy density solid biofuel from oil palm biomass and fast growing species available in Malaysia. *Industrial crops and products* 49: 768–774.

Ciolkosz, D., and Wallace, R. 2011. A review of torrefaction for bioenergy feedstock production. *Biofuels, Bioproducts and Biorefining* 5(3): 317–329.

Clark, J. H., and Deswarte, F. E. I. 2008. Introduction to Chemicals from Biomass. In: *Introduction to Chemicals from Biomass.* Chichester, UK: John Wiley & Sons.

Dai, X., Theppitak, S., and Yoshikawa, K. 2019. Pelletization of carbonized wood using organic binders with biomass gasification residue as additive. *Energy and Fuels* 33(1): 323–329.

Emadi, B., Iroba, K. L., and Tabil, L. G. 2017. Effect of polymer plastic binder on mechanical, storage and combustion characteristics of torrefied and pelletized herbaceous biomass. *Applied Energy* 198: 312–319.

Ghiasi, B., Kumar, L., Furubayashi, T., Lim, C. J., Bi, X., Kim, C. S., and Sokhansanj, S. 2014. Densified bio-coal from woodchips: Is it better to do torrefaction before or after densification? *Applied Energy* 134: 133–142.

Harun, N. Y., and Afzal, M. T. 2016. Effect of particle size on mechanical properties of pellets made from biomass blends. *Procedia Engineering* 148: 93–99.

Kumar, A. K., and Sharma, S. 2017. Recent updates on different methods of pretreatment of lignocellulosic feedstocks: A review. *Bioresources and Bioprocessing* 4(1): 1–19.

Kumar, L., Koukoulas, A. A., Mani, S., and Satyavolu, J. 2017. Integrating torrefaction in the wood pellet industry: A critical review. *Energy and Fuels* 31(1): 37–54.

Li, H., Liu, X., Legros, R., Bi, X. T., Lim, C. J., and Sokhansanj, S. 2012. Pelletization of torrefied sawdust and properties of torrefied pellets. *Applied Energy* 93: 680–685.

Li, Y., and Liu, H. 2000. High-pressure densification of wood residues to form an upgraded fuel. *Biomass and Bioenergy* 19(3): 177–186.

Liu, T., McConkey, B., Huffman, T., Smith, S., MacGregor, B., Yemshanov, D., and Kulshreshtha, S. 2014. Potential and impacts of renewable energy production from agricultural biomass in Canada. *Applied Energy* 130: 222–229.

Lu, D., Tabil, L. G., Wang, D., Wang, G., and Emami, S. 2014. Experimental trials to make wheat straw pellets with wood residue and binders. *Biomass and Bioenergy* 69: 287–296.

Lu, K. M., Lee, W. J., Chen, W. H., Liu, S. H., and Lin, T. C. 2012. Torrefaction and low temperature carbonization of oil palm fiber and eucalyptus in nitrogen and air atmospheres. *Bioresource Technology* 123: 98–105.

Mallory, E. 2013. Pelleting torrefied material. *Biomass Pelletization and Torrefaction Workshop*, Vancouver, Canada, pp. 18–20.

Mamvura, T. A., and Danha, G. 2020. Biomass torrefaction as an emerging technology to aid in energy production. *Heliyon* 6(3): 03531.

Mani, S., Tabil, L. G., and Sokhansanj, S. 2006. Effects of compressive force, particle size and moisture content on mechanical properties of biomass pellets from grasses. *Biomass and Bioenergy* 30: 648–654.

Massaro, M. M., Son, S. F., and Groven, L. J. 2014. Mechanical, pyrolysis, and combustion characterization of briquetted coal fines with municipal solid waste plastic (MSW) binders. *Fuel* 115: 62–69.

Medic, D., Darr, M., Potter, B., and Shah, A. 2010. Effect of torrefaction process parameters on biomass feedstock upgrading. *American Society of Agricultural and Biological Engineers,* Pittsburgh, Pennsylvania.

Mekonnen, T. H., Mussone, P. G., Stashko, N., Choi, P. Y., and Bressler, D. C. 2013. Recovery and characterization of proteinacious material recovered from thermal and alkaline hydrolyzed specified risk materials. *Process Biochemistry* 48: 885–892.

Meng, J., Park, J., Tilotta, D., and Park, S. 2012. The effect of torrefaction on the chemistry of fast-pyrolysis bio-oil. *Bioresource Technology* 111: 439–446.

Nanda, S., Gong, M., Hunter, H. N., Dalai, A. K., Gökalp, I., and Kozinski, J. A. 2017. An assessment of pinecone gasification in subcritical, near-critical and supercritical water. *Fuel Processing Technology* 168: 84–96.

Nanda, S., Mohanty, P., Pant, K. K., Naik, S., Kozinski, J. A., and Dalai, A. K. 2013. Characterization of North American lignocellulosic biomass and biochars in terms of their candidacy for alternate renewable fuels. *BioEnergy Research* 6: 663–677.

Natural Resources Canada. 2020. *Bioenergy and Bioproducts.* Government of Canada. www.nrcan.gc.ca/our-natural-resources/forests-forestry/forest-fact-book/bioenergy-bioproducts/21686 (accessed 21 December 2020)

Patel, M., Zhang, X., and Kumar, A. 2016. Techno-economic and lifecycle assessment on lignocellulosic biomass thermochemical conversion technologies: A review. *Renewable and Sustainable Energy Reviews* 53: 1486–1499.

Peng, J. H., Bi, X. T., Lim, C. J., Peng, H., Kim, C. S., Jia, D., and Zuo, H. 2015. Sawdust as an effective binder for making torrefied pellets. *Applied Energy* 157: 491–498.

Pirraglia, A., Gonzalez, R., Denig, J., and Saloni, D. 2013. Technical and economic modeling for the production of torrefied lignocellulosic biomass for the US densified fuel industry. *Bioenergy Research* 6(1): 263–275.

Poddar, S., Kamruzzaman, M., Sujan, S. M., Hossain, M., Jamal, M. S., Gafur, M. A., and Khanam, M. 2014. Effect of compression pressure on lignocellulosic biomass pellet to improve fuel properties: Higher heating value. *Fuel* 131: 43–48.

Ribeiro, J. M. C., Godina, R., Matias, J. C. D. O., and Nunes, L. J. R. 2018. Future perspectives of biomass torrefaction: Review of the current state-of-the-art and research development. *Sustainability* 10(7): 2323.

Samuelsson, R., Larsson, S. H., Thyrel, M., and Lestander, T. A. 2012. Moisture content and storage time influence the binding mechanisms in biofuel wood pellets. *Applied Energy* 99: 109–115.

Sarker, T. R., Azargohar, R., Dalai, A. K., and Meda, V. 2020. Physicochemical and fuel characteristics of torrefied agricultural residues for sustainable fuel production. *Energy and Fuels* 34: 14169–14181.

Sarker, T. R., Nanda, S., Dalai, A. K., and Meda, V. 2021. A review of torrefaction technology for upgrading lignocellulosic biomass to solid biofuels. *BioEnergy Research* 14: 645–669.

Satpathy, S. K., Tabil, L. G., Meda, V., Naik, S. N., and Prasad, R. 2014. Torrefaction of wheat and barley straw after microwave heating. *Fuel* 124: 269–278.

Shah, A., Baral, N. R., and Manandhar, A. 2016. Technoeconomic Analysis and Lifecycle Assessment of Bioenergy Systems. *Advances in Bioenergy* 1: 189–247.

Shah, A., Darr, M. J., Medic, D., Anex, R. P., Khanal, S., and Maski, D. 2012. Techno-economic analysis of a production-scale torrefaction system for cellulosic biomass upgrading. *Biofuels, Bioproducts and Biorefining* 6(1): 45–57.

Shang, L., Nielsen, N. P. K., Stelte, W., Dahl, J., Ahrenfeldt, J., Holm, J. K., Arnavat, M. P., Bach, L. S., and Henriksen, U. B. 2014. Lab and bench-scale pelletization of torrefied wood chips-process optimization and pellet quality. *Bioenergy Research* 7(1): 87–94.

Stelte, W., Holm, J. K., Sanadi, A. R., Barsberg, S., Ahrenfeldt, J., and Henriksen, U. B. 2011. Fuel pellets from biomass: The importance of the pelletizing pressure and its dependency on the processing conditions. *Fuel* 90(11): 3285–3290.

Stelte, W., Sanadi, A. R., Shang, L., Holm, J. K., Ahrenfeldt, J., and Henriksen, U. B. 2012. Recent developments in biomass pelletization – a review. *BioResources* 7(3): 4451–4490.

Svanberg, M., Olofsson, I., Flodén, J., and Nordin, A. 2013. Analysing biomass torrefaction supply chain costs. *Bioresource Technology* 142: 287–296.

Thrän, D., Witt, J., Schaubach, K., Kiel, J., Carbo, M., Maier, J., Ndibe, C., Koppejan, J., Alakangas, E., Majer, S., and Schipfer, F. 2016. Moving torrefaction towards market introduction – Technical improvements and economic-environmental assessment along the overall torrefaction supply chain through the SECTOR project. *Biomass and Bioenergy* 89: 184–200.

Tilay, A., Azargohar, R., Drisdelle, M., Dalai, A. K., and Kozinski, J. 2015. Canola meal moisture-resistant fuel pellets: Study on the effects of process variables and additives on the pellet quality and compression characteristics. *Industrial Crops and Products* 63: 337–348.

Tumuluru, J. S., Sokhansanj, S., Hess, J. R., Wright, C. T., and Boardman, R. D. 2011. A review on biomass torrefaction process and product properties for energy applications. *Industrial Biotechnology* 7: 384–401.

Tumuluru, J. S., Wright, C., Kenny, K., and Hess, R. 2010. A review on biomass densification technologies for energy application. U.S. Department of Energy. Doi: 10.2172/1016196.

Uslu, A., Faaij, A. P., and Bergman, P. C. 2008. Pre-treatment technologies, and their effect on international bioenergy supply chain logistics. Techno-economic evaluation of torrefaction, fast pyrolysis and pelletisation. *Energy* 33(8): 1206–1223.

Wang, Y., Wu, K., and Sun, Y. 2018. Effects of raw material particle size on the briquetting process of rice straw. *Journal of the Energy Institute* 91: 153–162.

Wu, Y. (2013). Systems analysis of integrated southern pine torrefaction and granulation technology (Doctoral dissertation, University of Georgia).

Xu, F., Linnebur, K., and Wang, D. 2014. Torrefaction of Conservation Reserve Program biomass: A techno-economic evaluation. *Industrial Crops and Products* 61: 382–387.

Zheng, A., Zhao, Z., Chang, S., Huang, Z., Wang, X., He, F., and Li, H. 2013. Effect of torrefaction on structure and fast pyrolysis behavior of corncobs. *Bioresource Technology* 128: 370–377.

7

Lignocellulosic Biomass Conversion to Syngas through Co-Gasification Approach

Minhaj Uddin Monir, Azrina Abd Aziz, Fatema Khatun,
Dai-Viet N. Vo and Nadzirah Mohd Mokhtar

CONTENTS

7.1 Introduction

Energy demand is rising globally due to unsustainable use of energy, urbanization, industrial extension, the fastest-growing population, and a decline in nonrenewable energy sources (Scheffran et al., 2020). Fossil fuels (e.g., oil, gas, coal) generate about 80% of energy to meet this energy demand, while only 10–15% of energy is supplied by wind, solar, biomass, geothermal, hydroelectric, etc. (Capuano, 2018; Singh and Sekhar, 2016). In spite of the continuous progress in petroleum exploration and exploitation innovation processes, the increased fuel production could not satisfy essential energy demand (Caineng et al., 2016). Shafiee and Topal (2009) estimated that oil reserves are expected to be reduced by 2040 and up to 2112 will be available for coal resources. It is therefore very crucial to find out about alternative, safe, and environmentally friendly energy sources. The fastest-growing biomass in this regard may be replacement sources of energy that are mostly available worldwide (Infield and Freris, 2020).

The sources of renewable and nonrenewable energy as shown in the flow chart (Figure 7.1) are compiled by Zabed et al. (2016) and Abo et al. (2019). The consumption of energy covers approximately 14% of the biomass resources as reported by Xu et al. (2018). They also mentioned that through certain conversion processes, biomass is transformed into bioenergy, biofuels, and other chemicals. Due to fuel safety and environmental conservation, this biomass is used for the production of syngas (Wyman, 2018). Carbon (C), hydrogen (H), and oxygen (O) are the primary elements of biomass that are disintegrated by combustion or decomposition (Basu, 2018). Owing to the presence of low sulfur content in biomass during biomass combustion, it produces lower air pollution than fossil

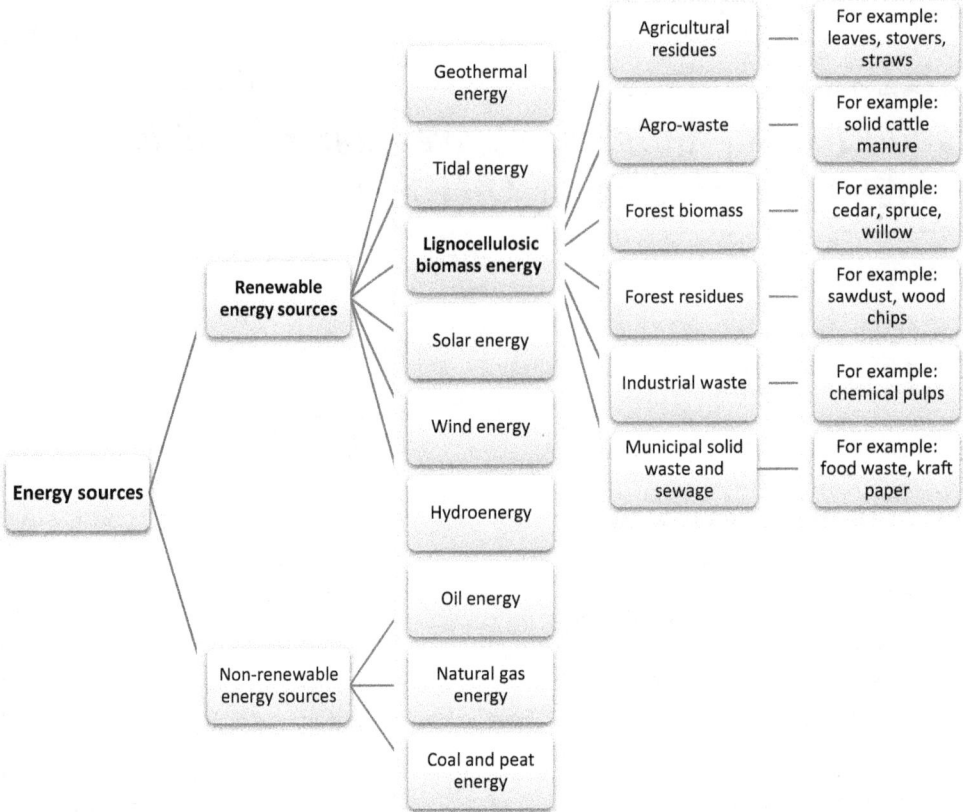

FIGURE 7.1   Sources of energy.

fuels (Monir et al., 2020a). Hence, the probability of the formation of sulfur dioxide (SO_2) and acid rain is very low.

Lignocellulose biomass is one of the major sources of energy made up of cellulose, hemicellulose, and lignin (Monir et al., 2018b). These are the most economical and fastest-growing bioenergy sources (Gaurav et al., 2017). This biomass is converted into bioenergy through three chemical processes: biochemical, thermochemical, and physiochemical. The thermochemical process is widely used for the conversion of biomass into biofuels and chemicals involving pyrolysis (300°C–700°C), liquefaction (140°C–200°C), and gasification (300°C–1000°C). The gasification process is typically used to turn biomass into syngas. This process is carried out with a controlled gasifying medium (e.g., air, oxygen, and/or steam) at high temperature by reacting with the biomass feedstock.

Co-gasification refers to the co-processing of two or more blends of feedstock, which is a new process enhancement strategy (Valdés et al., 2020). Co-gasification method has multiple advantages over the gasification of biomass or coal/charcoal (Mansur et al., 2020; Monir et al., 2020a). The benefits of this process are the reduction of sulfur and ash that cause reactor corrosion during coal gasification and environmental difficulties. In addition, the co-gasification process reduces the tar content in syngas (Monir et al., 2020d). By minimizing greenhouse gas emissions, this mechanism also protects the environment (Farzad et al., 2016). Some complex chemical reactions, including complete reaction (Eq. 7.1), partial combustion reaction (Eq. 7.2), total combustion reaction (Eq. 7.3), Boudouard equilibrium reaction (Eq. 7.4), water–gas shift reaction (Eq. 7.5), methanation reaction (Eq. 7.6), CO oxidation reaction (Eq. 7.7), H_2 oxidation reaction (Eq. 7.8), and H_2 oxidation reaction (Eq. 7.8), occur throughout the biomass gasification or co-gasification of biomass with other fuels (Monir et al., 2018b; Oh et al., 2018). This chapter aims to show a systematic method for biomass gasification and co-gasification process with other feedstocks such as coal, charcoal, petroleum coke, and their benefits. Therefore, in future, systemic co-gasification method and lignocellulosic-based bioenergy will be a sustainable substitute for fossil fuels.

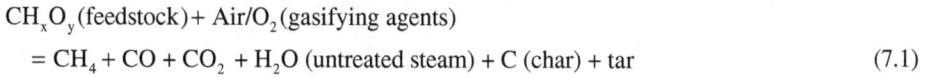

$$CH_xO_y\,(\text{feedstock}) + \text{Air}/O_2\,(\text{gasifying agents})$$
$$= CH_4 + CO + CO_2 + H_2O\ (\text{untreated steam}) + C\ (\text{char}) + \text{tar} \tag{7.1}$$

$$C + \frac{1}{2}O_2 \rightarrow CO \tag{7.2}$$

$$C + O_2 \rightarrow CO_2 \tag{7.3}$$

$$C + O_2 \rightarrow 2CO \tag{7.4}$$

$$C + H_2O \rightarrow CO + H_2 \tag{7.5}$$

$$C + 2H_2 \rightarrow CH_4 \tag{7.6}$$

$$CO + \frac{1}{2}O_2 \rightarrow CO_2 \tag{7.7}$$

$$H_2 + \frac{1}{2}O_2 \rightarrow H_2O \tag{7.8}$$

$$CH_4 + 2O_2 \rightarrow CO_2 + 2H_2O \tag{7.9}$$

$$CO + H_2O \rightarrow CO_2 + H_2 \tag{7.10}$$

$$CO + 3H_2 \rightarrow CH_4 + H_2O \tag{7.11}$$

7.2 Lignocellulosic Biomass

Lignocellulosic biomass refers to dry matter dependent on plants or plants that are converted by thermochemical conversion into fuel (Zhang and Brown, 2019). Globally, 11.4 billion tons of biomass is produced annually from agricultural land and forests of which 40% is produced by agricultural production, 30% by pasture production, 18% by timber production, and 12% by by-products reported by Zhu et al. (2019). A possible solution for the valorization of renewable energy resources is lignocellulose with other feedstocks for the production of syngas by co-gasification.

Agricultural residues, agro-waste, forest biomass, forest residues, industrial waste, municipal solid waste, and sewage are the most abundantly available lignocellulosic biomass (Figure 7.1). These are the most efficient sources of energy with safe goods and consumer value (Soccol et al., 2019). For lignocellulosic biomass conversion into liquid and gaseous product formation, single or mixed (thermochemical and biological) processes are involved (Claypool and Simmons, 2016; Monir et al., 2019; Monir et al., 2020c).

Cellulose, hemicellulose, and insoluble lignin are major ingredients in lignocellulosic biomass, depending on plant species, with a composition of 55%, 12%, and 25%, respectively (Musule et al., 2016). The chemical structure of cellulose, lignin, and hemicellulose cells is shown in Figure 7.2(d-e). The biomass cell wall is combined with solid polymer carbohydrate (cellulose) microfibers, and hemicellulose (dominant carbohydrate) impregnated with lignin is packed into the inner cell wall (Wang et al., 2017). The percentages of cellulose, hemicellulose and lignin in lignocellulosic biomass structures are shown in Table 7.1 and Figure 7.2.

The organic compound and the main structural feature of the biomass cell wall is cellulose, a genetic formulation of $(C_6H_{10}O_5)_n$. The existence of the hydroxyl (–OH) group makes it very rigid and fibrous. It is chemically stable and is insoluble in water. Mostly less dependent on the heating rate, the rate of decomposition at low temperatures is complex (Yu et al., 2017). It is usually hidden by hemicellulose, which forms a composite of cellulose–hemicellulose. A favorable condition for bioenergy production is

TABLE 7.1

The proportion of cellulose, hemicellulose, and lignin in lignocellulosic biomass

Lignocellulosic Biomass	Cellulose (wt.%)	Hemicellulose (wt.%)	Lignin (wt.%)	References
Agricultural residues	37–50	25–50	5–15	Limayem and Ricke (2012)
Rice straw	28–36	23–28	12–14	Rocha-Meneses et al. (2017)
Hardwood	45–47	25–40	20–25	Limayem and Ricke (2012)
Hardwood barks	22–40	20–38	30–55	Rocha-Meneses et al. (2017)
Softwood	40–45	25–29	30–60	Limayem and Ricke (2012)
Softwood barks	18–38	15–33	30–60	Rocha-Meneses et al. (2017)
Grasses	25–40	25–50	10–30	Rocha-Meneses et al. (2017)
Waste papers from chemical pulp	50–70	12–20	6–10	Limayem and Ricke (2012)
Bagasse	32–48	19–24	23–32	Rocha-Meneses et al. (2017)
Wheat straw	33–38	26–32	17–19	Rocha-Meneses et al. (2017)

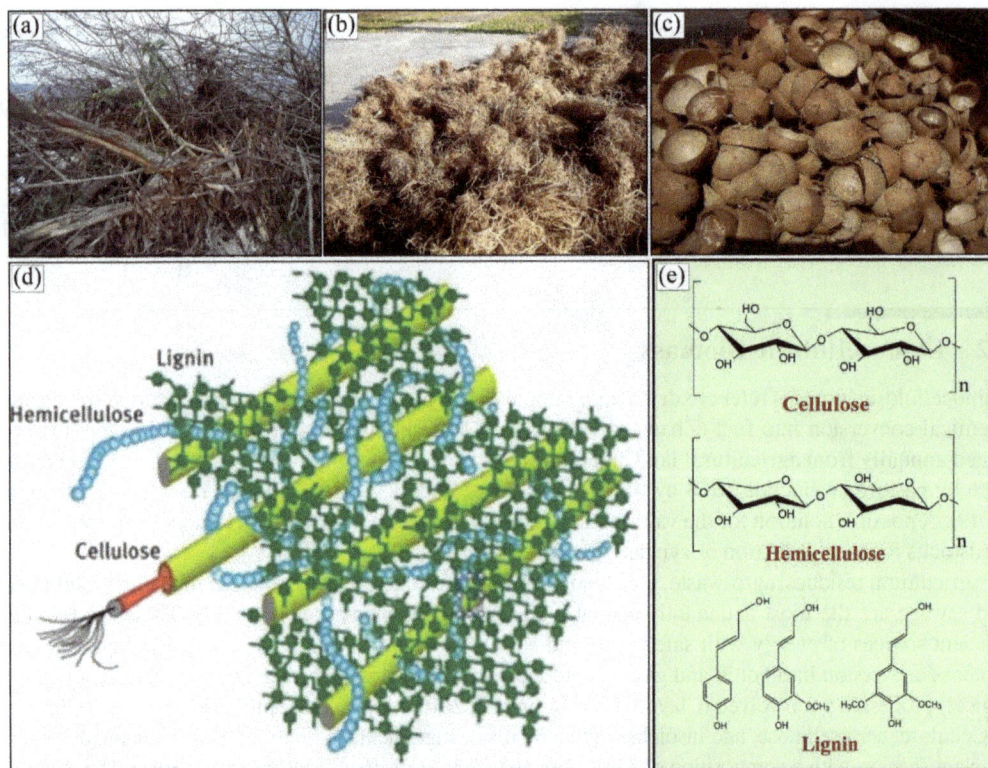

FIGURE 7.2 Lignocellulosic biomass: (a) forest residue, (b) empty fruit bunch of palm oil, (c) coconut shell, (d) and (e) structure of cellulose, hemicellulose, and lignin in plant cell wall.

the presence of a high cellulosic percentage in biomass (Koupaie et al., 2019). It has a higher productivity level of methane than hemicellulose and lignin (Ma et al., 2019).

Hemicellulose, with generic formula $(C_5H_8O_4)_n$ and including some hexose, pentose, sugar acid (uronic acids), methyl galacturonic acids, i.e., heteropolymers, is the second-highest abundant polymer of plant cell walls (Limayem and Ricke, 2012). Softwood hemicelluloses contain glucomannans, while there are xylans in hardwood hemicelluloses (Gírio et al., 2010). Chen (2014) estimated that 10–15 wt.% of the hemicellulose content was found in hardwood, 18–23 wt.% in softwood, and 20–25 wt.% in herbaceous plants. Hemicellulose is easier to digest than cellulose since it hydrolyzes at a higher rate (Ma et al., 2019).

Phenyl propionic alcohol is the primary composition of lignin. It is soluble in organic solvents but insoluble in water, and the traditional method is harder to degrade. It affects the biomass's digestibility. Cellulose and lignin polymers form the woody biomass of the forest. This stipulates plant structural support and provides resilience to microbial attacks (Ma et al., 2019). The highest lignin content is found in softwood barks (30–60 wt.%), while hardwood barks contain 30–55 wt.% and agricultural residues contain the lowest lignin content (3–15 wt.%) (Limayem and Ricke, 2012). The occurrence of high lignin in lignocellulosic biomass means that poor digestibility is caused by the enhanced removal of lignin (Ma et al., 2019).

7.3 Lignocellulosic Biomass-Based Energy

The energy that is derived from living organisms is based on biomass. Because of its intrinsic energy that comes directly from the sun (e.g., photo-electric, photochemical, thermal energy), lignocellulosic biomass is one of the most important renewable energy sources. In addition to processing biofuel and bioenergy, it also produces different chemicals from its cellulose, hemicellulose, and lignin components. Toklu (2017) recorded that around 77% of bioenergy is generated globally, while 87% is derived from wood biomass, 9% from crops, and 4% from solid and industrial municipal waste (Figure 7.3). During the years 2010 to 2035, the production of renewable energy for transport, heat demand, and electricity purposes (Ellabban et al., 2014) are shown graphically (Figure 7.4).

In the last few years (2010 to 2020), renewable energy production and usage in the transport, heating, and electricity sectors have been growing rapidly, and this trend may continue until 2035 (Figure 7.4). Ellabban et al. (2014) also suggested that the usage of conventional processes and new approaches have expanded exponentially by improving design and minimizing their costs. Biomass-based energy has not been used strictly for the exchange of oil, gas, or coal; however, it can disperse throughout the fields with minimal environmental effects. Bioenergy is produced from lignocellulosic biomass in terms of three different forms of products:

(i) Solid products: Biochar and torrefied biomass are the primary solid products of lignocellulosic biomass. Torrefaction (or bio-coal) is a thermal process resulting in higher fuel characteristics than raw biomass when the biomass is transformed into a coal-like material. Biochar (Figure 7.5c) is produced at reactor temperatures ranging from 300°C to 500°C (Soka and Oyekola, 2020) via the slow pyrolysis of biomass. Owing to the presence of similarities with

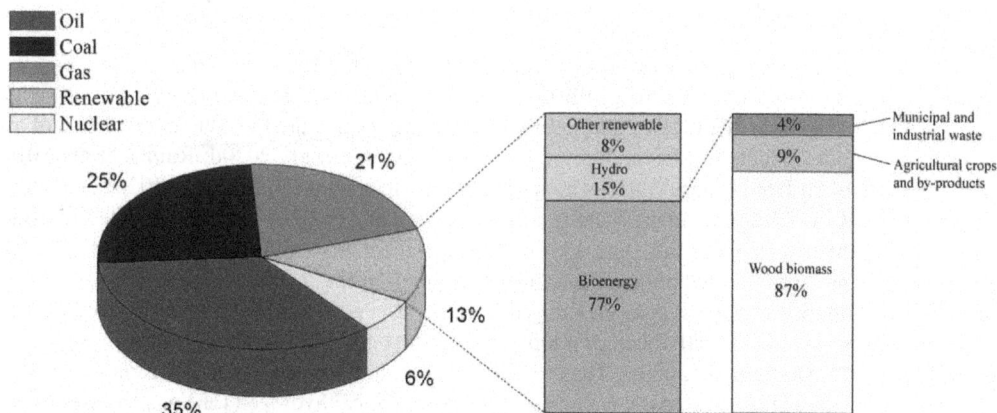

FIGURE 7.3 Worldwide share of bioenergy (data source: Toklu, 2017).

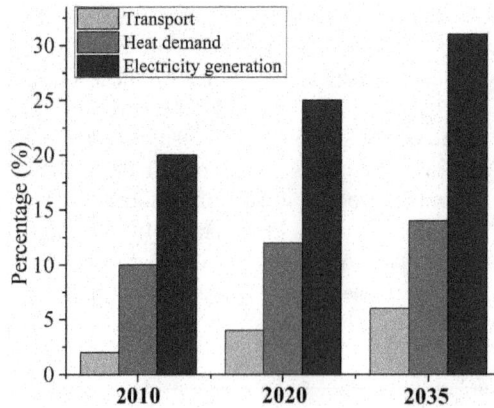

FIGURE 7.4 Generation of renewable energy for transport, heat, and electricity purposes from 2010 to 2035 (data source: Ellabban et al., 2014).

coal in its composition, it will be a good substitute for natural coal. Char can be the alternative to coal or peat for co-firing with lignocellulosic biomass due to the declining trend of fossil fuels. When it was co-gasified with an empty fruit bunch (EFB) of palm oil, forest residue, and coconut shell, the production rate of syngas increased significantly (Monir et al., 2017; Monir et al., 2018a; Monir et al., 2020a).

(ii) Liquid products: Bioethanol, methanol, biodiesel, vegetable oil, and pyrolysis oil are popular liquid products that are produced from biomass. Via syngas fermentation, biofuels (ethanol, methanol, hydrogen, dimethyl ether, etc.) are produced. Syngas is formed as a result of biomass gasification or co-gasification of two different feedstocks (Molino et al., 2018; Monir et al., 2020f). Biofuel is classified based on its form, origin, and processes as primary and secondary. Table 7.2 describes various types of biomass-based lignocellulosic biofuels and their parametric characteristics.

(iii) Gaseous products: Some chemical and thermochemical processes, including biogas (CH_4 and CO_2), producer gas (CO, H_2, CH_4, CO_2), syngas (CO, H_2, etc.), and methane (CH_4) generate the major gaseous products from lignocellulosic biomass. Syngas is one of the most essential products, and the addition of mixtures from other feedstocks increases its efficiency. Hydrogen (H_2), carbon monoxide (CO), methane (CH_4), carbon dioxide (CO_2), nitrogen (N_2), ethylene (C_2H_4), ethane (C_2H_6), ammonia (NH_3), hydrogen sulfide (H_2S), and tar (dark brown or black viscous hydrocarbon liquid) are obtained from specific feedstocks (Schmid et al., 2019). Through the thermal cracking process, tar impurities from syngas are reduced (Monir et al., 2020d).

During the thermochemical conversion process, it also included high molecular compounds that may be predictable with syngas. Ethane, sulfur dioxide, ammonia, benzene, phosphorus, nitrous oxide, hydrogen sulfide, methane, hydrogen cyanide, ethylene, acetylene, etc. are the solid components of ash and char, liquid co-products of tars, and gaseous products (Monir et al., 2020d; Ramos et al., 2018). For the removal of tar from syngas, various methods have been applied (Monir et al., 2020e). At a temperature of 900°C to 2000°C, thermal/heat treatment is available (Bell et al., 2019). Some important gasification parameters for feedstock type, gasifier type, gasification medium (air, oxygen, steam), and optimization of pressure and temperature parameters depend on the quality of syngas. Owing to the occurrence of impurities in the syngas cleaning technique, it is compulsory before it is used as fuel. Hydrogen-rich syngas has recently been developed using yeast and bacteria for bioethanol production (Monir et al., 2019; Monir et al., 2020c). They reported that treated syngas is more efficient for the production of bioethanol. During biomass gasification, the most critical hydrogen (H_2) gas (zero-emission fuel burned with oxygen) is developed (Dawood et al., 2020). Some pyrolysis, gasification, and steam-reforming thermochemical conversion processes are followed without and/or with chemical-looping

TABLE 7.2

Lignocellulosic Biomass-Based Biofuels and Their Characteristics

Parameters	First Generation Biofuels	Second Generation Biofuels	Third Generation Biofuels	References
Source	Food crops: sugarcane, maize, soybeans, wheat, cooking oil, etc.	Non-food crops: residues from forestry, urban waste, agricultural waste, vegetable oil waste, etc.	Algae and microbes containing seaweeds.	Oumer et al. (2018)
Manufacturing method	Fermentation process Starch/ Acetone–butanol–ethanol / Anaerobic; Transesterification of plant oils	Thermochemical processes (gasification, pyrolysis or torrefaction); Enzymatic hydrolysis	Biochemical or Thermochemical	Oumer et al. (2018)
Biofuel products	Bio-alcohol such as ethanol, propanol, butanol, biodiesel, green diesel.	Bioethanol/biobutanol/ biodiesel, methanol, dimethylfuran, fischer-Tropsch biodiesel (made from waste from paper and pulp manufacturing).	Bioethanol, Biohydrogen	Oumer et al. (2018)

technologies (Dawood et al., 2020; Monir et al., 2020b). Hydrogen gas is generated during biomass gasification with CO, CO_2, CH_4, etc. It is subsequently used when it is isolated from syngas as a safe and renewable fuel.

7.4 Non-Plant Feedstocks

For the conversion process of lignocellulosic biomass, the non-biomass feedstocks of coal, peat, charcoal, petroleum coke, etc. (Figure 7.5) are used. During conversion, these feedstocks help to increase the rate of gasification (Rana et al., 2019).

(i) Coal and peat:

Coal [Figure 7.5(a)] is a solid fossil fuel that tends to accumulate a significant volume of carbon-based material in swamps and peat bogs. It is a combustible sedimentary rock, black or brownish-black in color (Suárez-Ruiz et al., 2019). It primarily includes carbon, with the elements of hydrogen, nitrogen, oxygen, and sulfur missing. It is made from peat [(Figure 7.5(b)], laid down later on the top by the weight of rocks. Gastaldo et al. (2020) stated that forests are buried because of tectonic activity, including earthquakes and volcanic ashfall, and then peat and coal are formed by various geological parameters (such as time, temperature, pressure). Thus, the primitive type of coal that is formed by the geological process is peat. It is an essential hydrocarbon source that retains various forms of inorganic ions. Various methods for extracting hydrocarbons from coal and peat exist, such as gasification of coal/peat, liquefaction, and direct hydrocarbon conversion (Bailey et al., 2002). Carbon-containing coal particles are heated to 1225–1425°C (approximately) during thermal conversion (Mularski and Modliński, 2020). Energy-rich natural coal and peat are therefore valorized by co-combustion with lignocellulosic biomass, and syngas and other fuels are generated.

(ii) Petroleum coke: Petroleum coke (Figure 7.5d) is also a product of oil refining and is referred to as coke or petcoke. It is the final solid material rich in carbon and comprises a category of fuels known as coke. It consists of more than 80% carbon along with the other elements hydrogen, nitrogen, sulfur, etc. Because of the presence of a large portion of carbon, such as coal and peat, it is useful in industrial processes. During co-gasification with lignocellulosic biomass, the feedstock is also useful. It is used for the processing of cement, brick, lime, steel, glass, and fertilizers, as well as for commercial purposes.

FIGURE 7.5   Non-biomass feedstocks: (a) coal, (b) peat, (c) charcoal, and (d) petroleum coke.

7.5   Conversion of Lignocellulosic Biomass into Energy

There is considerable practical significance in converting lignocellulosic biomass (its components and raw materials) into energy. Different types of gasifiers perform the conversion process. The process of conversion includes mainly pyrolysis, gasification, co-gasification depending on temperature conversion parameters, gasifying agent, feedstock type, and conversion mechanism. The routes for the thermochemical and biochemical conversion of lignocellulosic biomass into biofuels are shown in Figure 7.6.

7.5.1   Pyrolysis

Pyrolysis is one of the thermochemical decomposition of materials for the conversion of energy resources. This process is carried out in the absence of oxygen (O_2), which occurs at a temperature of around 375–525°C under a certain pressure (0.5–1 MPa). Pyrolysis of biomass is essential for its combustion, gasification, and co-gasification processes (Dai et al., 2019). Different elemental compositions are decomposed in this process, and the chemical bonds in their molecules are broken down. After that, smaller molecules become fragments of the products and produce solid char, liquid oils, and gaseous compounds. During the pyrolysis phase, Basu (2010) defined four thermal stages: the drying stage occurs at a temperature of 100°C, the initial stage occurs at a temperature of 100–300°C, the intermediate stage occurs at a temperature higher than 200°C, and the final stage occurs at a temperature of 300–900°C. Charcoal is the primary outcome of slow pyrolysis and does not produce hydrogen due to slow pyrolysis. The gases produced from biomass pyrolysis are H_2, CO, CO_2, and CH_4 based on the nature of the feedstock content.

Pyrolysis processes and their functional groups are responsible for the production of cellulose, hemicellulose, and lignin, which are the main components of biomass as reported by Wang et al. (2017).

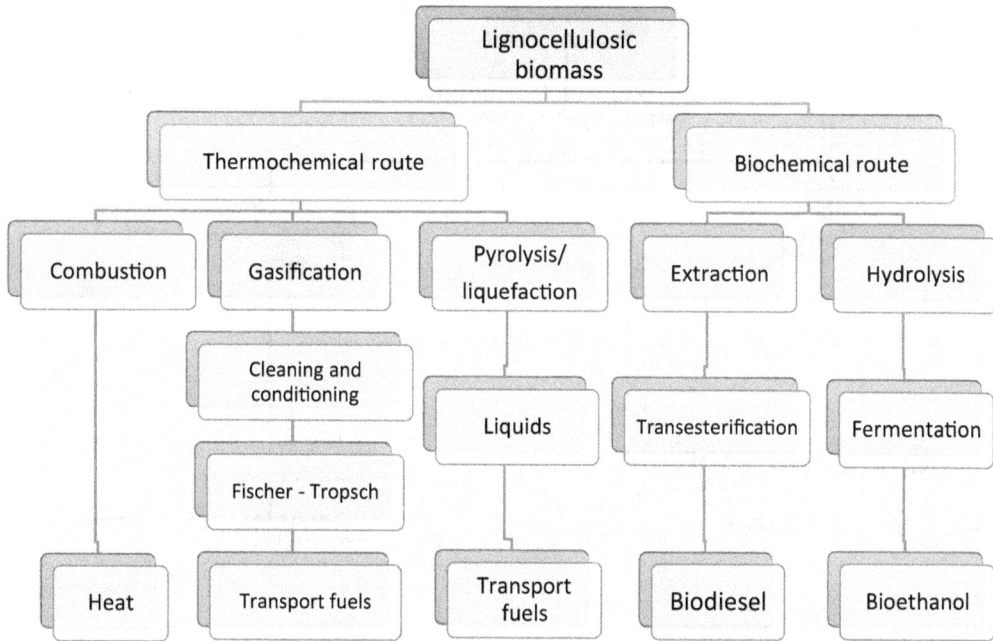

FIGURE 7.6 Lignocellulosic biomass conversion routes (thermochemical and biochemical) to biofuels.

Recent studies have illustrated that pyrolysis kinetics are directly related to thermochemical utilization of biomass. The pyrolysis response mechanism refers to the response rate and the degree of complexity of the response that can be predicted. Pyrolysis of lignin is more complex than cellulose and hemicellulose (Dhyani and Bhaskar, 2019). They also found that cellulose displays a sharp decay peak at a temperature of 350°C during pyrolysis, where lignin does not show a sharp decay peak. Overall reactions are as primary decomposition reactions at temperatures between 200°C and 400°C and secondary decomposition reactions at higher temperatures of 400°C.

7.5.2 Gasification

Gasification is one of the most promising thermochemical techniques that transform biomass or any carbonous material (e.g., carbon-based peat, coal, petcoke, etc.) into CO, H_2, CO_2, CH_4, and C_{2+} (Monir et al., 2018b) gaseous products in the oxygen-deficient setting. This process is performed by reacting at high temperatures (800–1000°C) with materials by partial oxidation reactions due to the presence of air/oxygen gasifying media. This includes undergoing a variety of processes including periodic drying, pyrolysis, oxidation, and reduction during energy conversion. The final products are developed in three phases: gaseous fuels, liquids (of low molecular weight), and solid carts (Monir et al., 2020a).

For the conversion of biomass, a fixed bed gasifier, a fluidized bed gasifier, and an entrained flow gasifier are used. Biomass is graded based on the physical configuration of the reactor. Among them, the downstream gasifier is widely used due to the conversion efficiency of biomass and low content of tar generated (Chaurasia, 2018). Gasifiers are selected on the basis of the form of feedstock, moisture content, fragment size, and gasifying medium (e.g., steam, oxygen or air). Syngas contains unwanted tars, impurities, and ash that cause problems when used as a fuel (Samiran et al., 2016; Singh and Sekhar, 2016). Gao et al. (2016) recorded relative errors in volumetric concentrations, lower heating value (1–13%), gas output (1–8%), cold gas efficiency (1–12%), and carbon conversion efficiency (1–11%). The key steps involved in gasification recorded by De et al. (2018) in an atmosphere of air and oxygen are shown in the following flow chart (Figure 7.7).

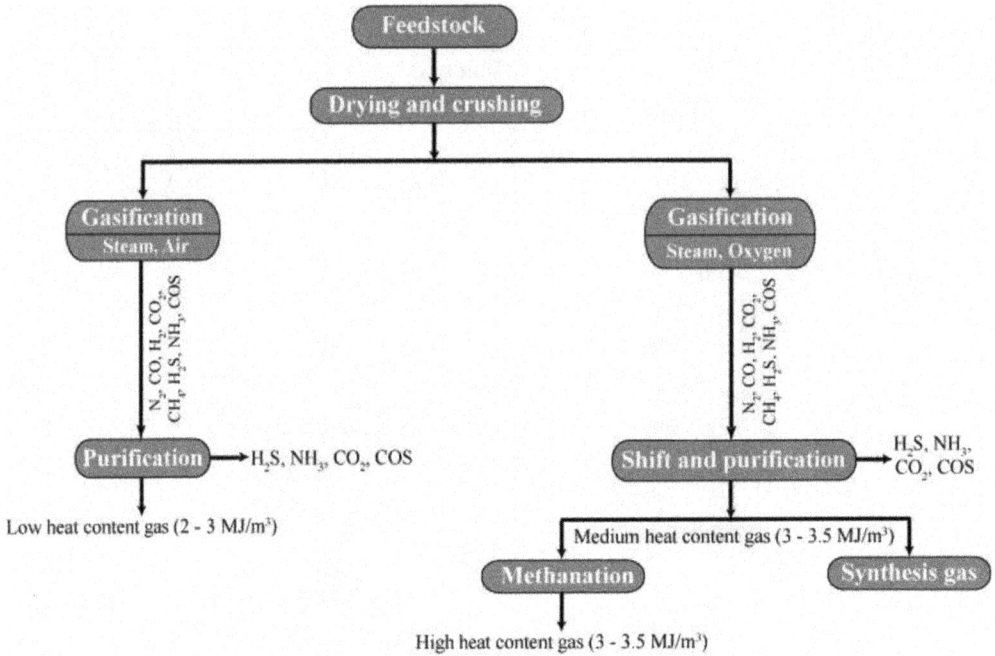

FIGURE 7.7 Steps involved in gasification in an air and oxygen ambiance (adapted from De et al., 2018).

During gasification, complex chemical reactions follow, and their mechanisms are shown in the literature (Monir et al., 2018b; Oh et al., 2018; Ren et al., 2020). At a temperature of 100–150°C, the moisture content is evaporated from biomass, and, at this time, partial hydrogen (H_2) is produced from steam (Aydin et al., 2018). This process is carried out in the biomass reactor drying zone. The feedstock materials are then tried for pyrolysis when the temperature of the reactor is increased from 200°C to 500°C (Monir et al., 2020a). Biomass feedstock materials are converted into volatile gases, and this time carbon is produced. Consequently, pyrolysis yields are moved to the hotter region called the oxidation region. Gasifying agents (e.g., air, steam, or oxygen) are added in the oxidation zone of the oxygen-deficient environment. When the reactor temperatures reach up to 1200°C, as reported by Aydin et al. (2018), the oxidation reactions take place. Gaseous products and solid char are then moved to the reduction zone (Monir et al., 2020a).

Co-gasification is referred to as thermochemical conversion of two or more mixtures of feedstocks (e.g., biomass, peat, petcoke, coal), that have many possibilities to ensure the quality of syngas to avoid GHG from polluting the atmosphere. Several researchers have recently carried out biomass and coal co-gasification (Mansur et al., 2020; Monir et al., 2018a; Monir et al., 2018b; Monir et al., 2020a) and listed the valorization of large quantities of combustible agricultural and wood waste in the fields of forestry, agriculture, and food processing. They also stated that biomass is less expensive than coal for co-firing. Monir et al. (2018a) analyzed the concentrations of H_2 and CO from EFB gasification relative to the co-gasification of EFB with charcoal in the downdraft gasifier. In their research, by using biomass through a co-gasification technique, attempts are made to minimize the use of fossil fuels.

During co-gasification, the fuel mixture is used, which due to physical and chemical changes reaches a high temperature. Due to the interaction of coal with biomass, the gasification reaction rate also shifts (Saw and Pang, 2013). In the petroleum field, CO_2, which is the co-product of syngas produced during co-gasification, may be used to boost the recovery of petroleum, which can reduce the content of greenhouse gases (Khatun et al., 2016; Mikulcic et al., 2016). An analysis of kernel shell (PKS) and polyethylene waste blend co-gasification has been performed by Moghadam et al. (2014). They optimized the temperature at 800°C with the P/B ratio in their study: 0.3 w/w and S/F. Patel et al. (2017) co-gasified based on mixtures of lignite and waste wood in a downdraft gasifier (10 kW) and also found that the

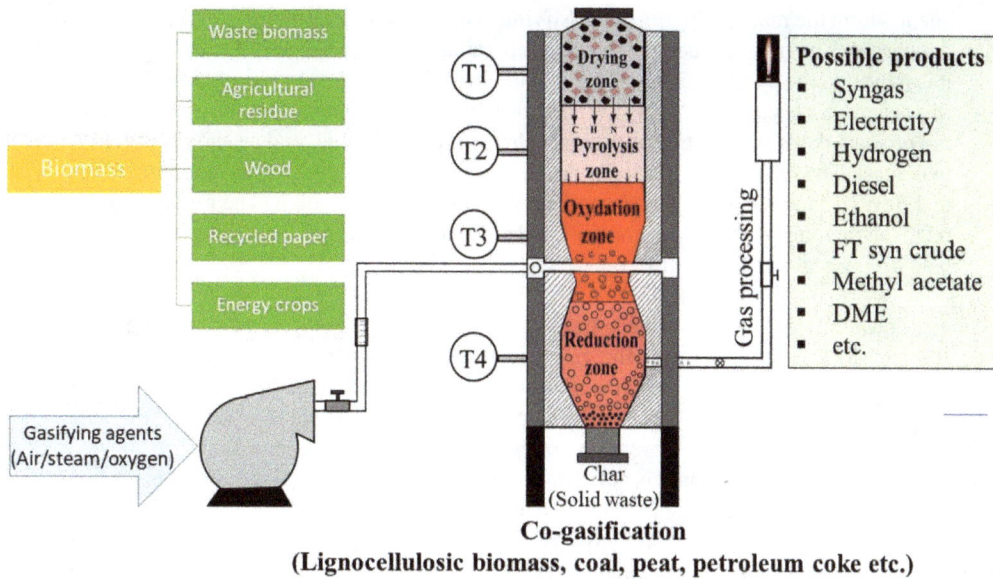

Co-gasification
(Lignocellulosic biomass, coal, peat, petroleum coke etc.)

FIGURE 7.8 Co-gasification process for biomass and coal.

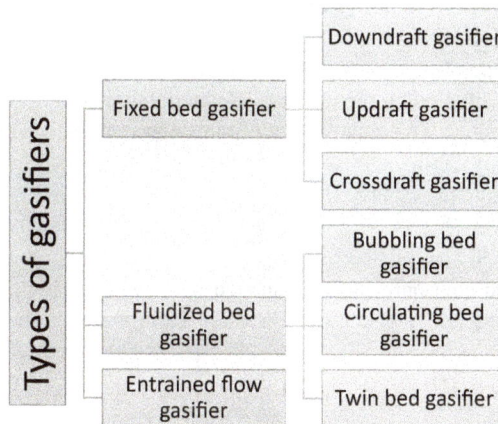

FIGURE 7.9 Types of gasifiers used for the conversion of biomass into syngas.

particle size of 22–25 mm is more appropriate for co-gasification. The conversion of lignocellulosic biomass along with other feedstocks (e.g., coal, peat, charcoal, petcoke) to potential products through the co-gasification process is shown in Figure 7.8.

Gasifiers are widely used for the thermal conversion of lignocellulosic biomass into syngas. These are classified based on the methods of gas and fuel interaction in the gasifiers. The types of gasifier are shown in Figure 7.9.

(i) Fixed bed gasifiers:

Three common fixed bed gasifiers are usually used for biomass conversion, which are knows as updraft, downdraft, and crossdraft (Figure 7.10). A schematic diagram of an updraft gasifier is shown in Figure 7.10a. It is also called a gasifier with a counter-current. This gasifier includes a fixed feedstock bed where the gasifying agent flows. Biomass feedstock is given at the top of the gasifier, and the gasifying agent is pumped out from the bottom of the reactor. Feedstock fragments slowly transfer from top to bottom because of gravity and react with the gasifying agent within the reactor. Finally, the gasifying agent pushes the counter up and feedstock flows downward. The moisture content is also evaporated from feedstocks in the drying zone. Then it is moved to the pyrolysis zone where volatile materials are evaporated from the feedstock. In this stage, char is yielded. During the char formation, water vapor is emitted upward along with the product syngas. In the oxidation zone, charcoal is combusted, and a gasifying agent comes in contact. The ash content is also produced as a liquid due to the high reactor temperature (up to 1200°C). Thermal efficiency an updraft gasifier is relatively high. The production of tar is relatively low due to low gas temperature due to the product syngas moving over a relatively cold drying region (Basu, 2018).

For the downdraft gasifier, the schematic diagram is shown in Figure 7.10(b). At the intermediate stage, the feedstock is fed at the top, the gasification medium (steam, oxygen, and/or air) is injected, and syngas is formed at the bottom. This gasifier is often referred to as the co-current gasifier due to the movement of feedstock and syngas in the same direction. As syngas is processed, the temperature is high, and its heat is typically recovered by preheating the gasifying agent. When tar moves over the hotbed of char, the gaseous product is also converted. Tiny quantities of water vapor evaporating from feedstock in the reduction zone react with the char.

The fragments of fuel travel to the pyrolysis zone, where char and gaseous products are produced. Then it passes downward into the oxidation region, where oxidation reactions take place in the reactor's throat. Fuel products are subsequently transferred downward and gaseous products with char are moved to the reduction zone and fuel syngas are produced. In this type of gasifier, tar production is relatively low as compared to updraft gasifier production.

The feedstock is fed into the crossdraft gasifier at the top of the gasifier [Figure 10(c)]. One side of the gasifier is injected with the gasifying medium (e.g., air or oxygen), while syngas is produced from the opposite side. Syngas is dried, pyrolyzed, and finally gasified in the reactor as the feedstock flows downward. The pyrolysis zone is situated on the top of the reduction/oxidation zone, and above the pyrolysis zone, the drying zone is situated. The temperature of the syngas produced is very high, influencing the composition of gases. Higher CO content was generated by the product syngas in this gasifier, while lower H_2 and CH_4 content were generated.

(ii) Fluidized bed gasifiers:

In an oxygen-deprived environment, the fluidized bed gasifier (Figure 7.11) provides biomass feedstock and standardized temperature-controlled facilities. For this gasifier, finely ground

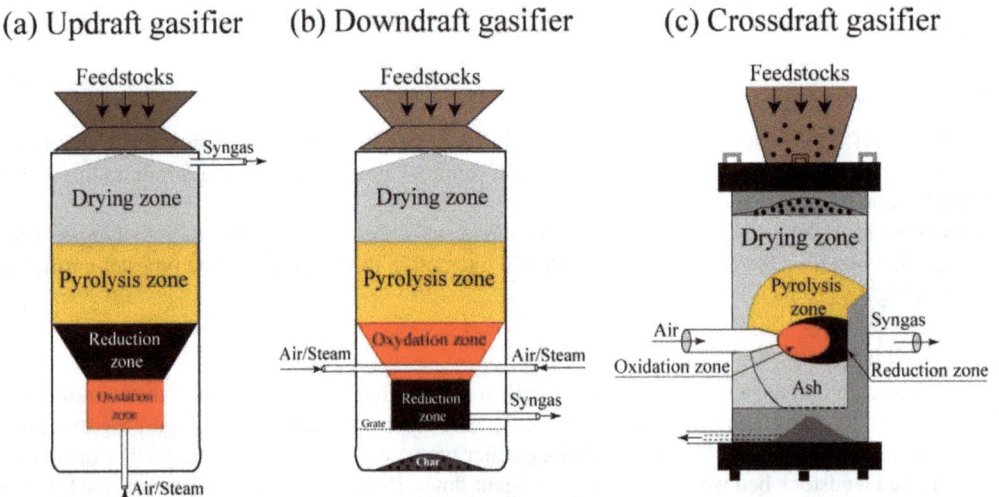

(a) Updraft gasifier   (b) Downdraft gasifier   (c) Crossdraft gasifier

FIGURE 7.10   Fixed bed gasifiers: (a) updraft gasifier, (b) downdraft gasifier, and (c) crossdraft gasifier.

feedstock samples are needed, and a distributor plate feeds gasifying agents (e.g., air, steam, or oxygen) (Sikarwar et al., 2016; Zhou et al., 2016). The drying zone, pyrolysis zone, oxidation zone, and reduction zone are not visible in any specific area of the gasifier, such as in fixed-bed gasifiers. This is a homogeneous type of reactor because of the complete reactions that occur within it.

Proper gas-solid mixing, high carbon conversion rate, flexibility, and low tar performance are the benefits of this gasifier. The rate of gasifier response is faster than fixed bed gasifiers because of intimate gas–solid touch and greater surface area of smaller feedstock particles. The most popular fluidized bed gasifiers are bubbling fluidized bed gasifiers [Figure 7.11(a)] and circulating fluidized bed gasifiers [Figure 7.11(b)].

The bubbling fluidized bed gasifier assembly schematic diagram is shown in Figure 7.11(a). The feedstock is applied, either from the top of the gasifier or deep within the bed. The cyclone separator is used for the separation of solid particles from syngas or solid particles isolated as fly ash. On the other side, fine ash falls back onto the bed and is emitted continuously through the gasifier's rim. A small quantity of tar just like that produced by an updraft gasifier is provided by this gasifier, while it produces a greater quantity than the downdraft gasifier.

Figure 7.11(b) displays the schematic diagram of a fluidized bed gasifier in circulation. This includes a riser, a separator for cyclones, and a loop-seal. The feedstock fuels are distributed through the riser height. From the cyclone separator, strong particles are isolated and are dropped down through the loop-seal and back to the foot of the riser. More intensive gas and particle mixing is needed in the rotating fluidized bed gasifier, which requires significant gas–solid interactions.

(iii) Entrained flow gasifier:

The feedstock materials and gasifying agents are fed from the top of an entrained flow gasifier (Figure 7.12). Feedstock and gasifying agents move simultaneously to the gasifier. The entire

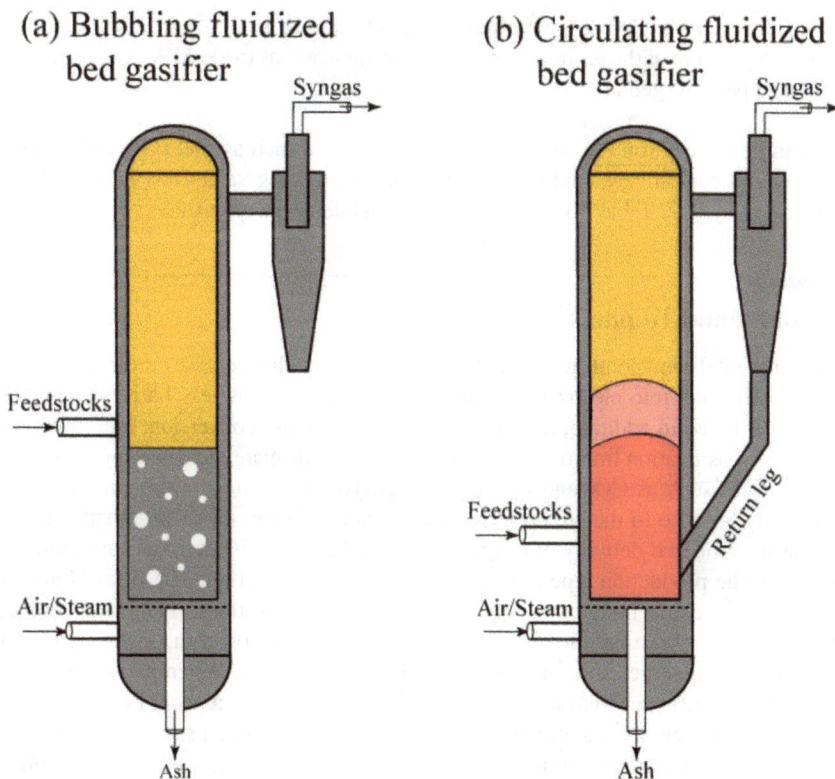

FIGURE 7.11 Fluidized bed gasifiers: (a) Bubbling fluidized bed gasifier, and (b) circulating fluidized bed gasifier.

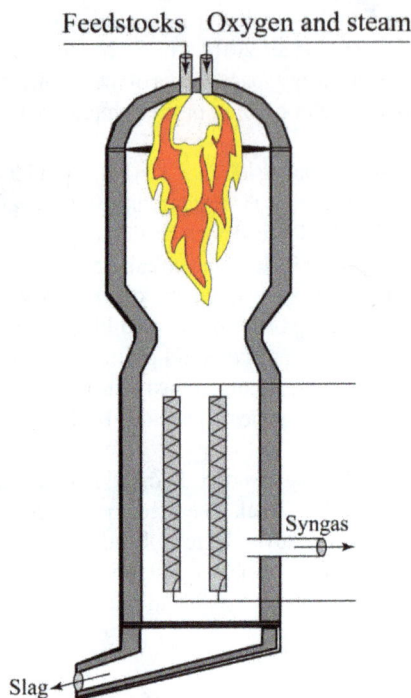

FIGURE 7.12   Schematic diagram of entrained flow gasifiers.

reaction occurs at a high rate and contains a lot of tar. The ash is isolated as slag due to the working temperature of the gasifier. For large-scale applications (>100 MW), this kind of gasifier is ideal, and pure oxygen operates under 2–5 MPa.

A gasifier can be selected on the basis of certain parameters such as fuel type, manufacturing cost, ease of operation, tar consistency, cold gas efficiency, higher heating value (HHV), lower heating value (LHV), and yield of syngas. Table 7.3 shows the merits and demerits of gasifiers.

7.6   Environmental Impacts

From an environmental viewpoint, it is found that various greenhouse gases are produced during the conversion of fossil fuels into electricity. Renewable energy contains less harmful compounds than the output of fossil fuels. In addition, as opposed to single biomass conversion, fuel mixtures produce low CO_2 gases. Co-gasification improves the efficiency of the ultimate yield of syngas. Syngas contain poisonous substances such as CO_2 and CO during biomass gasification, which are harmful not only to the human body but also to the environment. In addition, CO can mix with blood cell hemoglobin to inhibit the absorption and delivery of oxygen in human beings. It is therefore important to look out for any leakage of the production pipe during the process of gasification. The risk of burns should be reduced by having the entire body of the equipment isolated due to the formation of very hot syngas while running the process. In addition, due to the presence of hydrogen in syngas, if it is out of the air and combined at a specific ratio, there is a high risk of explosion. When by-product ash is mixed with agricultural land, it is very hazardous. A suitable disposal system should also be mandatory to minimize impacts on the environment. However, to prevent any unwanted environmental consequence, the disposal of tar obtained from biomass gasification should be done in a safe and controlled manner (Widjaya et al., 2018).

TABLE 7.3

Merits and Demerits of Gasifiers

Type of Gasifier		Merits	Demerits
Fixed Bed Gasifier	Downdraft	• Simple and easy control systems • Low entrainment of particulates • Low sensitivity to tar content and char dust • Maintenance cost is low • Simplicity and ease of operation	• Limited to small-scale applications • Inability to manage high moisture and ash content feedstock • Inability to work on many feedstocks that are unprocessed • Not practicable for the very small fuel particle size
	Updraft	• Good thermal efficiency • Small pressure drops • Slight preference towards slag formation	• High exposure to tar and fuel moisture content • Weak response capability with heavy gas loads • It takes a very long time to start an internal combustion (IC) engine
	Crossdraft	• Concise design height • Flexible gas production • Rapid response time to load	• Poor reduction in CO_2 • Particularly susceptible to the forming of slag • Limited to low ash materials
Fluidized Bed Gasifier		• Perfect gas to solid contact • Higher rate of heat and mass transfer • Excellent temperature control • Good heat transfer characteristics • A strong mixing degree • Large heat storage capacity	• The increased tar production rate • Incomplete carbon burn-out • High capital and operational costs • Weak response to load alterations • Complicated and expensive control systems • The scale-up choice is limited
Entrained Flow Gasifier		• Feedstock fed in slurry or dry form • Rapid feed conversion • Process characterized by high temperature and pressure	• High temperature increases the wear rate • Due to slurry feedstock water is injected into the gasifier

7.7 Conclusions

For the potential co-gasification approach, distinct lignocellulosic biomass and non-biomass feedstocks are discussed in this chapter. For the production of syngas, lignocellulosic biomass is a promising alternative as it could be converted into different energy forms while taking into account the environmental impact. The non-biomass feedstocks such as coal, peat, charcoal, and petcoke could be the potential co-gasifying feedstocks that will reinforce the gasification process. The selection of gasifiers for biomass conversion is very important for the production of syngas. Certain guidelines are summarized as follows:

(i) The gasification/co-gasification output and the control parameters of gasifying agents and unburned carbon are typically improved on industrial scales.

(ii) Theoretical gasification efficiency and process situations are different from those achievable in practice due to industrial constraints and biomass and non-biomass feedstock variability.

(iii) The composition of syngas varies from that of different gasifiers, as the ratio of the key components represents the equilibrium obtained in the existing circumstances inside the gasifier. It is also very important to pick the right gasifier to ensure the consistency of syngas through the co-gasification approach.

(iv) Emissions from various conversion processes are dependent on main variables such as feedstock type, structure, pre-processing, operating conditions, and emission control technology. Environmental protection is also a primary issue that must be taken care of during conversion.

REFERENCES

Abo, B. O., Gao, M., Wang, Y., Wu, C., Ma, H., and Wang, Q. 2019. Lignocellulosic biomass for bioethanol: An overview on pretreatment, hydrolysis and fermentation processes. *Reviews on Environmental Health* 34(1): 57–68.

Aydin, E. S., Yucel, O., and Sadikoglu, H. 2018. Numerical investigation of fixed-bed downdraft woody biomass gasification. In: *Exergetic, Energetic and Environmental Dimensions.* Elsevier, pp. 323–339.

Bailey, R. A., Clark, H. M., Ferris, J. P., Krause, S., and Strong, R. L. 2002. Petroleum, hydrocarbons, and coal. In: *Chemistry of the Environment.* 2nd Edition, (Eds.) R. A. Bailey, H. M. Clark, J. P. Ferris, S. Krause, and R. L. Strong. San Diego: Academic Press, pp. 147–192.

Basu, P. 2010. *Biomass Gasification and Pyrolysis: Practical Design and Theory.* Netherlands: Academic Press.

Basu, P. 2018. *Biomass Gasification, Pyrolysis and Torrefaction: Practical Design and Theory.* 3rd ed. London: Academic Press.

Bell, P. S., Ko, C.-W., Golab, J., Descales, B., and Eyraud, J. 2019. Apparatus and methods for tar removal from syngas. *Google Patents.*

Caineng, Z., Qun, Z., Guosheng, Z., and Bo, X. 2016. Energy revolution: From a fossil energy era to a new energy era. *Natural Gas Industry* 36(1): 1–10.

Capuano, L. 2018. International energy outlook 2018 (IEO2018). www.eia.gov/outlooks/ieo/executive_summary.php.

Chaurasia, A. 2018. Modeling of downdraft gasification process: Studies on particle geometries in thermally thick regime. *Energy* 142: 991–1009.

Chen, H. 2014. Chemical composition and structure of natural lignocellulose. In: *Biotechnology of Lignocellulose.* Netherlands: Springer, pp. 25–71.

Claypool, J. T., and Simmons, C. W. 2016. Hybrid thermochemical/biological processing: The economic hurdles and opportunities for biofuel production from bio-oil. *Renewable Energy* 96: 450–457.

Dai, G., Zhu, Y., Yang, J., Pan, Y., Wang, G., Reubroycharoen, P., and Wang, S. 2019. Mechanism study on the pyrolysis of the typical ether linkages in biomass. *Fuel* 249: 146–153.

Dawood, F., Anda, M., and Shafiullah, G. M. 2020. Hydrogen production for energy: An overview. *International Journal of Hydrogen Energy* 45(7): 3847–3869.

De, S., Agarwal, A. K., Moholkar, V., and Thallada, B. 2018. Coal and biomass gasification. *Energy, Environment, and Sustainability.* Singapore: Springer Nature.

Dhyani, V., and Bhaskar, T. 2019. Chapter 9 – Pyrolysis of biomass. In: *Biofuels: Alternative Feedstocks and Conversion Processes for the Production of Liquid and Gaseous Biofuels,* 2nd Edition, (Eds.) A. Pandey, C. Larroche, C.-G. Dussap, E. Gnansounou, S. K. Khanal, and S. Ricke. Netherland: Academic Press, pp. 217–244.

Ellabban, O., Abu-Rub, H., and Blaabjerg, F. 2014. Renewable energy resources: Current status, future prospects and their enabling technology. *Renewable and Sustainable Energy Reviews* 39: 748–764.

Farzad, S., Mandegari, M. A., and Görgens, J. F. 2016. A critical review on biomass gasification, co-gasification, and their environmental assessments. *Biofuel Research Journal* 3(4): 483–495.

Gao, X., Zhang, Y., Li, B., and Yu, X. 2016. Model development for biomass gasification in an entrained flow gasifier using intrinsic reaction rate submodel. *Energy Conversion and Management* 108: 120–131.

Gastaldo, R. A., Bamford, M., Calder, J., DiMichele, W. A., Iannuzzi, R., Jasper, A., Kerp, H., McLoughlin, S., Opluštil, S., and Pfefferkorn, H. W. 2020. The coal farms of the late Paleozoic. In: *Nature through Time.* Switzerland: Springer, pp. 317–343.

Gaurav, N., Sivasankari, S., Kiran, G. S., Ninawe, A., and Selvin, J. 2017. Utilization of bioresources for sustainable biofuels: A Review. *Renewable and Sustainable Energy Reviews* 73: 205–214.

Gírio, F. M., Fonseca, C., Carvalheiro, F., Duarte, L. C., Marques, S., and Bogel-Łukasik, R. 2010. Hemicelluloses for fuel ethanol: A review. *Bioresource Technology* 101(13): 4775–4800.

Infield, D., and Freris, L. 2020. *Renewable Energy in Power Systems.* Hoboken: John Wiley & Sons.

Khatun, F., Monir, M. M. U., Arham, S. M. N., and Wahid, Z. A. 2016. Implementation of carbon dioxide gas injection method for gas recovery at Rashidpur Gas Field, Bangladesh. *International Journal of Engineering Technology and Sciences* 5(1): 52–61.

Koupaie, E. H., Dahadha, S., Lakeh, A. B., Azizi, A., and Elbeshbishy, E. 2019. Enzymatic pretreatment of lignocellulosic biomass for enhanced biomethane production-A review. *Journal of Environmental Management* 233: 774–784.

Limayem, A., and Ricke, S. C. 2012. Lignocellulosic biomass for bioethanol production: Current perspectives, potential issues and future prospects. *Progress in Energy and Combustion Science* 38(4): 449–467.

Ma, S., Wang, H., Li, J., Fu, Y., and Zhu, W. 2019. Methane production performances of different compositions in lignocellulosic biomass through anaerobic digestion. *Energy* 189: 116190.

Mansur, F. Z., Faizal, C. K. M., Monir, M. U., Samad, N. A. F. A., Atnaw, S. M., and Sulaiman, S. A. 2020. Co-gasification between coal/sawdust and coal/wood pellet: A parametric study using response surface methodology. *International Journal of Hydrogen Energy* 45(32): 15963–15976.

Mikulcic, H., Klemeš, J. J., Vujanovic, M., Urbaniec, K., and Duic, N. 2016. Reducing greenhouse gasses emissions by fostering the deployment of alternative raw materials and energy sources in the cleaner cement manufacturing process. *Journal of Cleaner Production* 30: 1e14.

Moghadam, R. A., Yusup, S., Uemura, Y., Chin, B. L. F., Lam, H. L., Al Shoaibi, A. 2014. Syngas production from palm kernel shell and polyethylene waste blend in fluidized bed catalytic steam co-gasification process. *Energy* 75: 40–44.

Molino, A., Larocca, V., Chianese, S., and Musmarra, D. 2018. Biofuels production by biomass gasification: A review. *Energies* 11(4): 811.

Monir, M. U., Abd Aziz, A., Kristanti, R. A., and Yousuf, A. 2018a. Co-gasification of empty fruit bunch in a downdraft reactor: A pilot scale approach. *Bioresource Technology Reports* 1: 39–49.

Monir, M. U., Abd Aziz, A., Kristanti, R. A., and Yousuf, A. 2020a. Syngas Production from Co-gasification of Forest Residue and Charcoal in a Pilot Scale Downdraft Reactor. *Waste and Biomass Valorization* 11(2): 635–651.

Monir, M. U., Abd Aziz, A., Yousuf, A., and Alam, M. Z. 2019. Hydrogen-rich syngas fermentation for bioethanol production using Sacharomyces cerevisiea. *International Journal of Hydrogen Energy*: 1–9.

Monir, M. U., Aziz, A. A., Dai-Viet, N. Vo, and Khatun, F. 2020b. Enhanced Hydrogen Generation from Empty Fruit Bunches by Charcoal Addition into a Downdraft Gasifier. *Chemical Engineering & Technology* 43(4): 762–769.

Monir, M. U., Aziz, A. A., Khatun, F., and Yousuf, A. 2020c. Bioethanol production through syngas fermentation in a tar free bioreactor using Clostridium butyricum. *Renewable Energy* 157: 1116–1123.

Monir, M. U., Azrina, A. A., Kristanti, R. A., and Yousuf, A. 2018b. Gasification of lignocellulosic biomass to produce syngas in a 50 kW downdraft reactor. *Biomass and Bioenergy* 119: 335–345.

Monir, M. U., Khatun, F., Abd Aziz, A., and Vo, D.-V. N. 2020d. Thermal treatment of tar generated during co-gasification of coconut shell and charcoal. *Journal of Cleaner Production* 256: 1–9.

Monir, M. U., Khatun, F., Ramzilah, U. R., and Aziz, A. A. 2020e. Thermal effect on co-product tar produced with syngas through co-gasification of coconut shell and charcoal. *IOP Conference Series: Materials Science and Engineering* 736: 022007.

Monir, M. U., Yousuf, A., and Aziz, A. A. 2020f. Chapter 6 – Syngas fermentation to bioethanol. in: *Lignocellulosic Biomass to Liquid Biofuels*, (Eds.) A. Yousuf, D. Pirozzi, and F. Sannino. London: Academic Press, pp. 195–216.

Monir, M. U., Yousuf, A., Aziz, A. A., and Atnaw, S. M. 2017. Enhancing co-gasification of coconut shell by reusing char. *Indian Journal of Science and Technology* 10(6): 1–4.

Mularski, J., and Modliński, N. 2020. Entrained flow coal gasification process simulation with the emphasis on empirical devolatilization models optimization procedure. *Applied Thermal Engineering* 175: 115401.

Musule, R., Alarcón-Gutiérrez, E., Houbron, E. P., Bárcenas-Pazos, G. M., del Rosario Pineda-López, M., Domínguez, Z., Sánchez-Velásquez, L. R. 2016. Chemical composition of lignocellulosic biomass in the wood of Abies religiosa across an altitudinal gradient. *Journal of Wood Science* 62(6): 537–547.

Oh, G., Ra, H. W., Yoon, S. M., Mun, T. Y., Seo, M. W., Lee, J. G., and Yoon, S. J. 2018. Gasification of coal water mixture in an entrained-flow gasifier: Effect of air and oxygen mixing ratio. *Applied Thermal Engineering* 129: 657–664.

Oumer, A. N., Hasan, M. M., Baheta, A. T., Mamat, R., and Abdullah, A. A. 2018. Bio-based liquid fuels as a source of renewable energy: A review. *Renewable and Sustainable Energy Reviews* 88: 82–98.

Patel, V. R., Patel, D., Varia, N., and Patel, R. N. 2017. Co-gasification of lignite and waste wood in a pilot-scale (10 kWe) downdraft gasifier. *Energy* 119: 834–844.

Ramos, A., Monteiro, E., Silva, V., and Rouboa, A. 2018. Co-gasification and recent developments on waste-to-energy conversion: A review. *Renewable and Sustainable Energy Reviews* 81: 380–398.

Rana, R., Nanda, S., Maclennan, A., Hu, Y., Kozinski, J. A., and Dalai, A. K. 2019. Comparative evaluation for catalytic gasification of petroleum coke and asphaltene in subcritical and supercritical water. *Journal of Energy Chemistry* 31: 107–118.

Ren, J., Liu, Y.-L., Zhao, X.-Y., and Cao, J.-P. 2020. Methanation of syngas from biomass gasification: An overview. *International Journal of Hydrogen Energy* 45(7): 4223–4243.

Rocha-Meneses, L., Raud, M., Orupõld, K., and Kikas, T. 2017. Second-generation bioethanol production: A review of strategies for waste valorisation. *Agronomy Research* 15(3): 830–847.

Samiran, N. A., Jaafar, M. N. M., Ng, J.-H., Lam, S. S., and Chong, C. T. 2016. Progress in biomass gasification technique – With focus on Malaysian palm biomass for syngas production. *Renewable and Sustainable Energy Reviews* 62: 1047–1062.

Saw, W. L., and Pang, S. 2013. Co-gasification of blended lignite and wood pellets in a 100kW dual fluidised bed steam gasifier: The influence of lignite ratio on producer gas composition and tar content. *Fuel* 112: 117–124.

Scheffran, J., Felkers, M., and Froese, R. 2020. Economic growth and the global energy demand. *Green Energy to Sustainability: Strategies for Global Industries*: 1–44.

Schmid, J. C., Benedikt, F., Fuchs, J., Mauerhofer, A. M., Müller, S., and Hofbauer, H. 2019. Syngas for bio-refineries from thermochemical gasification of lignocellulosic fuels and residues – 5 years' experience with an advanced dual fluidized bed gasifier design. *Biomass Conversion and Biorefinery*: 1–38.

Shafiee, S., and Topal, E. 2009. When will fossil fuel reserves be diminished? *Energy Policy* 37(1): 181–189.

Sikarwar, V. S., Zhao, M., Clough, P., Yao, J., Zhong, X., Memon, M. Z., Shah, N., Anthony, E. J., and Fennell, P. S. 2016. An overview of advances in biomass gasification. *Energy & Environmental Science* 9(10): 2939–2977.

Singh, V. C. J., and Sekhar, S. J. 2016. Performance studies on a downdraft biomass gasifier with blends of coconut shell and rubber seed shell as feedstock. *Applied Thermal Engineering* 97: 22–27.

Soccol, C. R., Faraco, V., Karp, S. G., Vandenberghe, L. P., Thomaz-Soccol, V., Woiciechowski, A. L., and Pandey, A. 2019. Lignocellulosic bioethanol: Current status and future perspectives. In: *Biofuels: Alternative Feedstocks and Conversion Processes for the Production of Liquid and Gaseous Biofuels*. Netherland: Elsevier, pp. 331–354.

Soka, O., and Oyekola, O. 2020. A feasibility assessment of the production of char using the slow pyrolysis process. *Heliyon* 6(7): e04346.

Suárez-Ruiz, I., Diez, M. A., and Rubiera, F. 2019. Chapter 1 – Coal. In: *New Trends in Coal Conversion*, (Eds.) I. Suárez-Ruiz, M. A. Diez, and F. Rubiera. Duxford: Woodhead Publishing, pp. 1–30.

Toklu, E. 2017. Biomass energy potential and utilization in Turkey. *Renewable Energy* 107: 235–244.

Valdés, C. F., Marrugo, G. P., Chejne, F., Marin-Jaramillo, A., Franco-Ocampo, J., and Norena-Marin, L. 2020. Co-gasification and co-combustion of industrial solid waste mixtures and their implications on environmental emissions, as an alternative management. *Waste Management* 101: 54–65.

Wang, S. R., Dai, G. X., Yang, H. P., and Luo, Z. Y. 2017. Lignocellulosic biomass pyrolysis mechanism: A state-of-the-art review. *Progress in Energy and Combustion Science* 62: 33–86.

Widjaya, E. R., Chen, G., Bowtell, L., and Hills, C. 2018. Gasification of non-woody biomass: A literature review. *Renewable and Sustainable Energy Reviews* 89: 184–193.

Wyman, C. E. 2018. Ethanol production from lignocellulosic biomass: Overview. In: *Handbook on Bioethanol*. Boca Raton: Routledge, pp. 1–18.

Xu, C., Liao, B., Pang, S., Nazari, L., Mahmood, N., Tushar, M. S. H. K., Dutta, A., and Ray, M. B. 2018. Biomass energy. In: *Comprehensive Energy Systems*, (Ed.) I. Dincer. Oxford: Elsevier, pp. 770–794, January 1.

Yu, J., Paterson, N., Blamey, J., and Millan, M. 2017. Cellulose, xylan and lignin interactions during pyrolysis of lignocellulosic biomass. *Fuel* 191: 140–149.

Zabed, H., Sahu, J. N., Boyce, A. N., and Faruq, G. 2016. Fuel ethanol production from lignocellulosic biomass: An overview on feedstocks and technological approaches. *Renewable and Sustainable Energy Reviews* 66: 751–774.

Zhang, X., and Brown, R. C. 2019. Introduction to thermochemical processing of biomass into fuels, chemicals, and power. *Thermochemical Processing of Biomass: Conversion into Fuels, Chemicals and Power*: 1–16.

Zhou, C., Rosén, C., and Engvall, K. 2016. Biomass oxygen/steam gasification in a pressurized bubbling fluidized bed: Agglomeration behavior. *Applied Energy* 172: 230–250.

Zhu, H. L., Zhang, Y. S., Materazzi, M., Aranda, G., Brett, D. J. L., Shearing, P. R., and Manos, G. 2019. Co-gasification of beech-wood and polyethylene in a fluidized-bed reactor. *Fuel Processing Technology* 190: 29–37.

8

Glycerol: A Promising Green Source for Chemicals and Fuels

Thanh Khoa Phung, Khanh B. Vu, Quynh-Thy Song Nguyen, Khoa Dang Tong, Vy Anh Tran, Dai-Viet N. Vo, Hong Duc Pham

CONTENTS

8.1   Introduction

Glycerol is a simple polyol compound. It is an essential part of triglycerides (i.e., fats and oils) and phospholipids. The melting point and boiling point of pure glycerol are 17.8°C and 290°C, respectively. Glycerol is non-toxic in low concentrations. It is denser and viscous than water (density = 1.29 g/cm³), and it is easily soluble in water. Glycerol can be generated through transesterification, saponification, and hydrolysis reactions from fat and vegetable oils or from propylene through various synthesis processes (Tan et al., 2013). Glycerol is mainly produced from fat and vegetable oils because they are a green source and involve a cheap process.

Transesterification of fat and vegetable oils is a common process to produce biodiesel and a by-product glycerol (Sadaf et al., 2018). In the biodiesel process, fat and vegetable oils (triglycerides) react with alcohol, for example, methanol, via transesterification reaction in the presence of a catalyst to produce esters of fatty acid and alcohol in parallel to produce glycerol as a by-product. Due to the increase in demand for biodiesel in recent years, the production of biodiesel increased leading to an increase in the amount of glycerol. Nowadays, the biodiesel process is a major source of glycerol production (Tan et al., 2013).

Saponification reaction is a traditional reaction to produce soap (salts of fatty acid). In this reaction, triglycerides react with alkali to produce soap and glycerol. Indeed, the alkali breaks the ester bond of fatty acids and glycerol in triglycerides, releasing glycerol and salts of fatty acids. The formation of glycerol from saponification is usually in a large quantity (Tan et al., 2013). In the same manner as with saponification reaction, hydrolysis of oil and fats uses water instead of alkali to hydrolyze triglycerides into fatty acids and glycerol (Forero-Hernandez et al., 2020) During the reaction, the solution of this reaction separates into two phases comprising a light phase of fatty acid and a heavy phase of glycerol. This reaction contains impurities from fat or vegetable oils; however, the glycerol from hydrolysis comparatively is cleaner than that from biodiesel production and saponification reaction.

8.2   Purification of Glycerol

As aforementioned, crude glycerol contains many impurities arising from biodiesel production, saponification, and hydrolysis of triglycerides. The percentage of glycerol is around 60–70 wt.% in crude effluents from the biodiesel production process (Thompson and He, 2006). Glycerol solution can contain 35 wt.% and 15 wt.% impurities in the saponification and hydrolysis of triglycerides, respectively (Tan et al., 2013). Therefore, crude glycerol needs to be purified in order to be of use in the pharmaceuticals and cosmetics industries.

Glycerol purification is one of the most critical processes for increasing the glycerol value. Indeed, the highly pure glycerol can be applied in the pharmaceutical industry, while a low-quality glycerol can be used as waste material. Therefore, many purification techniques are used to convert crude glycerol into a cleaner form, including distillation, chemical treatment, filtration, ion-exchange (using resin), extraction, adsorption (using activated carbon), crystallization, and decantation. Different purification techniques are used for different crude glycerol, some techniques of which can be combined to obtain the highly pure glycerol (Tan et al., 2013). The common pathway to purify crude glycerol is to carry out the distillation to recover alcohol and remove salt by ion-exchange resin/adsorption process (Ardi et al., 2015).

Looking at the review of the available purification technologies of crude glycerol from biodiesel production, each technology has advantages and disadvantages (Muniru et al., 2019). Several technologies have shown high purity of purified glycerol (> 95%), ease of scale-up and simple process as advantages (Tianfeng et al., 2013; Muniru et al., 2016; Isahak et al., 2010; Manosak et al., 2011; Nanda et al., 2014a; Ooi et al., 2001). Hájek and Skopal (2010) showed a low glycerol purity (86%) and low glycerol yield. However, its advantages are simple to process and low maintenance cost. Therefore, developing new simple technology that enhances the yield and purity of glycerol and reduces the operation and maintenance costs is necessary for the purification of crude glycerol to be applied in further industries.

8.3   Glycerol Market and Applications

The demand for glycerol has risen in recent years and is expected to reach US $ 5 billion by 2024 as stated in a forecast growth report by Global Market Insights, Inc. (Hegde, 2018). The market for glycerol is likely to generate a demand of 5.5 million tons by 2027. The personal care and cosmetics segment emerged as a prominent end-use segment with a market share of approximately 37 vol.% in 2019. The Asia Pacific region is the fastest-growing region in glycerol use, while Europe is a top importer of glycerol (Research, 2020). Currently, the global glycerol market is shared by several big companies including DOW Chemicals, Solvay SA, Wilmar International, P&G Chemicals, ADM, BASF, and Godrej Industries.

Glycerol has many applications in food and beverage, pharmaceutical, nutraceutical, personal care, cosmetics, and value-added industries. The application areas of glycerol depend on its quality. The highly pure (> 90%) glycerol can be used in cosmetic and pharmaceuticals industries, and the low-quality

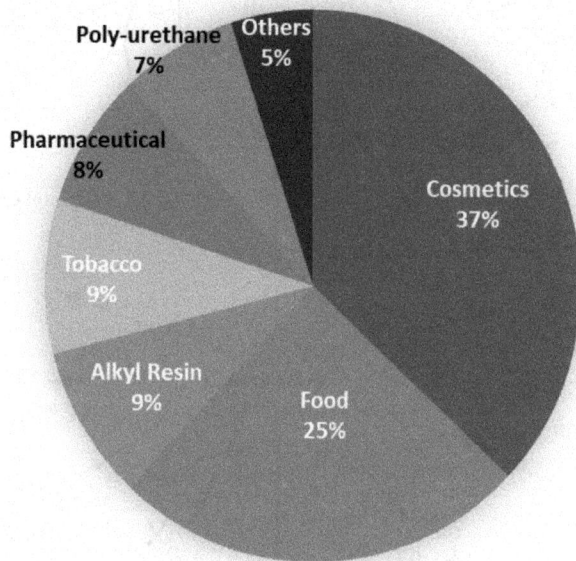

FIGURE 8.1 The percentage of application of glycerol in various industries. Reproduced with permission from Bagnato et al. (2017) (open-access article).

glycerol is used for other applications (Research, 2020). Glycerol is mostly used in cosmetics (37%) and food (25%), and the rest is used in making alkyl resin, tobacco, pharmaceuticals, polyurethane, and other products (Figure 8.1).

In the cosmetics industry, glycerol is one of the materials abundantly used after water and fragrance because it can be used as a moisture-control reagent, that is, for its ability in retaining moisture and emollient properties as well as preventing the drying and freezing of the cosmetics. Glycerol can be used in food because it does not contain a disagreeable flavor and is low in toxicity. Glycerol is used to enhance viscosity and texture, especially producing the intermolecular hydrogen bonds with water helping to preserve the shelf life of the product. In pharmaceutical industry, glycerol can increase the smoothness and taste of medicine. Indeed, glycerol makes it easy to swallow tablets and renders a sweet taste to cough lozenges.

Additionally, glycerol can be used in many industries such as polymer, tobacco, and biofuel, etc. (Bagnato et al., 2017). Glycerol is also used in the industrial production of antifreezes, textiles, waxes, paints, and resins (Sabbah et al., 2019). Glycerol is also used as an ingredient to prepare many valuable products including glycerol carbonate, 1,3-propanediol, acrolein, acetin, hydrogen, glycerol ethers, ketal, and lactic acid.

8.4 Conversion of Glycerol into Chemicals and Fuels

Glycerol can be converted into many useful products through different reaction processes. In this chapter, the preparation of several useful products (Figure 8.2) are discussed.

8.4.1 Glycerol Carbonate

Glycerol carbonate (GC) is an organic compound and is used in preparing polyurethanes, polycarbonates, polyamides, and polyesters (Nomanbhay et al., 2020). It can be used in paints and detergents as a solvent and an intermediate in textiles, cosmetics, plastics, and pharmaceutical industries (Ochoa-Gómez et al., 2009). Glycerol carbonate can be produced from glycerol through glycerolysis with urea, transesterification with dimethyl carbonate, and direct carboxylation with CO_2 (Figure 8.3).

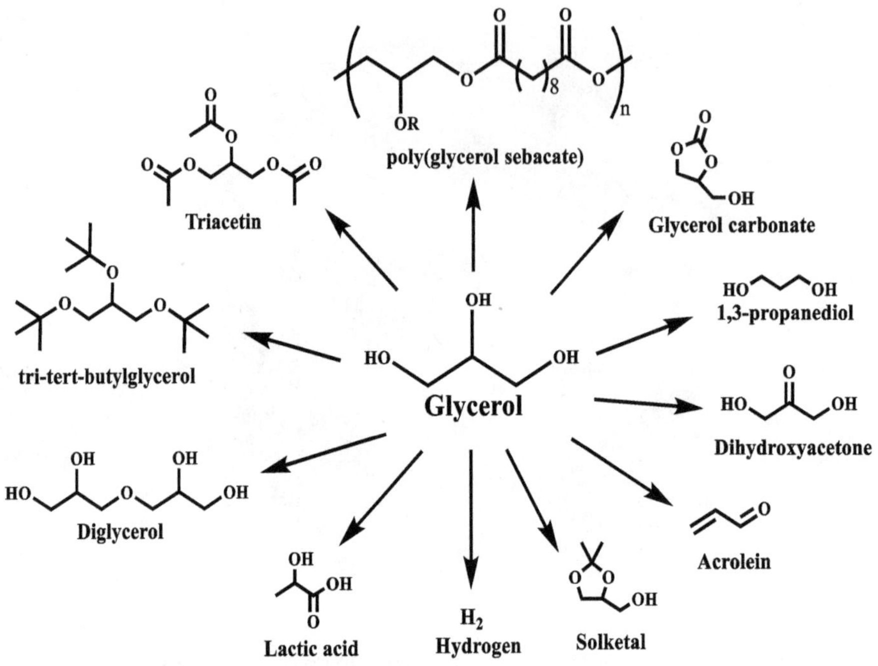

FIGURE 8.2   Useful products that can be produced from glycerol.

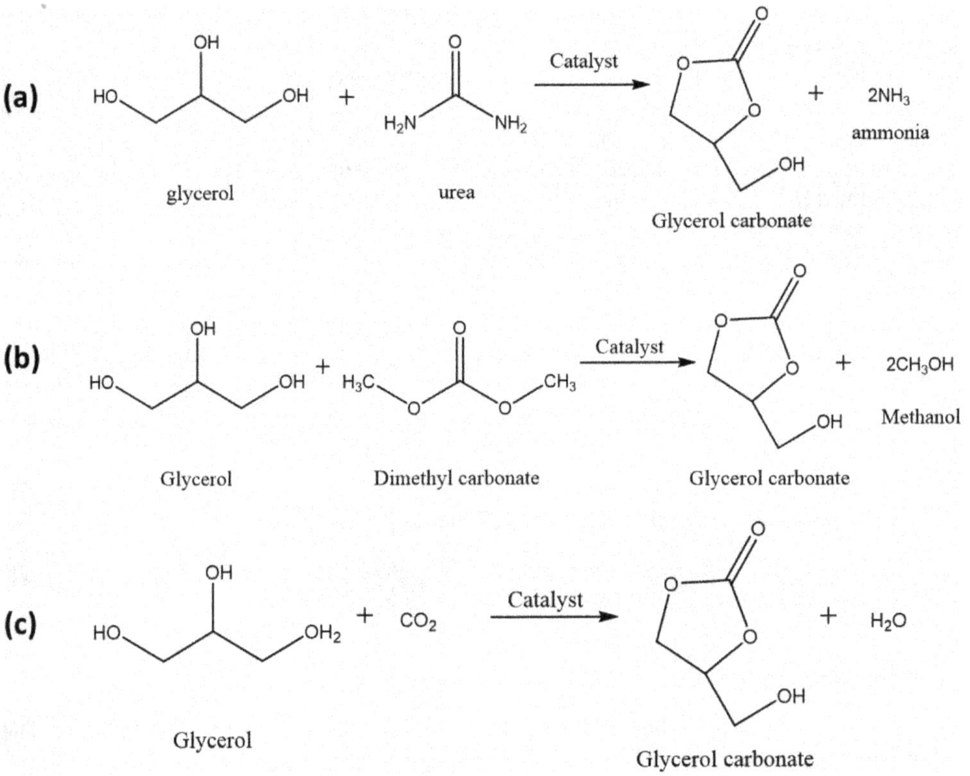

FIGURE 8.3   Glycerol carbonate is produced from glycerol through (a) glycerolysis with urea, (b) transesterification with dimethyl carbonate, and (c) direct carboxylation with carbon dioxide.

Homogeneous and heterogeneous catalysts can be employed to produce glycerol carbonate (Wu et al., 2017; Casiello et al., 2014). Homogeneous catalyst (copper (III) complexes) showed more than 92% selectivity of glycerol carbonate via oxidative carbonylation (Casiello et al., 2014). For heterogeneous catalysts, Wu et al. (2017) reported that the glycerol conversion was synergistically promoted by the well-dispersed NiO particles and the strong basic sites resulting from the replacement of Ce^{4+} by Ni^{2+} in the CeO_2 lattices. Zuhaimi et al. (2015) reported that the presence of Lewis acid sites (Ca^{2+}) and conjugate basic sites of SO_4^{2-} enhanced the glycerol carbonate yield (83.6%). Additionally, for the production of glycerol carbonate from glycerol and urea, a study by Climent et al. (2010) revealed that balanced bifunctional acid–base is the best for this reaction in which the carbonyl of the urea is activated on the Lewis acid sites, and the hydroxyl group of the glycerol is activated on the conjugated basic sites. Therefore, understanding the mechanism of the reaction is very important. For the reaction mechanism, there are several reaction mechanisms based on the reactants, which react with glycerol and the catalysts as well. Here is one of the reaction mechanisms for transesterifying glycerol with dimethyl carbonate (Figure 8.4) (Liu et al., 2015). Glycerol carbonate can be consummated due to the conversion into glycidol in the last step.

8.4.2 1,3-Propanediol

1,3-Propanediol (1,3-PD) is an organic liquid compound and soluble in water (Sullivan et al., 2018). It can be used in a variety of industrial products including cosmetics, lubricants, engine coolants, water-based inks, adhesives, coatings, antifreezes, and wood paints (Saxena et al., 2009). 1,3-PD can be fabricated from glycerol via hydrogenolysis as shown in Figure 8.5.

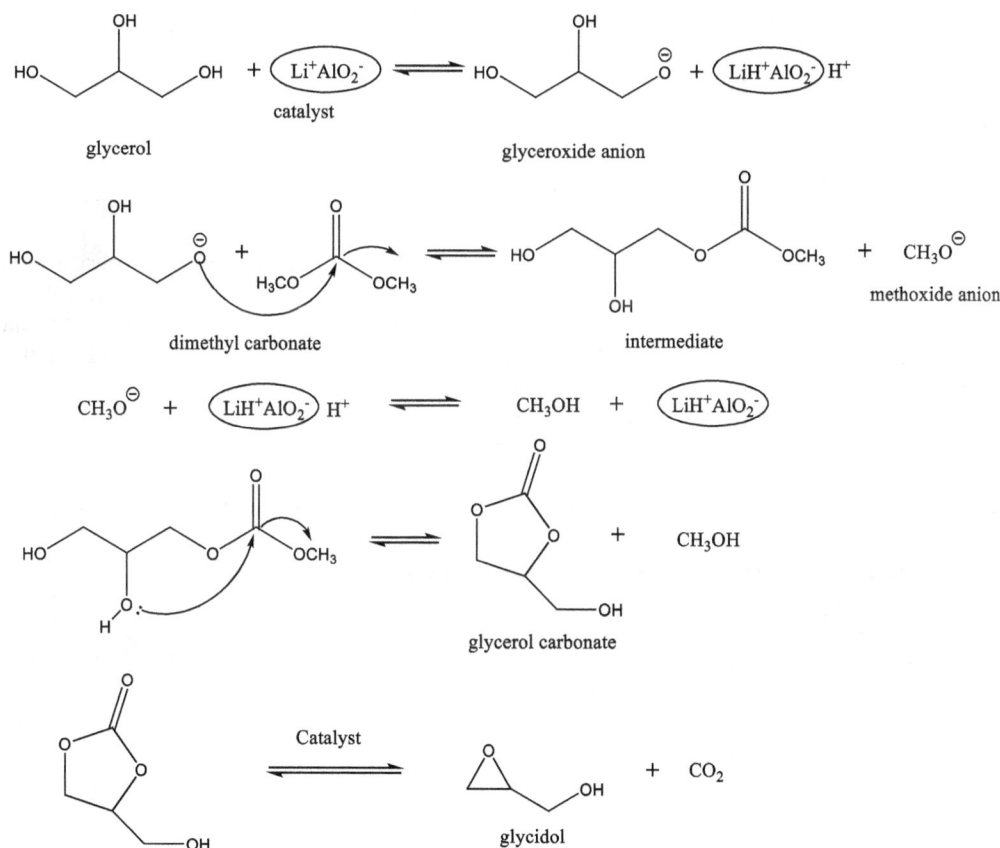

FIGURE 8.4 The reaction mechanism for transesterification of glycerol with dimethyl carbonate. Reproduced with permission from Liu et al. (2015).

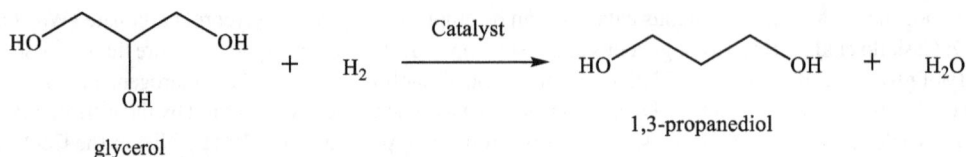

FIGURE 8.5 Scheme of 1,3-propanediol production by hydrogenolysis from glycerol.

Generally, noble catalysts are mostly used in hydrogenolysis reactions. In the case of 1,3-propanediol production by hydrogenolysis from glycerol, Pt is the metal catalyst mostly used (García-Fernández et al., 2017). Ru, Ni-Zr, Cu-Zr, and Zr are also used as active metal sites for this reaction (Kant et al., 2017; Zhou et al., 2017). The reaction conditions including reaction pressure, reaction temperature, and reaction time determine the yield of 1,3-propanediol. Besides that, the nature of the catalyst plays an important role in the preparation of 1,3-propanediol. Zhu et al. (2013a) reported that Brønsted acid sites favor the production of 1,3-propanediol from glycerol hydrogenolysis, while Lewis acid sites tend to produce 1,2-propanediol. The interaction of metal catalysts and supports is very important, as the high dispersion of metal catalysts, e.g., Pt, on the large surface area enhances the yield of 1,3-propanediol (Liu et al., 2012). García-Fernández et al. (2015) reported that the selection of 1,3-propanediol from glycerol hydrogenolysis depends not only on the acidity, but also on the interaction of Pt metal and support. They revealed that the closer proximity between metal catalyst (platinum) and support (tungsten oxides) at high metal catalyst content and the electronic interactions between acid and metallic sites could be responsible for the improvement of the yield of 1,3-propanediol. Moreover, the particle size of the catalyst also tailors the yield of hydrogenolysis glycerol into 1,3-propanediol (Deng et al., 2014).

8.4.3   Acrolein

Acrolein is the simplest unsaturated aldehyde and derived from dehydration of glycerol (Figure 8.6). Acrolein is an intermediate or a building block to acrylic acid, acrylic plastics, super absorbers, adhesives, and detergents (Yang et al., 2012).

It is well known that the process of dehydration requires an acid catalyst to remove water from alcohol (Phung and Busca, 2020). Many acid catalysts are employed to dehydrate glycerol into acrolein such as sulfuric acid-activated montmorillonite (Zhao et al., 2013), MFI zeolites (Possato et al., 2017; Jia et al., 2010), MUICaT-$_5$ (Yadav et al., 2013), nickel sulfate catalysts (Gu et al., 2013), and WO_3/TiO_2 (Akizuki and Oshima, 2012). Heteropoly acids are also used for this reaction, e.g., $Cs_{2.5}H_{0.5}PW_{12}O_{40}$ (CsPW) (Alhanash et al., 2010), $H_6P_2W_{18}O_{62}/MCM-41$, and $H_3PW_{12}O_{40}/MCM-41$ (Ma et al., 2017). Brønsted acid sites are the preferred active sites to get highly selective acrolein from glycerol (Possato et al., 2017). A strong acid catalyst, $NiSO_4$_350, proved to be a promising catalyst resulting in production of more than 70% selective acrolein at more than 90% glycerol conversion even after 10 h of reaction; however, the catalyst became deactivated due to oxidation and loss of sulfur (Gu et al., 2013). Among the strong acid catalysts, 80% selective acrolein at 94% glycerol conversion at 225°C could be achieved using the polyacid dodecatungstophosphoric acid (DTP) supported on hexagonal mesoporous silica (HMS) catalyst, with 20% w/w loading of DTP on HMS (Yadav et al., 2013).

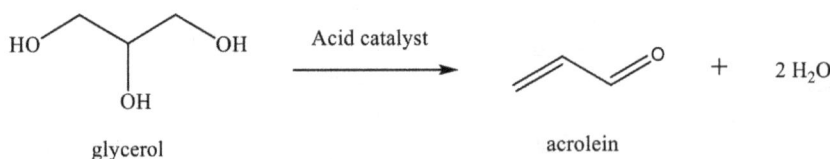

FIGURE 8.6 Scheme of conversion of glycerol to acrolein by dehydration.

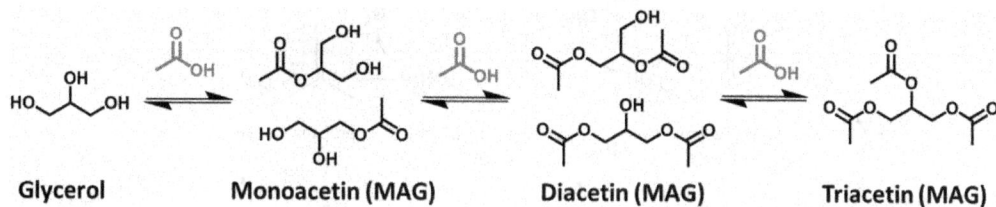

FIGURE 8.7 Reaction mechanism of glycerol acetylation with acetic acid catalyzed by acid.

8.4.4 Acetin

The esterification between glycerol and carboxylic acid with the utilization of catalysts produces acetins (Figure 8.7) (Dizoğlu and Sert, 2020). Acetin can also be produced from glycerol esterification with anhydride acid (Silva et al., 2010a). In Figure 8.7, the types of acetin produced include mono-, di-, and tri. Acetins have many applications in various industries: monoacetins and diacetins are used in cosmetics, medicines, and food industries, while triacetin can be used as fuel additives, solvents, and plasticizers (Nda-Umar et al., 2018).

Acid catalysts are commonly used for glycerol acetylation with acetic acid (Dizoğlu and Sert, 2020; Okoye et al., 2017). However, several zeolites showed poor performance due to the space hindrance of the pore channel for the diffusion of products (Gonçalves et al., 2008). In addition, in comparison with H-ZSM-5 and H-β zeolites, the mesoporous silica (Pr-SO_3H-SBA-15) functionalized with propyl sulfonic acid showed higher glycerol conversion (Dalla Costa et al., 2017). Using the Pr-SO_3H-SBA-15 catalyst, the conversion of glycerol can be enhanced by up to 96% with 87% diacetylglycerol and triacetylglycerol for 2.5 h. In the case of metalorganic frameworks (MOFs), Dizoğlu and Sert (2020) reported that the UiO-66/active carbon catalyst achieved triacetin production of 17.9% selectivity at acetic acid/glycerol molar ratio of 6. Moreover, heteropoly acids provided 100% glycerol conversion with 93.6% combined selectivity of di- and tri-acetin, respectively at 120°C, 4 h over HSiW/ZrO_2 catalyst (Zhu et al., 2013b). In particular, magnetic solid acid catalyst, Fe-Sn-Ti(SO_4^{2-})-400, can be considered a promising catalyst with 99% selectivity for triacetin at 100% glycerol conversion at 80°C for 30 min (Sun et al., 2016).

8.4.5 Steam Reforming of Glycerol into Hydrogen

Hydrogen can be produced from glycerol through steam reforming reaction (Bagnato et al., 2017; Roslan et al., 2020) or microbial method (Nanda et al., 2016; Sarma et al., 2012). In steam reforming reaction of glycerol, 1 mole of glycerol can produce 7 moles of hydrogen through the reaction.

$$C_3H_8O_3 + 3H_2O \rightarrow 7H_2 + 3CO_2 \qquad [\Delta H° = 129.41 \text{ kJ/mol}] \qquad (8.1)$$

Along with the main reaction (Eq. 8.1), glycerol steam reforming reaction also gives rise to secondary reactions – e.g., water–gas shift (WGS) (Eq. 8.2), methanation (Eq. 8.3), and decomposition of glycerol (Eq. 8.4).

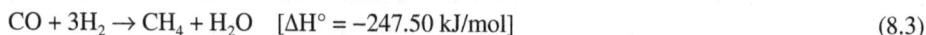

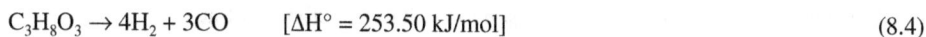

$$CO + H_2O \rightarrow CO_2 + H_2 \qquad [\Delta H° = -41.40 \text{ kJ/mol}] \qquad (8.2)$$

$$CO + 3H_2 \rightarrow CH_4 + H_2O \qquad [\Delta H° = -247.50 \text{ kJ/mol}] \qquad (8.3)$$

$$C_3H_8O_3 \rightarrow 4H_2 + 3CO \qquad [\Delta H° = 253.50 \text{ kJ/mol}] \qquad (8.4)$$

Nickel is one of the mostly used catalysts for glycerol steam reforming reaction (Yancheshmeh et al., 2020; Karakoc et al., 2019; Liu et al., 2018). Many studies focused on the effect of reaction conditions and the effect of metal addition and supports (Dieuzeide et al., 2016; Yurdakul et al., 2016). Dieuzeide et al. (2012) and Dieuzeide et al. (2016) studied the influence of Mg on Ni–Mg/Al_2O_3 catalysts, and they found that 3 wt.% of Mg showed higher Ni dispersion resulting in a high hydrogen yield. While Seung-hoon

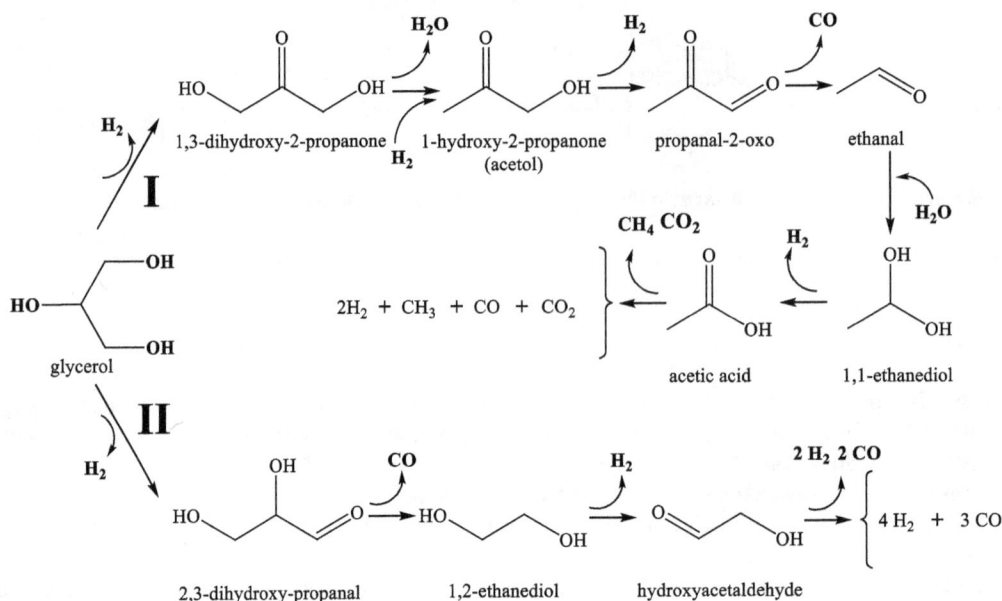

FIGURE 8.8 Reaction pathways for hydrogen production from glycerol steam reforming reaction. Reproduced with permission from Pompeo et al. (2010).

et al. (2014) showed that Sr was the best promoter for glycerol steam reforming reaction using Ni/Al$_2$O$_3$ catalyst. La$_2$O$_3$ was also found to improve the performance of glycerol steam reforming reaction (Kousi et al., 2016). Additionally, the addition of MgO, CeO$_2$ (Bobadilla et al. 2015), and Fe-Ce (Go et al. 2015) can reduce the coke formation at a high temperature.

Besides Ni, Pt is also a good candidate for glycerol steam reforming reaction (Pastor-Pérez et al., 2015; Sad et al., 2015). For Pt catalyst, the presence of promoters, such as Sn, showed a change in the catalytic activity and H$_2$ yield (Pastor-Pérez et al., 2015). Pt/SiO$_2$ showed the best catalyst among other supports such as Al$_2$O$_3$, MgO, SiO$_2$ and TiO$_2$ with higher H$_2$ yield (Sad et al., 2015). In the case of Pt catalyst, Pompeo et al. (2010) suggested the mechanism to produce H$_2$ with two reaction pathways (Figure 8.8). The first pathway includes dehydrogenation of glycerol, acetol dehydration, a second dehydrogenation, and then C–C breaking into ethanal followed by the formation of acetic acid with final decomposition into CO$_2$ and H$_2$. The second pathway contains dehydrogenation of glycerol followed by C–C bond breaking, a second dehydrogenation, then, the decomposition of hydroxyacetaldehyde into H$_2$ and CO.

Ruthenium catalyst is also used for glycerol steam reforming reaction (Hirai et al., 2005; Sundari and Vaidya, 2012), Hirai et al. (2005) found that Y$_2$O$_3$ is the best support among a compilation of MgO, Al$_2$O$_3$, SiO$_2$, Y$_2$O$_3$, ZrO$_2$, and CeO$_2$ for the production of hydrogen from glycerol. In the case of Co metallic, Co is commonly used as co-metallic, such as Ni-Co/Al$_2$O$_3$ (Sanchez and Comelli, 2014), Co-Ni-MgO/SBA-15 (Al-Salihi et al., 2020), and Co-Rh/CeZr (Araque et al., 2012), and Co can enhance the selective H$_2$ production related to the presence of its co-metallic. Moreover, perovskites were found to be promising candidates for glycerol steam reforming reaction (Ramesh et al., 2015; Surendar et al., 2016). Surender et al. (2016) found that the Pt-doped catalyst of LaCo$_{0.99}$Pt$_{0.01}$O$_3$ having the best hydrogen yield (78.4%) at 96% glycerol conversion. This superior catalytic activity of LaCo$_{0.99}$Pt$_{0.01}$O$_3$ is due to the high dispersion of the cobalt metal on the surface of the catalyst (Surendar et al., 2016).

8.4.6 Glycerol Ether

Glycerol ether including di- and tri-tert-butyl glycerol can be used as diesel additives (Manjunathan et al., 2016). Several studies have performed the etherification of glycerol with alcohol, e.g., with

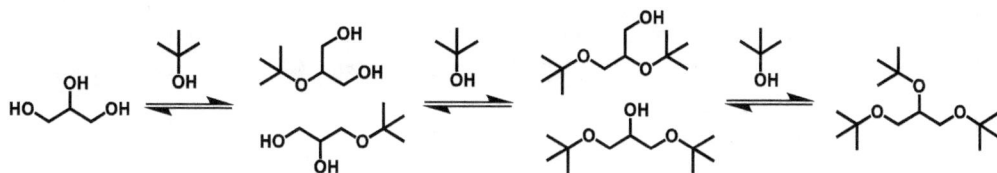

FIGURE 8.9 Scheme of the reaction pathway of glycerol etherification with tert-butanol

tert-butanol (Manjunathan et al., 2016; Srinivas et al., 2016; Viswanadham and Saxena, 2013). The etherification of glycerol with tert-butanol typically yields five major products including two mono-tert-butylglycerol ethers (monoethers), di-tert-butyl-glycerol ethers (diethers), and one tri-tert-butylglycerol (triether) (Figure 8.9).

Acid catalysts, including Amberlysts™ (Ozbay et al., 2013), Y zeolite (Srinivas et al., 2014), FAU, MOR, MFI, and β-zeolites (Pariente et al., 2009; Viswanadham and Saxena, 2013; González et al., 2014), tungstophosphoric acid (Srinivas et al., 2016), have been investigated. Amberlyst-15 gave very high glycerol conversion at 90–100°C due to the high Brønsted acid sites, but it is unstable at a higher temperature (>110°C) (Ozbay et al., 2013). Among zeolite catalysts, Pariente et al. (2009) found that β-zeolite with intermediate aluminum content (Si/Al = 25) is effective for glycerol etherification with ethanol. In addition, nano-β (N-BEA) zeolites (SiO_2/Al_2O_3 = 25) showed high selectivity of approximately 99% toward diethers and triethers and showed approximately 95% glycerol conversion (Viswanadham and Saxena, 2013).

8.4.7 Ketal

Ketal is a product of glycerol acetalization with a ketone. In the ketal family, the most significant compound is solketal ((2,2-dimethyl-1,3-dioxolan-4-yl)methanol) because of the wide applications of this material as fuel additives (Mota et al., 2010; Silva et al., 2010b), additives, and solvents for the pharmaceutical industry; as solvents for paint and ink industries; and as cleaning products (Garcia et al., 2014; Esteban et al., 2020).

Solketal can be produced from glycerol acetalization with acetone (Figure 8.10) under a mild reaction condition using many acid catalysts (Kowalska-Kuś et al., 2020; Talebian-Kiakalaieh and Tarighi, 2019; Bakuru et al., 2019). In glycerol acetalization reaction with acetone, water is a by-product, which can cover the active sites of catalyst reducing their activity (da Silva et al., 2009). The pore size of the catalyst is also an important factor for this reaction – e.g., the narrow pore structure of H-ZSM-5 hindered the diffusion of reactant and product, resulting in a low catalytic performance and solketal yield. H-Beta and HY zeolites with larger pore sizes can avoid the drawback of narrow-pore zeolites introducing a higher glycerol conversion. Many studies are focusing on using the β-zeolite catalyst for solketal production from glycerol (Venkatesha et al., 2016; Poly et al., 2019). The advantages of β-zeolite, because it possesses a large pore size (7.7 Å × 7.3 Å), make an easy diffusion of the product (Poly et al., 2019). Also, the increase in pore volume and a reduction in acidic density of β-zeolite can boost the solketal selectivity up to 100% (Venkatesha et al., 2016). Additionally, both the crystal size and the acidity of β-zeolite play crucial roles in the acetalization reaction of glycerol (Manjunathan et al., 2015; Poly et al., 2019) – i.e., the smaller crystal size of β-zeolite leading to higher conversion of glycerol and yield of solketal (Manjunathan et al., 2015).

In the presence of water and NaCl impurities, $HR/Y-W_{20}$ proved to be a good catalyst for the conversion of crude glycerol. Recently, Rahaman et al. (2020) reported that the OTS (n-octadecyl trichlorosilane)-grafted HY catalyst improved the hydrophobicity of HY zeolite resulting in a high catalytic activity (89%) compared with that of HY (28%) at 30°C after 60 min.

For the reaction mechanism, there are two possible reaction mechanisms. The first reaction mechanism includes three steps (Figure 8.11): (step 1) the adsorbed acetone reacts with glycerol to form hemiacetal, (step 2) the formation of water (the rate-limiting step), and (step 3) the formation solketal.

The second possible reaction mechanism occurs on the surface of Lewis acid (Figure 8.12). First, the metallic sites serving as Lewis acid attack the carbonyl functional group of acetone. Then, the primary

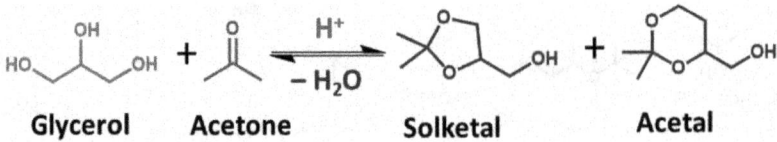

FIGURE 8.10   Scheme of glycerol acetalization reaction with acetone.

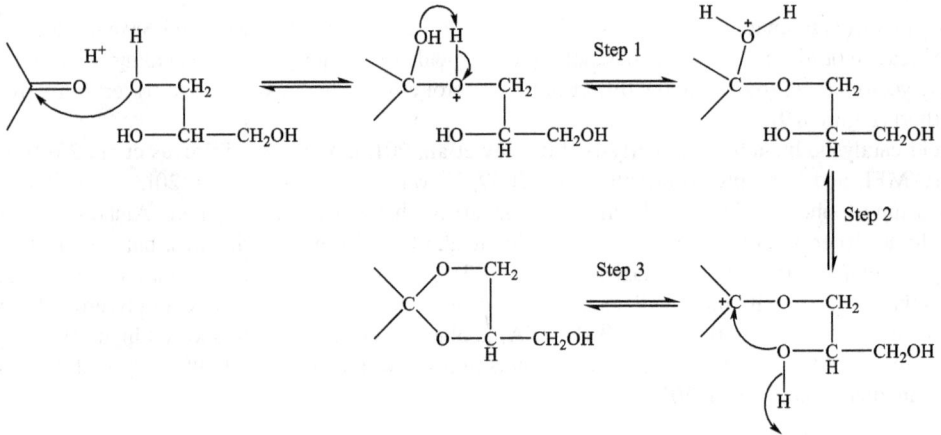

FIGURE 8.11   The proposed mechanism for ketalization reaction between glycerol and acetone. Reproduced with permission from Nanda et al. (2014b).

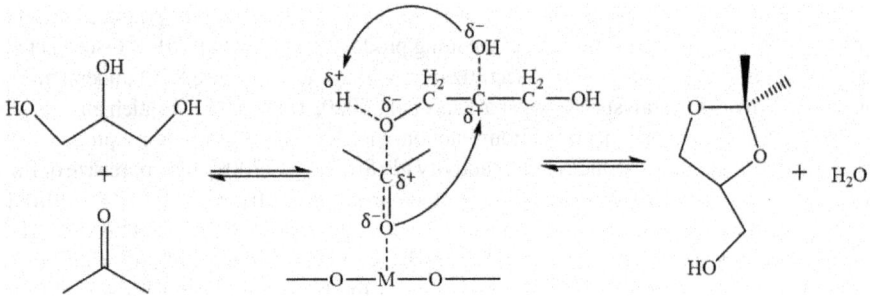

FIGURE 8.12   The mechanism of Solketal production from glycerol acetalization with acetone in the presence of Lewis acid catalysts. Reproduced with permission from Li et al. (2012).

alcoholic functional group of glycerol attacks the carbon atom of the carbonyl. In the end, the five-membered-ring solketal is produced via the C–O bond formation between the β-carbon of the glycerol and the carbonyl oxygen atom (Li et al., 2012).

8.4.8   Lactic Acid

Lactic acid ($CH_3CH(OH)COOH$) is an organic acid and an important material of use in many industries (Razali and Abdullah, 2017), especially in the polymer industry. Lactic acid can be produced from glycerol via a catalytic selective oxidation reaction or catalytic hydrothermal reaction (Razali and Abdullah, 2017). Lactic acid preparation from glycerol selective oxidation is a clean process (Crotti and Farnetti, 2015). Many catalysts were used for selective oxidation of glycerol such as Pd/C, Pt/C (Arcanjo et al., 2017), $AlPMo_{12}O_{40}$, $CrPMo_{12}O_{40}$ (Tao et al., 2016), HPMo/C (Tao et al., 2015), $Au\text{-}Pt/TiO_2$ (Shen et al.,

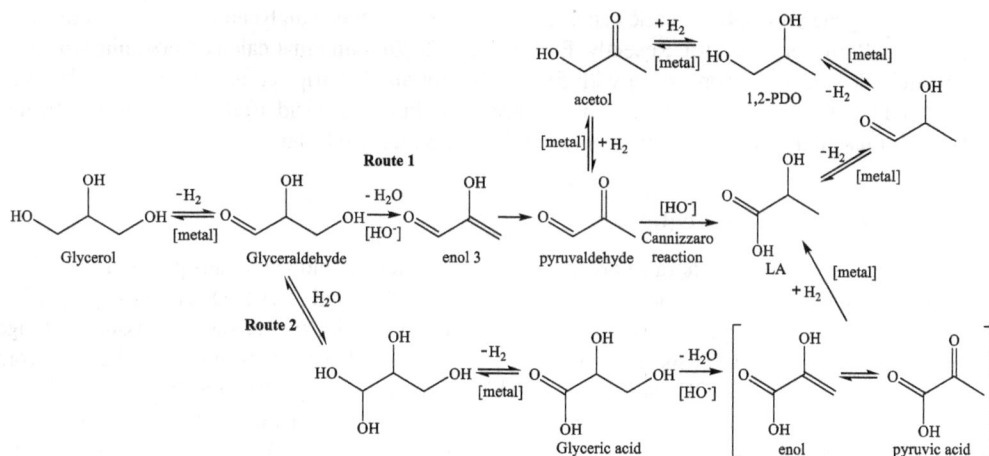

FIGURE 8.13 The mechanism of lactic production from glycerol via hydrothermal reaction. Reproduced with permission from Ftouni et al. (2015).

2010), and $AuPd/TiO_2$ (Xu et al., 2013). Among them, HPMo/C showed a high lactic selectivity (92%) at a high glycerol conversion (98%) (Tao et al., 2015). Also, $AlPMo_{12}O_{40}$ gave a good lactic selectivity (90.5) at glycerol conversion (93.7%) due to the contribution of bifunctional sites as Lewis acidity and redox reaction (Tao et al., 2016). Additionally, the yield of lactic acid is dependent on the particle size and distribution of metal, support, and the acidity of catalyst (Razali and Abdullah, 2017).

In the case of the hydrothermal reaction of glycerol, Ftouni et al. (2015) suggested the reaction mechanism for the conversion of glycerol into lactic acid (LA) starting with a dehydrogenation step as shown in Figure 8.13. In this reaction, both homogeneous and heterogeneous catalysts are used and show promising results (Shen et al., 2019; Ramírez-López et al., 2010). Copper proved to be an excellent candidate for this reaction with high lactic acid selectivity at a high glycerol conversion, for example, 92.5% lactic acid selectivity (S) and 88.6% glycerol conversion (X) over Cu/Al_2O_3 (Moreira et al., 2016); S = 90% and X = 95.4% over Cu/MgO (Moreira et al., 2016), S = 90% and X = 91% over Cu/HAP (Yin et al., 2016), and S = 89.1% and X = 92.5% over Cu_2O (Shen et al., 2017).

8.4.9 Short-Chain Polyglycerol

Diglycerols and triglycerols are products derived from glycerol through oligomerization of glycerol. They have many applications in the food industry, cosmetics, and polymer production (Martin and Richter, 2011; Chong et al., 2020). In cosmetics, diglycerol improves flavor and fragrance impact and longevity in toothpaste, mouthwashes, and deodorant sticks. It also helps to reduce the evaporation of menthol and enhances the transparency of emulsions due to its high refractive index than glycerol in the aqueous–oil mixture. In the food industry, diglycerol and polyglycerol can be used in the production of emulsifiers or food additives – for example, polyglycerol polyricinoleates act as emulsifiers in chocolate products. Additionally, diglycerol can esterify with fatty acid to form emulsifiers, which have good flavor and lipophilic/hydrophilic properties used in bakery products.

Oligomerization of glycerol can be carried out in the presence of acid/base homogeneous and heterogeneous catalysts (Sayoud et al., 2015; Barros et al., 2020). The base catalysts such as metal oxides, K_2CO_3, NaOH, and Cs-MCM41 can be used, while zeolite, H_2SO_4, cation exchange resins are used as acid catalysts (Sayoud et al., 2015). Generally, acid catalysts can cause dehydration (a side reaction) of glycerol or derived glycerol. Whereas, basic catalysts require high temperature (>220°C) to oligomerize glycerol, but high-temperature favor dehydration and dehydrogenation as well (Sayoud et al., 2015). To avoid the drawback of homogeneous catalysts, heterogeneous ones can be used for oligomerization reaction of glycerol.

Among heterogeneous catalysts, basic catalysts are favored to produce diglycerols and triglycerols than acid catalysts (Barros et al., 2020). Recently, Barros et al. (2020) found that calcined dolomite can show approximately 80% glycerol conversion with 51% diglycerol and 3% triglycerol selectivities. The same authors found that Ca and Mg metals can give a constant of diglycerol yield after two cycles, confirming that the rate of reaction is slow at the first hour, and the rate is increased later.

8.4.10 Poly(Glycerol Sebacate)

Poly(glycerol sebacate) (PGS) is a transparent, colorless polyester and biodegradable polymer. It is also a soft elastomeric material with nonlinear stress–strain behavior (Rai et al., 2012). Due to those properties, PGS has many biomedical applications in tissue engineering of cardiac tissue, vascular tissue, cartilage tissue, retinal tissue, and nerve tissue and can be utilized in drug delivery. PGS can be synthesized from sebacic acid ($HOOC(CH_2)_8COOH$) and glycerol. This reaction occurs in two steps including the pre-polycondensation step and cross-linking (Wang et al., 2002). Both the steps worked at low temperatures (~120°C), but they needed low pressure (40 mTorr) and long reaction time (5 h for the first step and 48 h for the second step). According to the application of PGS, the synthesis of PGS has followed five criteria as follows:

(i) PGS must undergo hydrolytic degradation to minimize the variation in degradation kinetics caused by enzymatic degradation.
(ii) The structure should contain hydrolyzable ester bonds.
(iii) PGS needs to contain a low degree of cross-linking.
(iv) Cross-link chemical bonds need to be hydrolyzable and identical to those in the backbone to minimize the possibility of heterogeneous degradation.
(v) Materials for synthesis are not toxic. It should have hydroxyl groups for cross-linking to form hydrogen bonding.

8.4.11 Dihydroxyacetone Production by Aerobic Glycerol Oxidation

Dihydroxyacetone (DHA) is a valuable product and has many applications for the synthesis of fine chemicals used in pharmaceutical, chemical, and biological products and in sunless tanning lotions (cosmetics) (Walgode et al., 2020). DHA can be produced via selective oxidation of glycerol at the secondary hydroxyl group. In aerobic glycerol oxidation reactions, noble metals are the most active catalysts used. It should be noted that DHA is not stable in basic conditions and just stable in acid conditions (Katryniok et al., 2011); therefore, the reaction system must have the base-free condition. Under base-free conditions, Au/ZnO, Au/CuO, Pt–Bi/mesoporous materials, and Pt–Sb/mesoporous materials are the most common catalysts in the oxidation of glycerol into DHA (Walgode et al., 2020). In the case of under basic media, Rh/AC showed 28% selectivity of DHA with full glycerol conversion (Rodrigues et al., 2011). Besides noble metals, transition metals (e.g., Ni, Cu, Co, CeO_2, ZrO_2, etc.) are also studied to replace noble metals to reduce operational costs. In the synthesis of DHA, the increase of oxygen pressure (1.5–10 bar) and reaction temperature (40–100°C) enhances the selectivity of DHA.

8.5 Conclusions

This chapter provided an overview of the source, purification, application, and several conversion processes of glycerol. The products from these processes can be used as end-used products or intermediate/building blocks for further conversion into high-value products such as polymers, detergents, surfactants, and solvents. In addition, the use of purified glycerol in cosmetics, food and pharmaceuticals also contributes significantly to the glycerol industry. In short, green glycerol from fats and vegetable oils can be a promising source for chemicals and fuels in the future to reduce the dependency on fossil resources and mitigate environmental pollution.

Acknowledgments

This research was funded by the Vietnam National Foundation for Science and Technology Development (NAFOSTED) under grant number 104.05–2019.39.

REFERENCES

Akizuki, M., and Oshima, Y. 2012. Kinetics of glycerol dehydration with WO_3/TiO_2 in supercritical water. *Industrial & Engineering Chemistry Research* 51(38): 12253–12257.

Alhanash, A., Kozhevnikova, E. F., and Kozhevnikov, I. V. 2010. Gas-phase dehydration of glycerol to acrolein catalysed by caesium heteropoly salt. *Applied Catalysis A: General* 378(1): 11–18.

Al-Salihi, S., Abrokwah, R., Dade, W., Deshmane, V., Hossain, T., and Kuila, D. 2020. Renewable hydrogen from glycerol steam reforming using Co – Ni – MgO based SBA-15 nanocatalysts. *International Journal of Hydrogen Energy* 45: 14183–14198

Araque, M., Martínez T, L. M., Vargas, J. C., Centeno, M. A., and Roger, A. C. 2012. Effect of the active metals on the selective H_2 production in glycerol steam reforming. *Applied Catalysis B: Environmental* 125: 556–566.

Arcanjo, M. R. A., Silva, I. J., Rodríguez-Castellón, E., Infantes-Molina, A., and Vieira, R. S. 2017. Conversion of glycerol into lactic acid using Pd or Pt supported on carbon as catalyst. *Catalysis Today* 279: 317–326.

Ardi, M. S., Aroua, M. K., and Hashim, N. A. 2015. Progress, prospect and challenges in glycerol purification process: A review. *Renewable and Sustainable Energy Reviews* 42: 1164–1173.

Bagnato, G., Iulianelli, A., Sanna, A., and Basile, A. 2017. Glycerol production and transformation: A critical review with particular emphasis on glycerol reforming reaction for producing hydrogen in conventional and membrane reactors. *Membranes* 7(2).

Bakuru, V. R., Churipard, S. R., Maradur, S. P., and Kalidindi, S. B. 2019. Exploring the Brønsted acidity of UiO-66 (Zr, Ce, Hf) metal – organic frameworks for efficient solketal synthesis from glycerol acetalization. *Dalton Transactions* 48(3): 843–847.

Barros, F. J. S., Cecilia, J. A., Moreno-Tost, R., de Oliveira, M. F., Rodríguez-Castellón, E., Luna, F. M. T., and Vieira, R. S. 2020. Glycerol oligomerization using low cost dolomite catalyst. *Waste and Biomass Valorization* 11(4): 1499–1512.

Bobadilla, L. F., Penkova, A., Álvarez, A., Domínguez, M. I., Romero-Sarria, F., Centeno, M. A., and Odriozola, J. A. 2015. Glycerol steam reforming on bimetallic $NiSn/CeO_2–MgO–Al_2O_3$ catalysts: Influence of the support, reaction parameters and deactivation/regeneration processes. *Applied Catalysis A: General* 492: 38–47.

Casiello, M., Monopoli, A., Cotugno, P., Milella, A., Dell'Anna, M. M., Ciminale, F., and Nacci, A. 2014. Copper(II) chloride-catalyzed oxidative carbonylation of glycerol to glycerol carbonate. *Journal of Molecular Catalysis A: Chemical* 381: 99–106.

Chong, C. C., Aqsha, A., Ayoub, M., Sajid, M., Abdullah, A. Z., Yusup, S., and Abdullah, B. 2020. A review over the role of catalysts for selective short-chain polyglycerol production from biodiesel derived waste glycerol. *Environmental Technology & Innovation* 19: 100859.

Climent, M. J., Corma, A., De Frutos, P., Iborra, S., Noy, M., Velty, A., and Concepción, P. 2010. Chemicals from biomass: Synthesis of glycerol carbonate by transesterification and carbonylation with urea with hydrotalcite catalysts. The role of acid – base pairs. *Journal of Catalysis* 269(1): 140–149.

Crotti, C., and Farnetti, E. 2015. Selective oxidation of glycerol catalyzed by iron complexes. *Journal of Molecular Catalysis A: Chemical* 396: 353–359.

Dalla Costa, B. O., Decolatti, H. P., Legnoverde, M. S., and Querini, C. A. 2017. Influence of acidic properties of different solid acid catalysts for glycerol acetylation. *Catalysis Today* 289: 222–230.

da Silva, C. X. A., Gonçalves, V. L. C., and Mota, C. J. A. 2009. Water-tolerant zeolitecatalyst for the acetalisation of glycerol. *Green Chemistry* 11(1): 38–41.

Deng, C., Duan, X., Zhou, J., Chen, D., Zhou, X., and Yuan, W. 2014. Size effects of Pt-Re bimetallic catalysts for glycerol hydrogenolysis. *Catalysis Today* 234: 208–214.

Dieuzeide, M. L., Iannibelli, V., Jobbagy, M., and Amadeo, N. 2012. Steam reforming of glycerol over Ni/ Mg/γ-Al$_2$O$_3$ catalysts. Effect of calcination temperatures. *International Journal of Hydrogen Energy* 37(19): 14926–14930.

Dieuzeide, M. L., Laborde, M., Amadeo, N., Cannilla, C., Bonura, G., and Frusteri, F. 2016. Hydrogen production by glycerol steam reforming: How Mg doping affects the catalytic behaviour of Ni/Al$_2$O$_3$ catalysts. *International Journal of Hydrogen Energy* 41(1): 157–166.

Dizoğlu, G., and Sert, E. 2020. Fuel additive synthesis by acetylation of glycerol using activated carbon/UiO-66 composite materials. *Fuel* 281: 118584.

Esteban, J., Vorholt, A. J., and Leitner, W. 2020. An overview of the biphasic dehydration of sugars to 5-hydroxymethylfurfural and furfural: A rational selection of solvents using COSMO-RS and selection guides. *Green Chemistry* 22(7): 2097–2128.

Forero-Hernandez, H., Jones, M. N., Sarup, B., Jensen, A. D., Abildskov, J., and Sin, G. 2020. Comprehensive development, uncertainty and sensitivity analysis of a model for the hydrolysis of rapeseed oil. *Computers & Chemical Engineering* 133: 106631.

Ftouni, J., Villandier, N., Auneau, F., Besson, M., Djakovitch, L., and Pinel, C. 2015. From glycerol to lactic acid under inert conditions in the presence of platinum-based catalysts: The influence of support. *Catalysis Today* 257: 267–273.

Garcia, J. I., Garcia-Marin, H., and Pires, E. 2014. Glycerol based solvents: Synthesis, properties and applications. *Green Chemistry* 16(3): 1007–1033.

García-Fernández, S., Gandarias, I., Requies, J., Güemez, M. B., Bennici, S., Auroux, A., and Arias, P. L. 2015. New approaches to the Pt/WOx/Al$_2$O$_3$ catalytic system behavior for the selective glycerol hydrogenolysis to 1,3-propanediol. *Journal of catalysis* 323: 65–75.

García-Fernández, S., Gandarias, I., Requies, J., Soulimani, F., Arias, P. L., and Weckhuysen, B. M. 2017. The role of tungsten oxide in the selective hydrogenolysis of glycerol to 1,3-propanediol over Pt/WOx/Al$_2$O$_3$. *Applied Catalysis B: Environmental* 204: 260–272.

Go, Y.-J., Go, G.-S., Lee, H.-J., Moon, D.-J., Park, N.-C., and Kim, Y.-C. 2015. The relation between carbon deposition and hydrogen production in glycerol steam reforming. *International Journal of Hydrogen Energy* 40(35): 11840–11847.

Gonçalves, V. L. C., Pinto, B. P., Silva, J. C., and Mota, C. J. A. 2008. Acetylation of glycerol catalyzed by different solid acids. *Catalysis Today* 133–135: 673–677.

González, M. D., Salagre, P., Linares, M., García, R., Serrano, D., and Cesteros, Y. 2014. Effect of hierarchical porosity and fluorination on the catalytic properties of zeolite beta for glycerol etherification. *Applied Catalysis A* 473: 75–82.

Gu, Y., Liu, S., Li, C., and Cui, Q. 2013. Selective conversion of glycerol to acrolein over supported nickel sulfate catalysts. *Journal of catalysis* 301: 93–102.

Hájek, M., and Skopal, F. 2010. Treatment of glycerol phase formed by biodiesel production. *Bioresource Technology* 101(9): 3242–3245.

Hegde, A. 2018. Glycerol Market to Hit $5 bn by 2024. Global Market Insights, Inc. https://www.globenewswire.com/fr/news-release/2018/11/13/1650184/0/en/Glycerol-Market-to-hit-5-bn-by-2024-Global-Market-Insights-Inc.html (accessed 1 December 2020)

Hirai, T., N.-o. Ikenaga, Miyake, T., and Suzuki, T. 2005. Production of hydrogen by steam reforming of glycerin on ruthenium catalyst. *Energy & Fuels* 19(4): 1761–1762.

Isahak, W. N. R. W., Ismail, M., Yarmo, M., Jahim, J., and Salimon, J. 2010. Purification of crude glycerol from transesterification RBD palm oil over homogeneous and heterogeneous catalysts for the biolubricant preparation. *Journal of Applied Sciences* 10(21): 2590–2595.

Jia, C.-J., Liu, Y., Schmidt, W., Lu, A.-H., and Schüth, F. 2010. Small-sized HZSM-5 zeolite as highly active catalyst for gas phase dehydration of glycerol to acrolein. *Journal of Catalysis* 269(1): 71–79.

Kant, A., He, Y., Jawad, A., Li, X., Rezaei, F., Smith, J. D., and Rownaghi, A. A. 2017. Hydrogenolysis of glycerol over Ni, Cu, Zn, and Zr supported on H-beta. *Chemical Engineering Journal* 317: 1–8.

Karakoc, O. P., Kibar, M., Akin, A., and Yildiz, M. 2019. Nickel-based catalysts for hydrogen production by steam reforming of glycerol. *International Journal of Environmental Science and Technology* 16(9): 5117–5124.

Katryniok, B., Kimura, H., Skrzyńska, E., Girardon, J.-S., Fongarland, P., Capron, M., Ducoulombier, R., Mimura, N., Paul, S., and Dumeignil, F. 2011. Selective catalytic oxidation of glycerol: Perspectives for high value chemicals. *Green Chemistry* 13(8): 1960–1979.

Kousi, K., Chourdakis, N., Matralis, H., Kontarides, D., Papadopoulou, C., and Verykios, X. 2016. Glycerol steam reforming over modified Ni-based catalysts. *Applied Catalysis A: General* 518: 129–141.

Kowalska-Kuś, J., Held, A., and Nowińska, K. 2020. A continuous-flow process for the acetalization of crude glycerol with acetone on zeolite catalysts. *Chemical Engineering Journal* 401: 126143.

Li, L., Korányi, T. I., Sels, B. F., and Pescarmona, P. P. 2012. Highly-efficient conversion of glycerol to solketal over heterogeneous Lewis acid catalysts. *Green Chemistry* 14(6): 1611–1619.

Liu, L., Zhang, Y., Wang, A., and Zhang, T. 2012. Mesoporous WO_3 supported Pt catalyst for hydrogenolysis of glycerol to 1,3-propanediol. *Chinese Journal of Catalysis* 33(7): 1257–1261.

Liu, S., Yan, Z., Zhang, Y., Wang, R., Luo, S.-Z., Jing, F., and Chu, W. 2018. Carbon nanotubes supported nickel as the highly efficient catalyst for hydrogen production through glycerol steam reforming. *ACS Sustainable Chemistry & Engineering* 6(11): 14403–14413.

Liu, Z., Wang, J., Kang, M., Yin, N., Wang, X., Tan, Y., and Zhu, Y. 2015. Structure-activity correlations of $LiNO_3/Mg_4AlO_{5.5}$ catalysts for glycerol carbonate synthesis from glycerol and dimethyl carbonate. *Journal of Industrial and Engineering Chemistry* 21: 394–399.

Ma, T., Ding, J., Shao, R., Xu, W., and Yun, Z. 2017. Dehydration of glycerol to acrolein over Wells – Dawson and Keggin type phosphotungstic acids supported on MCM-41 catalysts. *Chemical Engineering Journal* 316: 797–806.

Manjunathan, P., Kumar, M., Churipard, S. R., Sivasankaran, S., Shanbhag, G. V., and Maradur, S. P. 2016. Catalytic etherification of glycerol to tert-butyl glycerol ethers using tert-butanol over sulfonic acid functionalized mesoporous polymer. *RSC Advances* 6(86): 82654–82660.

Manjunathan, P., Maradur, S. P., Halgeri, A. B., and Shanbhag, G. V. 2015. Room temperature synthesis of solketal from acetalization of glycerol with acetone: Effect of crystallite size and the role of acidity of beta zeolite. *Journal of Molecular Catalysis A: Chemical* 396: 47–54.

Manosak, R., Limpattayanate, S., and Hunsom, M. 2011. Sequential-refining of crude glycerol derived from waste used-oil methyl ester plant via a combined process of chemical and adsorption. *Fuel Processing Technology* 92(1): 92–99.

Martin, A., and Richter, M. 2011. Oligomerization of glycerol – a critical review. *European Journal of Lipid Science and Technology* 113(1): 100–117.

Moreira, A. B. F., Bruno, A. M., Souza, M. M. V. M., and Manfro, R. L. 2016. Continuous production of lactic acid from glycerol in alkaline medium using supported copper catalysts. *Fuel Processing Technology* 144: 170–180.

Mota, C. J. A., da Silva, C. X. A., Rosenbach, N., Costa, J., and da Silva, F. 2010. Glycerin derivatives as fuel additives: The addition of glycerol/acetone ketal (solketal) in gasolines. *Energy Fuel* 24(4): 2733–2736.

Muniru, O. S., Ezeanyanaso, C. S., Akubueze, E. U., Igwe, C. C., and Elemo, G. N. 2019. Review of different purification techniques for crude glycerol from biodiesel production. *Journal of Energy Research and Reviews*:1–6.

Muniru, O. S., Ezeanyanaso, C. S., Fagbemigun, T., Akubueze, E., Oyewole, A., Okunola, O., Asieba, G., Shifatu, A., Igwe, C., and Elemo, G. 2016. Valorization of biodiesel production: Focus on crude glycerine refining/purification. *Journal of Scientific Research and Reports* 11: 1–8.

Nanda, M. R., Yuan, Z., Qin, W., Ghaziaskar, H. S., Poirier, M.-A., and Xu, C. C. 2014b. Thermodynamic and kinetic studies of a catalytic process to convert glycerol into solketal as an oxygenated fuel additive. *Fuel* 117: 470–477.

Nanda, M. R., Yuan, Z., Qin, W., Poirier, M., and Chunbao, X. 2014a. Purification of crude glycerol using acidification: Effects of acid types and product characterization. *Austin Journal of Chemical Engineering* 1(1): 1–7.

Nanda, M. R., Yuan, Z., Qin, W., and Xu, C. 2016. Recent advancements in catalytic conversion of glycerol into propylene glycol: A review. *Catalysis Reviews* 58(3): 309–336.

Nda-Umar, U., Ramli, I., Taufiq-Yap, Y., and Muhamad, E. 2018. An overview of recent research in the conversion of glycerol into biofuels, fuel additives and other bio-based chemicals. *Catalysts* 9(1): 15.

Nomanbhay, S., Ong, M. Y., Chew, K. W., Show, P.-L., Lam, M. K., and Chen, W.-H. 2020. Organic carbonate production utilizing crude glycerol derived as by-product of biodiesel production: A Review. *Energies* 13(6): 1483.

Ochoa-Gómez, J. R., Gómez-Jiménez-Aberasturi, O., Maestro-Madurga, B., Pesquera-Rodríguez, A., Ramírez-López, C., Lorenzo-Ibarreta, L., Torrecilla-Soria, J., and Villarán-Velasco, M. C. 2009. Synthesis of glycerol carbonate from glycerol and dimethyl carbonate by transesterification: Catalyst screening and reaction optimization. *Applied Catalysis A: General* 366(2): 315–324.

Okoye, P. U., Abdullah, A. Z., and Hameed, B. H. 2017. Synthesis of oxygenated fuel additives via glycerol esterification with acetic acid over bio-derived carbon catalyst. *Fuel* 209: 538–544.

Ooi, T., Yong, K., Dzulkefly, K., Wan Yunus, W., and Hazimah, A. 2001. Crude glycerine recovery from glycerol residue waste from a palm kernel oil methyl ester plant. *Journal of Oil Palm Research* 13(2): 16–22.

Ozbay, N., Oktar, N., Dogu, G., and Dogu, T. 2013. Activity comparison of different solid acid catalysts in etherification of glycerol with tert-butyl alcohol in flow and batch reactors. *Topics in Catalysis* 56(18–20): 1790–1803.

Pariente, S., Tanchoux, N., and Fajula, F. 2009. Etherification of glycerol with ethanol over solid acid catalysts. *Green Chemistry* 11(8): 1256–1261.

Pastor-Pérez, L., Merlo, A., Buitrago-Sierra, R., Casella, M., and Sepúlveda-Escribano, A. 2015. Bimetallic PtSn/C catalysts obtained via SOMC/M for glycerol steam reforming. *Journal of Colloid and Interface Science* 459: 160–166.

Phung, T. K., and Busca, G. 2020. Selective bioethanol conversion to chemicals and fuels via advanced catalytic approaches. In: *Biorefinery of Alternative Resources: Targeting Green Fuels and Platform Chemicals*, (Eds.) S. Nanda, D. V. N. Vo, and P. K. Sarangi. Singapore: Springer Nature, pp. 75–103.

Poly, S. S., Jamil, M. A. R., Touchy, A. S., Yasumura, S., Siddiki, S. M. A. H., Toyao, T., Maeno, Z., and Shimizu, K.-I. 2019. Acetalization of glycerol with ketones and aldehydes catalyzed by high silica Hβ zeolite. *Molecular Catalysis* 479: 110608.

Pompeo, F., Santori, G., and Nichio, N. N. 2010. Hydrogen and/or syngas from steam reforming of glycerol. Study of platinum catalysts. *International Journal of Hydrogen Energy* 35(17): 8912–8920.

Possato, L. G., Chaves, T. F., Cassinelli, W. H., Pulcinelli, S. H., Santilli, C. V., and Martins, L. 2017. The multiple benefits of glycerol conversion to acrolein and acrylic acid catalyzed by vanadium oxides supported on micro-mesoporous MFI zeolites. *Catalysis Today* 289: 20–28.

Rahaman, M. S., Phung, T. K., Hossain, M. A., Chowdhury, E., Tulaphol, S., Lalvani, S. B., O'Toole, M., Willing, G. A., Jasinski, J. B., Crocker, M., and Sathitsuksanoh, N. 2020. Hydrophobic functionalization of HY zeolites for efficient conversion of glycerol to solketal. *Applied Catalysis A: General* 592: 117369.

Rai, R., Tallawi, M., Grigore, A., and Boccaccini, A. R. 2012. Synthesis, properties and biomedical applications of poly(glycerol sebacate) (PGS): A review. *Progress in Polymer Science* 37(8): 1051–1078.

Ramesh, S., Yang, E.-H., Jung, J.-S., and Moon, D. J. 2015. Copper decorated perovskite an efficient catalyst for low temperature hydrogen production by steam reforming of glycerol. *International Journal of Hydrogen Energy* 40(35): 11428–11435.

Ramírez-López, C. A., Ochoa-Gómez, J. R., Fernández-Santos, M. a., Gómez-Jiménez-Aberasturi, O., Alonso-Vicario, A., and Torrecilla-Soria, J. 2010. Synthesis of lactic acid by alkaline hydrothermal conversion of glycerol at high glycerol concentration. *Industrial & Engineering Chemistry Research* 49(14): 6270–6278.

Razali, N., and Abdullah, A. Z. 2017. Production of lactic acid from glycerol via chemical conversion using solid catalyst: A review. *Applied Catalysis A: General* 543: 234–246.

Research, G. V. 2020. Glycerol market growth & trends. In: *Glycerol Market Size Worth $3.5 Billion By 2027 | CAGR 4.0%*. grandviewresearch.com.

Rodrigues, E. G., Carabineiro, S. A. C., Chen, X., Delgado, J. J., Figueiredo, J. L., Pereira, M. F. R., and Órfão, J. J. M. 2011. Selective oxidation of glycerol catalyzed by Rh/Activated Carbon: Importance of support surface chemistry. *Catalysis Letters* 141(3): 420–431.

Roslan, N. A., Abidin, S. Z., Ideris, A., and Vo, D.-V. N. 2020. A review on glycerol reforming processes over Ni-based catalyst for hydrogen and syngas productions. *International Journal of Hydrogen Energy* 45(36): 18466–18489.

Sabbah, M., Di Pierro, P., Cammarota, M., Dell'Olmo, E., Arciello, A., and Porta, R. 2019. Development and properties of new chitosan-based films plasticized with spermidine and/or glycerol. *Food Hydrocolloids* 87: 245–252.

Sad, M. E., Duarte, H. A., Vignatti, C., Padro, C., and Apesteguía, C. R. 2015. Steam reforming of glycerol: Hydrogen production optimization. *International Journal of Hydrogen Energy* 40(18): 6097–6106.

Sadaf, S., Iqbal, J., Ullah, I., Bhatti, H. N., Nouren, S., Nisar, J., and Iqbal, M. 2018. Biodiesel production from waste cooking oil: An efficient technique to convert waste into biodiesel. *Sustainable Cities and Society* 41: 220–226.

Sanchez, E. A., and Comelli, R. A. 2014. Hydrogen production by glycerol steam-reforming over nickel and nickel-cobalt impregnated on alumina. *International Journal of Hydrogen Energy* 39(16): 8650–8655.

Sarma, S. J., Brar, S. K., Sydney, E. B., Le Bihan, Y., Buelna, G., and Soccol, C. R. 2012. Microbial hydrogen production by bioconversion of crude glycerol: A review. *International Journal of Hydrogen Energy* 37(8): 6473–6490.

Saxena, R. K., Anand, P., Saran, S., and Isar, J. 2009. Microbial production of 1,3-propanediol: Recent developments and emerging opportunities. *Biotechnology Advances* 27(6): 895–913.

Sayoud, N., De Oliveira Vigier, K., Cucu, T., De Meulenaer, B., Fan, Z., Lai, J., Clacens, J.-M., Liebens, A., and Jérôme, F. 2015. Homogeneously-acid catalyzed oligomerization of glycerol. *Green Chemistry* 17(8): 4307–4314.

Seung-hoon, K., Jung, J.-s., Yang, E.-h., Lee, K.-Y., and Moon, D. J. 2014. Hydrogen production by steam reforming of biomass-derived glycerol over Ni-based catalysts. *Catalysis Today* 228: 145–151.

Shen, L., Yin, H., Yin, H., Liu, S., and Wang, A. 2017. Conversion of glycerol to lactic acid catalyzed by different-sized Cu_2O nanoparticles in NaOH aqueous solution. *Journal of Nanoscience and Nanotechnology* 17(1): 780–787.

Shen, L., Zhou, X., Zhang, C., Yin, H., Wang, A., and Wang, C. 2019. Functional characterization of bimetallic CuPdx nanoparticles in hydrothermal conversion of glycerol to lactic acid. *Journal of Food Biochemistry* 43(8): e12931.

Shen, Y., Zhang, S., Li, H., Ren, Y., and Liu, H. 2010. Efficient synthesis of lactic acid by aerobic oxidation of glycerol on Au – Pt/TiO_2 catalysts. *Chemistry – A European Journal* 16(25): 7368–7371.

Silva, L. N., Gonçalves, V. L. C., and Mota, C. J. A. 2010a. Catalytic acetylation of glycerol with acetic anhydride. *Catalysis Communications* 11(12): 1036–1039.

Silva, P. H. R., Gonçalves, V. L. C., and Mota, C. J. A. 2010b. Glycerol acetals as anti-freezing additives for biodiesel. *Bioresource Technology* 101(15): 6225–6229.

Srinivas, M., Raveendra, G., Parameswaram, G., and Prasad, P. S. S., and Lingaiah, N. 2016. Cesium exchanged tungstophosphoric acid supported on tin oxide: An efficient solid acid catalyst for etherification of glycerol with tert-butanol to synthesize biofuel additives. *Journal of Molecular Catalysis A: Chemical* 413: 7–14.

Srinivas, M., Sree, R., Raveendra, G., Kumar, C. R., Prasad, P. S. S., and Lingaiah, N. 2014. Selective etherification of glycerol with tert-butanol over 12-tungstophosphoric acid catalysts supported on Y-zeolite. *Indian Journal of Chemistry* 53A:524–529.

Sullivan, C. J., Kuenz, A., and Vorlop, K. D. 2018. Propanediols. In *Ullmann's Encyclopedia of Industrial Chemistry*: 1–15.

Sun, J., Tong, X., Yu, L., and Wan, J. 2016. An efficient and sustainable production of triacetin from the acetylation of glycerol using magnetic solid acid catalysts under mild conditions. *Catalysis Today* 264: 115–122.

Sundari, R., and Vaidya, P. D. 2012. Reaction kinetics of glycerol steam reforming using a Ru/Al_2O_3 catalyst. *Energy & Fuels* 26(7): 4195–4204.

Surendar, M., Sagar, T., Raveendra, G., Kumar, M. A., Lingaiah, N., Rao, K. R., and Prasad, P. S. 2016. Pt doped $LaCoO_3$ perovskite: A precursor for a highly efficient catalyst for hydrogen production from glycerol. *International Journal of Hydrogen Energy* 41(4): 2285–2297.

Talebian-Kiakalaieh, A., and Tarighi, S. 2019. Hierarchical faujasite zeolite-supported heteropoly acid catalyst for acetalization of crude-glycerol to fuel additives. *Journal of Industrial and Engineering Chemistry* 79: 452–464.

Tan, H. W., Abdul Aziz, A. R., and Aroua, M. K. 2013. Glycerol production and its applications as a raw material: A review. *Renewable and Sustainable Energy Reviews* 27: 118–127.

Tao, M., Yi, X., Delidovich, I., Palkovits, R., Shi, J., and Wang, X. 2015. Hetropolyacid-catalyzed oxidation of glycerol into lactic acid under mild base-free conditions. *ChemSusChem* 8(24): 4195–4201.

Tao, M., Zhang, D., Deng, X., Li, X., Shi, J., and Wang, X. 2016. Lewis-acid-promoted catalytic cascade conversion of glycerol to lactic acid by polyoxometalates. *Chemical Communications* 52(16): 3332–3335.

Thompson, J. C., and He, B. B. 2006. Characterization of crude glycerol from biodiesel production from multiple feedstocks. *Applied Engineering in Agriculture* 22(2): 261–265.

Tianfeng, C., Huipeng, L., Hua, Z., and Kejian, L. 2013. Purification of crude glycerol from waste cooking oil based biodiesel production by orthogonal test method. *China Petroleum Processing and Petrochemical Technology* 15(1).

Venkatesha, N. J., Bhat, Y. S., and Jai Prakash, B. S. 2016. Dealuminated BEA zeolite for selective synthesis of five-membered cyclic acetal from glycerol under ambient conditions. *RSC Advance* 6(23): 18824–18833.

Viswanadham, N., and Saxena, S. K. 2013. Etherification of glycerol for improved production of oxygenates. *Fuel* 103: 980–986.

Walgode, P. M., Faria, R. P., and Rodrigues, A. E. 2020. A review of aerobic glycerol oxidation processes using heterogeneous catalysts: A sustainable pathway for the production of dihydroxyacetone. *Catalysis Reviews*: 1–90.

Wang, Y., Ameer, G. A., Sheppard, B. J., and Langer, R. 2002. A tough biodegradable elastomer. *Nature Biotechnology* 20(6): 602–606.

Wu, Y., Song, X., Cai, F., and Xiao, G. 2017. Synthesis of glycerol carbonate from glycerol and diethyl carbonate over Ce-NiO catalyst: The role of multiphase Ni. *Journal of Alloys and Compounds* 720: 360–368.

Xu, J., Zhang, H., Zhao, Y., Yu, B., Chen, S., Li, Y., Hao, L., and Liu, Z. 2013. Selective oxidation of glycerol to lactic acid under acidic conditions using AuPd/TiO$_2$ catalyst. *Green Chemistry* 15(6): 1520–1525.

Yadav, G. D., Sharma, R. V., and Katole, S. O. 2013. Selective dehydration of glycerol to acrolein: Development of efficient and robust solid acid catalyst MUICaT-5. *Industrial & Engineering Chemistry Research* 52(30): 10133–10144.

Yancheshmeh, M. S., Sahraei, O. A., Aissaoui, M., and Iliuta, M. C. 2020. A novel synthesis of NiAl$_2$O$_4$ spinel from a Ni-Al mixed-metal alkoxide as a highly efficient catalyst for hydrogen production by glycerol steam reforming. *Applied Catalysis B: Environmental* 265: 118535.

Yang, F., Hanna, M. A., and Sun, R. 2012. Value-added uses for crude glycerol – a byproduct of biodiesel production. *Biotechnology for Biofuels* 5(1): 1–10.

Yin, H., Zhang, C., Yin, H., Gao, D., Shen, L., and Wang, A. 2016. Hydrothermal conversion of glycerol to lactic acid catalyzed by Cu/hydroxyapatite, Cu/MgO, and Cu/ZrO$_2$ and reaction kinetics. *Chemical Engineering Journal* 288: 332–343.

Yurdakul, M., Ayas, N., Bizkarra, K., El Doukkali, M., and Cambra, J. F. 2016. Preparation of Ni-based catalysts to produce hydrogen from glycerol by steam reforming process. *International Journal of Hydrogen Energy* 41(19): 8084–8091.

Zhao, H., Zhou, C. H., Wu, L. M., Lou, J. Y., Li, N., Yang, H. M., Tong, D. S., and Yu, W. H. 2013. Catalytic dehydration of glycerol to acrolein over sulfuric acid-activated montmorillonite catalysts. *Applied Clay Science* 74: 154–162.

Zhou, W., Zhao, Y., Wang, S., and Ma, X. 2017. The effect of metal properties on the reaction routes of glycerol hydrogenolysis over platinum and ruthenium catalysts. *Catalysis Today* 298: 2–8.

Zhu, S., Qiu, Y., Zhu, Y., Hao, S., Zheng, H., and Li, Y. 2013a. Hydrogenolysis of glycerol to 1,3-propanediol over bifunctional catalysts containing Pt and heteropolyacids. *Catalysis Today* 212: 120–126.

Zhu, S., Zhu, Y., Gao, X., Mo, T., Zhu, Y., and Li, Y. 2013b. Production of bioadditives from glycerol esterification over zirconia supported heteropolyacids. *Bioresource Technology* 130: 45–51.

Zuhaimi, N. A. S., Indran, V. P., Deraman, M. A., Mudrikah, N. F., Maniam, G. P., Taufiq-Yap, Y. H., and Rahim, M. H. Ab. 2015. Reusable gypsum based catalyst for synthesis of glycerol carbonate from glycerol and urea. *Applied Catalysis A: General* 502: 312–319.

9

Effect of Substrates on the Performance of Microbial Fuel Cell for Sustainable Energy Production

M. Amirul Islam, Ahasanul Karim, Fuad Ameen

CONTENTS

9.1 Introduction

Microbial fuel cell (MFC) is a promising technique for harvesting electricity by utilizing substrates as electron donor (Yu et al., 2019). Using the MFC technology, direct electricity can be generated from electroactive microorganisms by employing an external electrical circuit (Rabaey et al., 2007). Among the several bioelectrochemical devices, MFCs can spontaneously generate electricity by converting substrates or biomasses through the metabolic activity of microbes (Yu et al., 2019). Therefore, this technology is considered a sustainable technology to fulfill growing energy demands particularly by using different wastewaters as nutrients (Lu et al., 2009). Generally, MFC comprises an anodic compartment and a cathodic compartment while both compartments are separated via an electron exchange membrane (Ye et al., 2019). In the anodic chamber, oxidative conversions occur through microbial catalysis, whereas the chemical reduction process occurs in the cathodic chamber via the oxidoreduction process (Jaiswal et al., 2020). Both anode and cathode are connected together through an external circuit with an external resistor. The detailed concept and working principle of MFC have been described in the subsequent section.

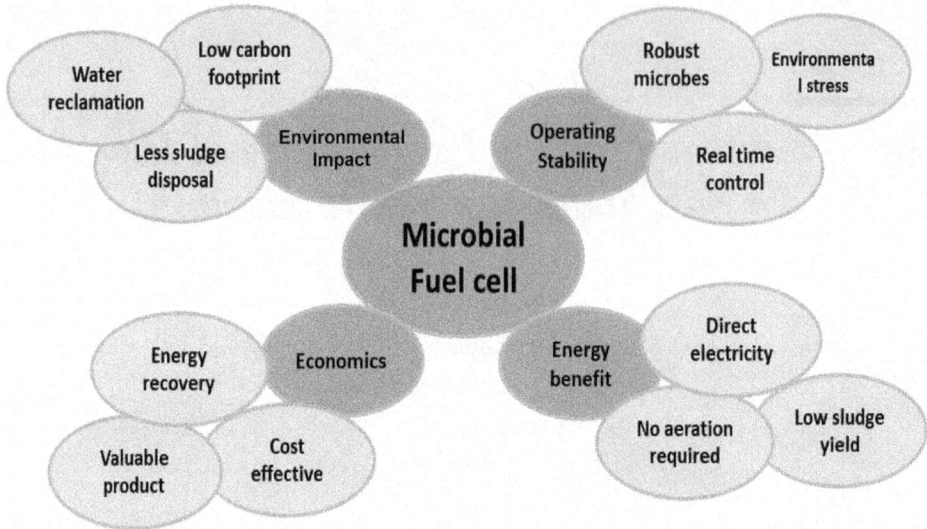

FIGURE 9.1 Potential benefits of microbial fuel cells compared to traditional wastewater treatment in the context of environment, energy, and sustainability.

In the past decade, MFC technology has been established as a new source of bioenergy (Kumar et al., 2019). Besides, the MFCs have been considered as a replacement for traditional anaerobic digester for treating wastewater (Jaiswal et al., 2020) since the wastewater treated by anaerobic treatment cannot meet the environmental wastewater standard (Cheng et al., 2010). Moreover, the anaerobic treatment is itself an energy-intensive technique to treat wastewater. The potential benefits of MFC compared to traditional anaerobic wastewater treatment in the context of energy and environment are illustrated in Figure 9.1. Several factors such as electrode materials, microbial community composition, inoculum (Pham et al., 2009), cathodic reaction performance (Rismani-Yazdi et al., 2008) have been widely studied including few recent developments in MFCs. However, a comprehensive study on the substrates, which can substantially influence the performance of MFCs, is still a bottleneck.

The substrates play a crucial role in any biochemical process especially for bioreactors as they are the only source of carbon or nutrient for the system (Kumar et al., 2019). The economic viability and efficiency of converting organic wastes, particularly wastewaters into bioenergy, substantially are governed by the nature and characteristics of the waste substrates or wastewaters (Wang et al., 2020). More importantly, the chemical structure and the components, which can be transformed into fuels or products, are of foremost interest for choosing substrates in the bioelectrochemical systems (BES), especially, for the MFCs. Notably, the composition of the substrate simultaneously influences the integral microbial community consortia in the anodic biofilm, together with the Coulombic efficiency (CE) and power density (PD) of the MFCs (Chae et al., 2009). In this chapter, we have discussed the major substrates that have been employed in the MFCs to date. Besides, the impacts of the substrates on the performance of MFCs have also been covered in this chapter.

9.2 Concept and Working Principles of Microbial Fuel Cells

An MFC can be defined as a bioelectrochemical device in which the chemical energy stored in organic components of wastewater can be converted into electricity through the metabolism of microorganisms (Logan et al., 2006; Pant et al., 2012; Baranitharan et al., 2015). The electro-active microbes that can oxidize substrates are the key components of MFC, thereby differentiating MFCs from the conventional fuel cells, especially chemical fuel cells (Logan and Regan, 2006; Lovley, 2006).

A typical double-chambered MFC is shown in Figure 9.2. In the anode of MFCs, several organic and inorganic substrates are metabolized by electrogenic microorganisms to produce charge or

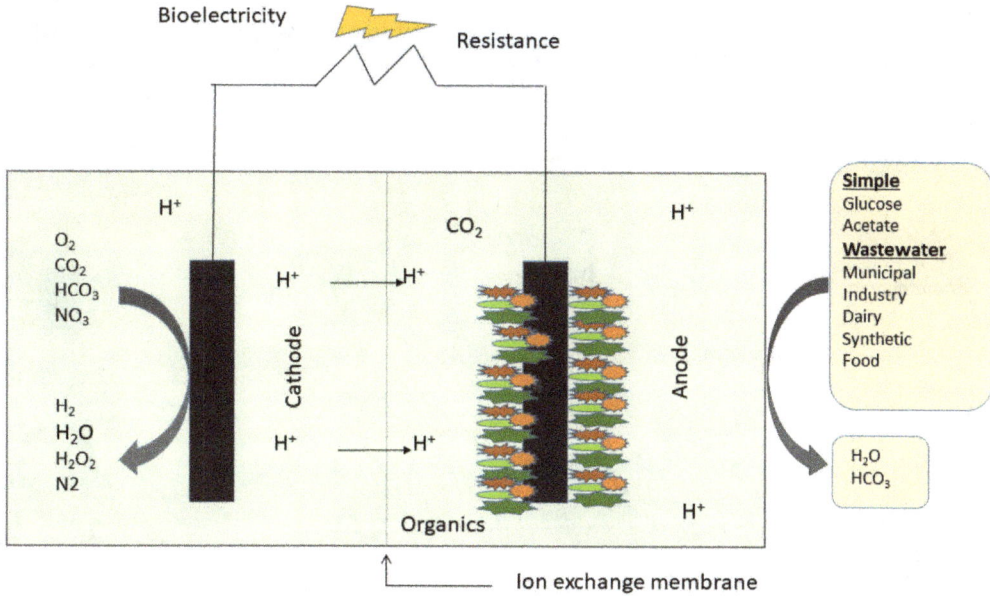

FIGURE 9.2 Schematic of the functional principle of a typical dual-chamber microbial fuel cell.

TABLE 9.1

Typical Electron Donors and Acceptors in a Microbial Fuel Cell

Electrode	Electron Donor/Acceptors	Reactions
Anode	Glucose	$C_6H_{12}O_6 + H_2O \rightarrow 6CO_2 + 24H^+ + 24e^-$
	Acetate	$C_2H_3O_2^- + 4H_2O \rightarrow 2HCO_3^- + 9H^+ + 8e^-$
	Glycerol	$C_3H_8O_3 + 6H_2O \rightarrow 3HCO_3^- + 17H^+ + 14e^-$
	Citrate	$C_4H_8O_2 + 2H_2O \rightarrow 2C_2H_4O_2 + 4H^+ + 4e^- C_6H_5O_73^- + 11H_2O \rightarrow 6H_2CO_3$ $+15H^+ + 18e^-$
	Malate	$C_4H_5O_5^- + 7H_2O \rightarrow 4H_2CO_3 + 11H^+ + 12e^-$
	Sulfide	$HS- \rightarrow S0 + H+ + 2e^-$
Cathode	Nitrate	$2NO_3^- + 12H^+ + 10e^- \rightarrow N_2 + 6H_2O$
	Oxygen	$+ 4e^- + 4H+ \rightarrow 2H_2O$
	Nitrite	$NO_2^- + 2e^- + 2H^+ \rightarrow N_2 + H_2O$
	Manganese dioxide	$MnO_2 + H^+ + e^- \rightarrow MnO + OH^-(s)$
	Permanganate	$MnO_4^- + 4 H^+ + 3e^- \rightarrow MnO_2 + 2H_2O$
	Iron (III)	$Fe_3^+ + e^- \rightarrow Fe_2^+$
	Potassium persulfate	$S_2O_2^- + 2e^- \rightarrow 2SO_4^{2-}$
	Copper (II)	$4 Cu_2^+ + 8e^- \rightarrow 4 Cu(s)$
	Ferricyanide	$Fe (CN)_6^{3-} + e^- \rightarrow Fe (CN)_6^{4-}$

electrons that reach the anode via an electron transport system (ETS) while the oxidation–reduction process occurs in the cathode of MFCs (Umar et al., 2020). The conventional electron acceptors and donors used in MFCs have been shown in Table 9.1. Generally, microorganisms can transfer electrons either indirectly from membrane to solid surface through an electron-shuttling mediator or directly by a membrane-bound cytochrome. It has been reported that microorganisms can directly transfer electrons through the conductive pilli (Reguera et al., 2006; Strycharz-Glaven et al., 2011; Renslow et al., 2013).

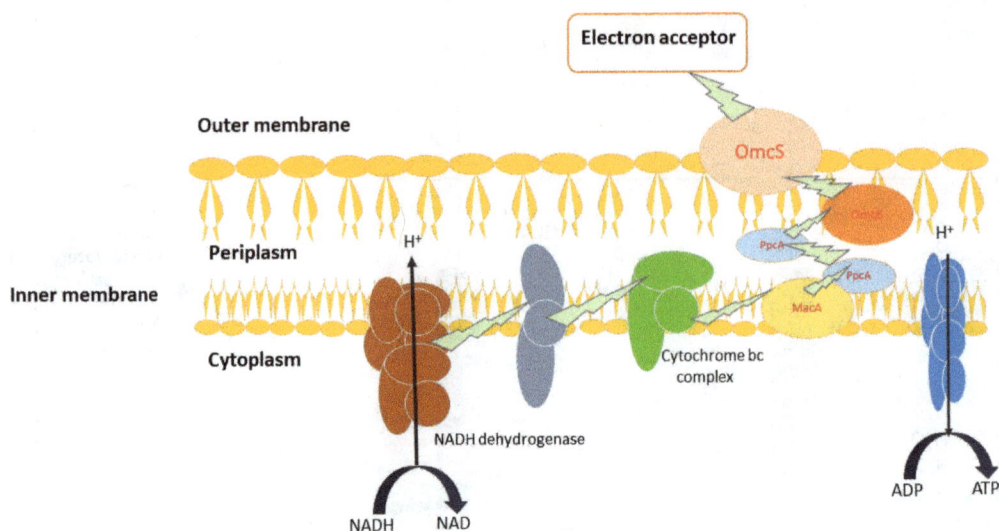

FIGURE 9.3 Protein involved in electron transfer mechanism of exoelectrogenic bacteria.

The understanding of microbial metabolism, particularly bacterial electron transfer mechanism, is a crucial part of enhancing the performance of MFCs. Electricity generation in the MFC technology substantially depends on the metabolic activity of electroactive microorganisms and the biocatalytic reaction rate in the anode of MFCs. This is because the organic substrates are oxidized by the electrogenic bacteria for generating electricity (Du et al., 2007).

The electrons are transferred from the donor microbes into the final electron acceptor. The whole mechanism is completed within a microbial cell utilizing several electron-conducting proteins and enzyme co-factors such as ubiquinone, cytochromes, NADH dehydrogenase as illustrated in Figure 9.3. The energy generated during the electron transfer mechanism of bacteria concomitantly pushes the proton to the cellular periplasms.

9.3 Substrates Used in the Microbial Fuel Cells

Substrates are considered as one of the crucial biological factors, which are associated with the amount of power produced in the MFCs (Liu et al., 2009). The organic substrates ranging from simple sugars (e.g., glucose, acetate, sucrose, etc.) to complex organics (e.g., amino acids, proteins and wastewater) and their mixtures can be used as possible nutrient sources for electroactive microorganisms for electricity generation (Singhania et al., 2013; Tong et al., 2013; Pandey et al., 2016).

Cheng and Logan (2007) reported that the acetate-fed MFC obtained the highest colonic efficiency of up to 98% along with a maximum power density of 115 W/m³ using mixed culture inoculum. Besides, several wastewaters such as palm oil mill effluent (POME) (Islam et al., 2017; Islam et al., 2020), dairy wastewater (Karim and Aider, 2020a, Marassi et al., 2020), petroleum refinery wastewater (Sevda and Abu-Reesh, 2019), and synthetic wastewater (Leung, 2020) have been used as complex substrates in the MFCs. In most cases, the intention behind the usage of wastewaters is to remove pollutants or micropollutants from wastewaters before discharging them into the environment.

9.3.1 Acetate

Acetate is widely used as a nutrient or carbon source in the anode of MFCs (Bond et al., 2002). As a simple substrate, electroactive microorganisms can easily metabolize acetate. Notably, acetate is the final product of numerous metabolic pathways (Biffinger et al., 2008). It has been reported that acetate-fed

MFCs obtained higher power generation compared to glucose-fed MFCs (Biffinger et al., 2008; Xing et al., 2009). Liu et al. (2005) observed that the acetate-fed single-chamber MFC obtained the power generation of 506 mW/m² which was approximately 66% higher than that of the butyrate-fed MFCs (305 mW/m²).

Chae et al. (2009) evaluated the power generation of four diverse substrates where the acetate-fed MFC obtained the maximum power generation with CE of 72.3% while the glucose-operated MFC achieved the lowest power generation with CE of 15%. In addition, it is reported that the acetate-fed MFC obtained a higher power generation even compared to the wastewater-fed MFCs. Liu et al. (2009) observed that the MFC fed by acetate substrate and acetate-acclimatized microbial consortia obtained two-times higher power generation in comparison of MFC operated with protein-enriched wastewater. However, a wide range of microbial community composition was observed in the anode biofilm for the protein-enriched wastewater compared to the acetate substrate.

9.3.2 Glucose

Glucose is also widely used as a substrate of MFCs. In most cases, pure culture inoculated MFCs used glucose as a substrate rather than mixed culture MFCs. Rabaey et al. (2003) operated MFCs fed with glucose and achieved the highest power production of 216 mW/m³ while ferric cyanide was used as an oxidizing agent. In another study, Kim et al. (2000) observed that the power generated by MFCs inoculated with *Proteus vulgaris* was substantially influenced by the carbon sources such as glucose and galactose.

In a different study, Hu (2008) compared the performance of MFCs operated with glucose and anaerobic sludge as fuel. According to Chae et al. (2009), the lowest CE was achieved by a glucose-fed MFC due to lower electron transfer efficiency between electroactive bacteria and electrode. The lower CE could be attributed to the fact that methanogenic and fermentative bacteria that are not associated with the production of electricity in the MFC system can easily consume glucose. Therefore, it is recommended to operate MFCs using a combination of simple substrates (i.e., glucose and acetate) and complex substrate.

9.3.3 Synthetic Wastewater

Several synthetic or chemical wastewaters with known composition have been used in MFCs as the conductivity, pH, and other parameters can easily be controlled. Mohan et al. (2008) operated MFCs fed with various loadings of synthetic wastewater to obtain an ideal loading rate. Aldrovandi et al. (2009) operated synthetic wastewater-fed MFCs with the supplementation of additional media, namely cysteine, for producing a significant amount of electron-shuttling mediators through the faster growth of bacteria, and in this way, they augmented power generation of MFCs within a very short period (Aldrovandi et al., 2009).

Rodrigo et al. (2009) employed MFCs using synthetic wastewater substrate for investigating the impact of wastewater components on the power generation of MFCs. Interestingly, they observed that the MFCs run with similar organic contaminants and identical organic loadings showed different performances. In addition, they also observed that MFCs with low biodegradable waste components were found to enhance electricity generation.

9.3.4 Brewery Wastewater

In general, brewery wastewater is nontoxic and contains high chemical oxygen demand (COD) in the presence of higher organic contents mainly consisting of protein and starch components (Guo et al., 2020). Among the wastewater substrates, the brewery wastewater is the most proffered substrate in MFCs by researchers due to its higher organic contents (Guo et al., 2020). It is considered as a suitable substrate for electricity production through the MFCs due to the presence of a lower amount of inhibitory substances (Feng et al., 2008; Wang et al., 2008).

Generally, brewery wastewater contains the COD ranging from 3,000 to 5,000 mg/L which is significantly (~10 times) higher than the other wastewaters, particularly domestic wastewater (Vijayaraghavan et al., 2006). It contains high concentration of carbohydrates and a lower concentration of ammonium–nitrogen; therefore, it could be considered as an efficient substrate for MFCs. Feng et al. (2008) obtained a maximum power density of 528 mW/m² from single-chambered air-cathode MFC while brewery wastewater was used as substrate.

9.3.5 Starch-Processing Wastewater

Starch-processing wastewater (SPW) is organic-rich wastewater that predominantly contains starch (~1500–2600 mg/L), sugars (~0.65–1.18%), protein (~0.12–0.15%), and carbohydrates (~2300–3500 mg/L) (Jin et al., 1998). It has been reported that the starch-processing wastewater could be used as a substrate for enriching an electroactive microbial community. Lu et al. (2009) conducted MFCs using starch processing wastewater and obtained a power density and maximum voltage of 239.4 mW/m² and 490.8 mV, respectively. Alternatively, 7 percent of CE was achieved which could be attributed to the low oxygen diffusion in the anode of MFCs.

9.3.6 Dye Wastewater

Around 7×10^5 tons of dye wastewater are produced per year, which predominantly contain recalcitrant organic molecules. The removal of these detrimental organics from dye wastewater is essential because they are associated with severe environmental problems (Pant et al., 2007). Generally, several pretreatment techniques such as physical, chemical, and electrochemical treatments are being implemented before discharging into the land (Parmar and Shukla, 2019; Karim et al., 2020). Recently, the MFC technique has been used to bioremediate dye wastewaters through the generation of electricity. Sun et al. (2009) used dye wastewater as a substrate of MFCs for removing contaminants from wastewater. In another study, Kalathil et al. (2012) operated granular-activated single-chambered MFCs for treating dye wastewater.

9.3.7 Landfill Leachates

Landfill leachates are highly contaminated effluents that predominantly contain dissolved inorganic macro-components, organic matter, xenobiotic organic compounds, and heavy metals (Kjeldsen et al., 2002). Habermann and Pommer (1991) treated landfill effluent using biofuel cell for removing contaminants from the landfill leachates wastewater. Zhang et al. (2008) designed a single-chambered air-cathode MFC for simultaneous generation of electricity and treatment of landfill leachates wastewater. Gálvez et al. (2009) conducted three different sets of MFCs, which were related to each other through the fluid flow for generating electricity and concurrent leachate treatment.

9.3.8 Palm Oil Mill Effluent

Palm oil mill effluent (POME) is one of the agricultural-based industrial wastewaters that are produced from the palm oil seeds. POME is a viscous brown liquid or generally yellowish wastewater. The POME is produced during the extraction of palm oil seeds, which are considered as one of the predominant and high-yielding crops in some tropical countries such as Thailand, Malaysia, and Indonesia (Karim et al., 2019). The production of 1 ton of crude palm oil usually results in the generation of around 2.5 tons of POME. The fresh POME is hot (~80–90°C) and possesses a very high content of COD in the range of approximately 50,000–100,000 mg/L and biochemical oxygen demand (BOD) in the range of approximately 25,000–54,000 mg/L), which is 100 times more polluting than domestic sewage (Islam et al., 2018b; Karim et al., 2019).

The direct discharge of POME in the environment can result in environmental hazards due to its organic pollutants. Therefore, the palm oil industries face vast challenges for complying with environmental standards due to the huge demand for crude palm oil all over the world (Cheng et al., 2009).

In this text, MFCs can be considered as a sustainable technology for generating energy from POME through its bioremediation (Logan and Rabaey, 2012; Ivanov et al., 2013). Islam et al. (2018a) operated POME-fed MFCs inoculated with *Klebsiella variicola* and *Pseudomonas aeruginosa* co-culture where they obtained maximum power generation of 14.8 W/m^3. In their separate study (Islam et al., 2020), the POME-fed MFC achieved a maximum power density of 11.8 W/m^3 while *Klebsiella variicola* and *Bacillus cereus* were used as inoculum. In a different study, Nor et al. (2015) achieved the maximum power density of 85.1 mW/m^2 from POME-mediated MFCs while *Pseudomonas aeruginosa* was used as inoculum. The result of these studies suggests that the POME could be effectively utilized as substrate of MFCs for bioelectricity generation.

9.3.9 Lignocellulosic Biomass

The lignocellulosic biomass (LCB) is abundantly available in nature which is mainly generated from agricultural waste, and it is considered a promising and cost-effective feedstock for energy production (Huang et al., 2008). However, most of the electroactive microorganisms cannot directly metabolize lignocellulosic biomass due to their complex structure (Liang et al., 2020). Hence, the LCB needs to be transformed into simple sugars such as monosaccharides or further low-weight sugars to be easily catalyzed by bacteria (Ren et al., 2007). It has been reported that the combination of exoelectrogenic and cellulolytic bacteria could achieve a significant amount of power generation in LCB-fed MFCs (Rezaei et al., 2009b).

Zuo et al. (2006) converted corn stover into soluble sugars via acid or steam-exploded hydrolysis and subsequently operated MFCs with such simple sugars. Wang et al. (2009b) used corn stover substrate as a source of nutrient for power production using a single-chamber MFC and observed lower power generation compared to the glucose. It is important to note that microorganisms cannot directly ferment the pentoses which are one of the major constituents of lignocellulose hydrolysates. Huang and Angelidaki (2008) have reported that the xylose (typical pentose)-fed MFC obtained the maximum power generation of 69 mW/m^2, which was relatively lower than the glucose-fed MFC (97 mW/m^2). These results suggested that xylose is harder to metabolize compared to the glucose.

9.3.10 Cellulose and Chitin

Cellulose and chitin are abundantly available and cheap bio-polymeric materials that can be utilized for electricity production. The cellulose and chitin are considered renewable substrates and are mainly found in municipal and industrial wastewaters as organic matter (Rezaei et al., 2009a). Very few studies have reported the uses of cellulose and chitin directly as substrates in the MFCs. In most cases, the cellulose and chitin have been hydrolyzed before using them as a substrate in the MFCs. Rismani-Yazdi et al. (2007) conducted cellulose-fed MFCs and obtained a maximum power generation of 55 mW/m^2. In another study, Ren et al. (2008) obtained the maximum power of 153 mW/m^2 while carboxymethyl cellulose was used as a nutrient source in MFCs. Rezaei et al. (2009a) used different sizes of chitin particles for generating electricity using MFCs and achieved the maximum power generation of 176 mW/m^2 from largest particles (0.78 mm). Thus, these substrates can be considered as nutrient sources of electrogenic microorganisms in bioelectrochemical systems.

A comprehensive review of substrates used in MFC technology is presented in Table 9.2. We note that it is difficult to compare the performance of MFCs as reported in the literature, owing to different surface area, operating conditions, electrodes, and microorganisms. Additionally, several researchers have used different units to demonstrate the performances.

9.4 Current Challenges of Microbial Fuel Cells

In the past decade, significant research efforts have been given to improve power generation in MFCs to make the system practically applicable to at least run small devices. However, scaling up is still challenging due to several limitations such as the cost of ion exchange membranes and the bio-fouling of membrane and electrodes, which substantially increase the resistance of the system that limits the power

TABLE 9.2

An Overview of Various Types of Materials Used as Substrates in the Microbial Fuel Cell

Substrate Category	Type of Substrate	Concentration of Substrate	Microbes Used as Inoculum	Working Volume (mL)	Anode	Maximum Open-Circuit Voltage (V)	Coulombic Efficiency (%)	COD Removal (%)	Maximum Power Density (mW/m²)	Reference
Carbohydrates	D-xylose	0.008 M	Mixed culture	12	Carbon-cloth	0.38	31	95	2330	Catal et al. (2008a)
	D-glucose	0.0067 M	Mixed bacterial culture	12	Carbon-cloth	0.39	28	93	2160	Catal et al. (2008a)
	D-fructose	0.0067 M	Mixed culture of bacteria	12	Carbon-cloth	0.31	23	88	1810	Catal et al. (2008a)
	Sucrose	0.1 g/L	Anaerobic sludge	–	Carbon-fiber veil	–	4	94	1.79 W/m³	Beecroft et al. (2012)
Sugar acids	D-glucuronic acid	6.7 mM	Mixed bacterial culture	12	Carbon-cloth	0.44	24	89	2770	Catal et al. (2008a)
	D-gluconic acid	0.0067 M	Mixed bacterial culture	12	Carbon-cloth	0.28	30	93	2050	Catal et al. (2008a)
Polyalcohols	Galactitol	0.0067 M	Mixed culture of bacteria	12	Carbon-cloth	0.34	13	90	2650	Catal et al. (2008b)
	Arabitol	0.008 M	Mixed culture	12	Carbon-cloth	0.26	25	91	2030	Catal et al. (2008b)
	Xylitol	0.008 M	Mixed culture	12	Carbon-cloth	0.29	21	91	2110	Catal et al. (2008b)
Amino acids	L-serine	0.002 M	Domestic wastewater	28	Carbon-cloth	0.37	20	93	768	Yang et al. (2012)
	L-aspartic acid	0.0015 M	Domestic wastewater	28	Carbon-cloth	0.25	25	94	601	Yang et al. (2012)
	L-glutamic acid	0.0012 M	Domestic wastewater	28	Carbon-cloth	0.33	~28	95	686	Yang et al. (2012)
	L-histidine	0.001 M	Domestic wastewater	28	Carbon-cloth	0.42	25	93	718	Yang et al. (2012)
Organic acids	Lactic acid	1 g/L	Domestic wastewater	28	Graphite-fiber brushes	0.4	13.4	–	739	Kiely et al. (2011)
	Acetic acid	1 g/L	Domestic wastewater	28	Graphite fiber brushes	0.48	19.9	–	835	Kiely et al. (2011)
	Acetate	0.8 g COD/L	Domestic wastewater	28	Toray carbon-paper	0.79	7	>99	506	Liu et al. (2005)
	Formate	0.002 M	Fluid of anaerobic digester	5	Graphite	–	6.5	83	–	Ha et al. (2008)

(Continued)

TABLE 9.2 (Continued)

Substrate Category	Type of Substrate	Concentration of Substrate	Microbes Used as Inoculum	Working Volume (mL)	Anode	Maximum Open-Circuit Voltage (V)	Coulombic Efficiency (%)	COD Removal (%)	Maximum Power Density (mW/m²)	Reference
	Propionate	0.005 M	Anaerobic sludge	–	Graphite	0.8	31.5	–	115.6	De Cárcer et al. (2011)
	Lactate	0.002 M	Anaerobic cultures with soil	25	Graphite	0.75	12.5	–	320	Futamata et al. (2013)
Alcohols	Ethanol	0.0015 M	Anaerobic sludge	28	Toray carbon paper	0.75	10	–	488	Kim et al. (2007)
	Glycerol	0.50 mL/L	*Bacillus subtilis*	500	CarbonCloth	0.56	23.1	–	600	Nimje et al. (2011)
Wastewaters	Dairy industrial wastewater	53.22 kg COD/ m³/d	Activated sludge	2	Graphite-plate	0.86	37.2	90.5	621.1	Mansoorian et al. (2016)
	Palm oil mill effluent sludge	2.68 g COD/L	*Pseudomonas aeruginosa* ZH1	100	Carbon-graphite	–	–	3	451.3	Nor et al. (2015)
	Petroleum refinery wastewater	0.25 g COD/L	Activated sludge	400	Carbon-rod	0.31	–	64	330.4 mW/cm³	Guo et al. (2016)
	Rice straw	1.00 g/L	Mixed culture bacteria	160	Carbon-paper	0.71	37	–	190	Gurung and Oh (2015)
	Red wine lees wastewater	10.10 g COD/L	Denitrification tank wastewater	28	Graphite-brush	0.34	9	27	111	Sciarria et al. (2015)
	Retting wastewater (Coconut husk)	2.69 g COD/L/d	Retting wastewater	600	Graphite-sheet	0.88	0.8–8	32	254	Jayashree et al. (2015)
	Toxic refractory organic pesticide, hexachlorobenzene	0.040 g/kg	Anaerobic sludge with soil	140	GAC	0.33	–	71.2	77.5	Cao et al. (2015)
	White wine less wastewater	6.40 g COD/L	Denitrification Reservoir wastewater	28	Graphite-Brushes	0.42	15	90	262	Sciarria et al. (2015)
Miscellaneous	Pyridine	0.12 g COD/L	Anaerobic sludge	250	Carbon-paper	0.52	<8	86	142.1	Hu et al. (2011)
	Phenol	600 mL/L	Anaerobic sludge	330	Carbon-felt	0.63	3.7	88.9	31.3	Song et al. (2014)
	Furfural	0.0067 M	Aerobic and anaerobic sludge	–	Carbon-cloth	0.42	30.3	68	361	Sulonen et al. (2015)

Abbreviations: PTFE, polytetrafluoroethylene; Pt, Platinum; GAC, granular activated carbon; COD, chemical oxygen demand.

generation as well as practical applicability of MFCs (Hu, 2008; Baranitharan et al., 2015). Nevertheless, the treatment of wastewaters through MFCs is considered a promising technology compared to the conventional aerobic and anaerobic digestion techniques. The wastewater, especially domestic wastewater, which contains a high amount of organic matter requires at least 10 times higher energy as given by the conventional techniques to treat it (Aiken et al., 2019). In this context, the MFC technique can successfully treat such wastewater by generating significant electrical energies. However, the start-up time is considered as a major drawback of MFCs which can differ from 4 to 180 days due to the influence of inoculum compositions, electrode materials, substrate compositions, operating conditions, reactor design, etc. (Wang et al., 2009a). The survivability of microorganisms in wastewater-fed MFCs is another big concern for generating electricity and bioremediating harsh wastewater through this technology (You et al., 2009).

Several initiatives can be taken to overcome the current limitations in MFC technology. The research community has realized that the power generation from MFC is still quite low for targeted applications (Lovley, 2008; Borole et al., 2011). Moreover, the cost of materials particularly cathode catalyst, which is used as an oxidizing agent in the cathode of single-chamber MFC, is another concern for upscaling the system. In this context, the open-air bio-cathode could be an alternative solution to make it cost-effective and sustainable (Clauwaert et al., 2007; Shehab et al., 2013). Some studies have shown that the high-cost catalysts, especially platinum, could be replaced by other low-cost catalysts by achieving a similar performance (Zhang et al., 2009a; Ghasemi et al., 2013).

Zhang et al. (2009b) used MnO_2 as a catalyst in the cathode of MFCs while Selembo et al. (2009) used nickel alloys and stainless steel as a substitute for platinum. Besides, the substrate is another crucial factor for enhancing the power generation of MFCs. In this paper, we have covered several substrates that have been widely used in MFCs as nutrients or carbon sources. However, there are some other possible substrates, which can enhance the performance of the system. For instance, the vegetable and fruit wastewater (Damian, 2019), whey mill wastewater (Karim and Aider, 2020b), livestock industry wastewater (Bolognesi et al., 2020), and so forth are also considered as potential substrates in the MFC. In addition, the effluent of distilleries-based cane-molasses enriched with organic matters (Pant and Adholeya, 2007) can also be regarded as an efficient substrate in MFCs.

Several studies have shown that the recalcitrant compounds could also be treated by the bio-anode of MFCs. Notably, while contaminants, especially wastewaters, are used to serve as oxidizing agents or electron acceptors in the cathode of MFC, the benefit in terms of the environmental perspective of MFCs would be significantly enhanced. Few studies have shown that microorganisms can denitrify wastewater substrate by supplementing electrons in the cathode of MFCs (Jia et al., 2008; Virdis et al., 2010).

Another interesting finding has shown that the value-added platform chemicals such as polyhydroxybutyrate, polyhydroxyalkanoates, and 1, 3 propanediol could be produced by the MFC systems. Zhu and Ni (2009) produced H_2O_2 using carbon felt in the cathode of MFC. In another study, Di Lorenzo et al. (2009) developed a biosensor from monitoring BOD of artificial wastewater by utilizing single-chambered air cathode MFC, and they observed a relationship between current output and the concentration of COD.

9.5   Conclusions

We have summarized the predominant substrates used in MFCs for treatment and power generation purposes. However, several new substrates, particularly lesser used substrates, in MFCs have not been discussed in this chapter. In earlier studies, simple substrates such as glucose and acetate were widely used; however, in recent years, some unconventional substrates have been used as anode nutrients for utilizing waste biomass along with their treatment. The electricity generation from renewable waste biomass by utilizing the MFC technique has been regarded as a great development compared to the traditional nonrenewable biofuels. It is expected that the MFC technology would be designed for accommodating a wide range of substrates to make it a source of sustainable bioenergy. The major conclusions that can be drawn are:

(i) The use of complex substrates, particularly wastewaters, creates complexity in MFCs due to their high organic load and inhibiting agents. Hence, it is required to establish an electrochemically active and diverse microbial community consortium to effectively utilize the wastewater substrates and improve system performances.

(ii) The power output of the systems in both cases of electric power and an electric current is still very low and far from practical applications. Hence, more technological advancement, especially designing low-cost materials is required.

(iii) The lower rate of substrate conversion is another concern for the system. In this context, a suitable mathematical and statistical model can be designed to maximize the substrate conversion efficiency in the anode of MFCs.

(iv) The incorporation of MFCs with conventional wastewater treatment techniques could be the best feasible option for the industry.

REFERENCES

Aiken, D. C., Curtis, T. P., and Heidrich, E. S. 2019. Avenues to the financial viability of microbial electrolysis cells [MEC] for domestic wastewater treatment and hydrogen production. *International Journal of Hydrogen Energy* 44(5): 2426–2434.

Aldrovandi, A., Marsili, E., Stante, L., Paganin, P., Tabacchioni, S., and Giordano, A. 2009. Sustainable power production in a membrane-less and mediator-less synthetic wastewater microbial fuel cell. *Bioresource Technology* 100(13): 3252–3260.

Baranitharan, E., Khan, M. R., Prasad, D., Teo, W. F. A., Tan, G. Y. A., and Jose, R. 2015. Effect of biofilm formation on the performance of microbial fuel cell for the treatment of palm oil mill effluent. *Bioprocess and Biosystems Engineering* 38(1): 15–24.

Beecroft, N. J., Zhao, F., Varcoe, J. R., Slade, R. C., Thumser, A. E., and Avignone-Rossa, C. 2012. Dynamic changes in the microbial community composition in microbial fuel cells fed with sucrose. *Applied Microbiology and Biotechnology* 93(1): 423–437.

Biffinger, J. C., Byrd, J. N., Dudley, B. L., and Ringeisen, B. R. 2008. Oxygen exposure promotes fuel diversity for *Shewanella oneidensis* microbial fuel cells. *Biosensors and Bioelectronics* 23(6): 820–826.

Bolognesi, S., Cecconet, D., and Capodaglio, A. G. 2020. Agro-industrial wastewater treatment in microbial fuel cells. In: *Integrated Microbial Fuel Cells for Wastewater Treatment*. Elsevier, pp. 93–133.

Bond, D. R., Holmes, D. E., Tender, L. M., and Lovley, D. R. 2002. Electrode-reducing microorganisms that harvest energy from marine sediments. *Science* 295(5554): 483–485.

Borole, A. P., Reguera, G., Ringeisen, B., Wang, Z.-W., Feng, Y., and Kim, B. H. 2011. Electroactive biofilms: Current status and future research needs. *Energy & Environmental Science* 4(12): 4813–4834.

Cao, X., Song, H.-l., Yu, C.-y., and Li, X.-n. 2015. Simultaneous degradation of toxic refractory organic pesticide and bioelectricity generation using a soil microbial fuel cell. *Bioresource Technology* 189: 87–93.

Catal, T., Li, K., Bermek, H., and Liu, H. 2008a. Electricity production from twelve monosaccharides using microbial fuel cells. *Journal of Power Sources* 175(1): 196–200.

Catal, T., Xu, S., Li, K., Bermek, H., and Liu, H. 2008b. Electricity generation from polyalcohols in single-chamber microbial fuel cells. *Biosensors and Bioelectronics* 24(4): 849–854.

Chae, K.-J., Choi, M.-J., Lee, J.-W., Kim, K.-Y., and Kim, I. S. 2009. Effect of different substrates on the performance, bacterial diversity, and bacterial viability in microbial fuel cells. *Bioresource Technology* 100(14): 3518–3525.

Cheng, J., Zhu, X., Ni, J., and Borthwick, A. 2010. Palm oil mill effluent treatment using a two-stage microbial fuel cells system integrated with immobilized biological aerated filters. *Bioresource Technology* 101(8): 2729–2734.

Cheng, S., and Logan, B. E. 2007. Ammonia treatment of carbon cloth anodes to enhance power generation of microbial fuel cells. *Electrochemistry Communications* 9(3): 492–496.

Cheng, S., Xing, D., Call, D. F., and Logan, B. E. 2009. Direct biological conversion of electrical current into methane by electromethanogenesis. *Environmental Science & Technology* 43(10): 3953–3958.

Clauwaert, P., Van der Ha, D., Boon, N., Verbeken, K., Verhaege, M., Rabaey, K., and Verstraete, W. 2007. Open air biocathode enables effective electricity generation with microbial fuel cells. *Environmental Science & Technology* 41(21): 7564–7569.

Damian, C. 2019. Assessment of waste and wastewater treatment from fruits and vegetables processing industry in *Romania. International Multidisciplinary Scientific GeoConference: SGEM* 19(6.3): 119–124.

De Cárcer, D. A., Ha, P. T., Jang, J. K., and Chang, I. S. 2011. Microbial community differences between propionate-fed microbial fuel cell systems under open and closed circuit conditions. *Applied Microbiology and Biotechnology* 89(3): 605–612.

Di Lorenzo, M., Curtis, T. P., Head, I. M., and Scott, K. 2009. A single-chamber microbial fuel cell as a biosensor for wastewaters. *Water Research* 43(13): 3145–3154.

Du, Z., Li, H., and Gu, T. 2007. A state of the art review on microbial fuel cells: A promising technology for wastewater treatment and bioenergy. *Biotechnology Advances* 25(5): 464–482.

Feng, Y., Wang, X., Logan, B. E., and Lee. H. 2008. Brewery wastewater treatment using air-cathode microbial fuel cells. *Applied Microbiology and Biotechnology* 78(5): 873–880.

Futamata, H., Bretschger, O., Cheung, A., Kan, J., Owen, R., and Nealson. K. H. 2013. Adaptation of soil microbes during establishment of microbial fuel cell consortium fed with lactate. *Journal of Bioscience and Bioengineering* 115(1): 58–63.

Gálvez, A., Greenman, J., and Ieropoulos. I. 2009. Landfill leachate treatment with microbial fuel cells; scale-up through plurality. *Bioresource Technology* 100(21): 5085–5091.

Ghasemi, M., Daud, W. R. W., Rahimnejad, M., Rezayi, M., Fatemi, A., Jafari, Y., Somalu, M., and Manzour. A. 2013. Copper-phthalocyanine and nickel nanoparticles as novel cathode catalysts in microbial fuel cells. *International Journal of Hydrogen Energy* 38(22): 9533–9540.

Guo, F., Luo, H., Shi, Z., Wu, Y., and Liu, H. 2020. Substrate salinity: A critical factor regulating the performance of microbial fuel cells, a review. *Science of The Total Environment* 143021.

Guo, X., Zhan, Y., Chen, C., Cai, B., Wang, Y., and Guo, S. 2016. Influence of packing material characteristics on the performance of microbial fuel cells using petroleum refinery wastewater as fuel. *Renewable Energy* 87: 437–444.

Gurung, A., and Oh, S.-E. 2015. Rice straw as a potential biomass for generation of bioelectrical energy using microbial fuel cells (MFCs). *Energy Sources, Part A: Recovery, Utilization, and Environmental Effects* 37(24): 2625–2631.

Ha, P. T., Tae, B., and Chang, I. S. 2008. Performance and bacterial consortium of microbial fuel cell fed with formate. *Energy & Fuels* 22(1): 164–168.

Habermann, W., and Pommer, E. 1991. Biological fuel cells with sulphide storage capacity. *Applied Microbiology and Biotechnology* 35(1): 128–133.

Hu, W.-J., Niu, C.-G., Wang, Y., Zeng, G.-M., and Wu, Z. 2011. Nitrogenous heterocyclic compounds degradation in the microbial fuel cells. *Process Safety and Environmental Protection* 89(2): 133–140.

Hu, Z. 2008. Electricity generation by a baffle-chamber membraneless microbial fuel cell. *Journal of Power Sources* 179(1): 27–33.

Huang, L., and Angelidaki, I. 2008. Effect of humic acids on electricity generation integrated with xylose degradation in microbial fuel cells. *Biotechnology and Bioengineering* 100(3): 413–422.

Huang, L., Zeng, R. J., and Angelidaki, I. 2008. Electricity production from xylose using a mediator-less microbial fuel cell. *Bioresource Technology* 99(10): 4178–4184.

Islam, M. A., Ethiraj, B., Cheng, C. K., Yousuf, A., and Khan, M. M. R. 2018a. An insight of synergy between *Pseudomonas aeruginosa* and *Klebsiella variicola* in a microbial fuel cell. *ACS Sustainable Chemistry & Engineering* 6(3): 4130–4137.

Islam, M. A., Karim, A., Mishra, P., Dubowski, J. J., Yousuf, A., Sarmin, S., and Khan, M. M. R. 2020. Microbial synergistic interactions enhanced power generation in co-culture driven microbial fuel cell. *Science of The Total Environment* 140138.

Islam, M. A., Karim, A., Woon, C. W., Ethiraj, B., Cheng, C. K., Yousuf, A., and Khan, M. M. R. 2017. Augmentation of air cathode microbial fuel cell performance using wild type *Klebsiella variicola*. *RSC Advances* 7(8): 4798–4805.

Islam, M. A., Yousuf, A., Karim, A., Pirozzi, D., Khan, M. R., and Ab Wahid, Z. 2018b. Bioremediation of palm oil mill effluent and lipid production by *Lipomyces starkeyi*: A combined approach. *Journal of Cleaner Production* 172: 1779–1787.

Ivanov, I., Ren, L., Siegert, M., and Logan, B. E. 2013. A quantitative method to evaluate microbial electrolysis cell effectiveness for energy recovery and wastewater treatment. *International Journal of Hydrogen Energy* 38(30): 13135–13142.

Jaiswal, K. K., Kumar, V., Vlaskin, M., Sharma, N., Rautela, I., Nanda, M., Arora, N., Singh, A., and Chauhan, P. 2020. Microalgae fuel cell for wastewater treatment: Recent advances and challenges. *Journal of Water Process Engineering* 38: 101549.

Jayashree, C., Sweta, S., Arulazhagan, P., Yeom, I., Iqbal, M., and Banu, J. R. 2015. Electricity generation from retting wastewater consisting of recalcitrant compounds using continuous upflow microbial fuel cell. *Biotechnology and Bioprocess Engineering* 20(4): 753–759.

Jia, Y.-H., H.-Tran, T., Kim, D.-H., Oh, S.-J., Park, D.-H., Zhang, R.-H., and Ahn, D.-H. 2008. Simultaneous organics removal and bio-electrochemical denitrification in microbial fuel cells. *Bioprocess and Biosystems Engineering* 31(4): 315–321.

Jin, B., Van Leeuwen, H., Patel, B., and Yu, Q. 1998. Utilisation of starch processing wastewater for production of microbial biomass protein and fungal α-amylase by *Aspergillus oryzae. Bioresource Technology* 66(3): 201–206.

Kalathil, S., Lee, J., and Cho, M. H. 2012. Efficient decolorization of real dye wastewater and bioelectricity generation using a novel single chamber biocathode-microbial fuel cell. *Bioresource Technology* 119: 22–27.

Karim, A., and Aider, M. 2020a. Contribution to the process development for lactulose production through complete valorization of whey permeate by using electro-activation technology versus a chemical isomerization process. *ACS Omega* 5: 28831–28843.

Karim, A., and Aider, M. 2020b. Sustainable valorization of whey by electroactivation technology for in situ isomerization of lactose into lactulose: Comparison between electroactivation and chemical processes at equivalent solution alkalinity. *ACS Omega* 5(14): 8380–8392.

Karim, A., Gerliani, N., and Aïder, M. 2020. *Kluyveromyces marxianus*: An emerging yeast cell factory for applications in food and biotechnology. *International Journal of Food Microbiology* 108818.

Karim, A., Islam, M. A., Yousuf, A., Khan, M. M. R., and Faizal, C. K. M. 2019. Microbial lipid accumulation through bioremediation of palm oil mill wastewater by *Bacillus cereus. ACS Sustainable Chemistry & Engineering* 7(17): 14500–14508.

Kiely, P. D., Rader, G., Regan, J. M., and Logan, B. E. 2011. Long-term cathode performance and the microbial communities that develop in microbial fuel cells fed different fermentation endproducts. *Bioresource Technology* 102(1): 361–366.

Kim, J. R., Jung, S. H., Regan, J. M., and Logan, B. E. 2007. Electricity generation and microbial community analysis of alcohol powered microbial fuel cells. *Bioresource Technology* 98(13): 2568–2577.

Kim, N., Choi, Y., Jung, S., and Kim, S. 2000. Effect of initial carbon sources on the performance of microbial fuel cells containing *Proteus vulgaris. Biotechnology and Bioengineering* 70(1): 109–114.

Kjeldsen, P., Barlaz, M. A., Rooker, A. P., Baun, A., Ledin, A., and Christensen, T. H. 2002. Present and long-term composition of MSW landfill leachate: A review. *Critical reviews in Environmental Science and Technology* 32(4): 297–336.

Kumar, S. S., Kumar, V., Kumar, R., Malyan, S. K., and Pugazhendhi, A. 2019. Microbial fuel cells as a sustainable platform technology for bioenergy, biosensing, environmental monitoring, and other low power device applications. *Fuel* 255: 115682.

Leung, D. H. L. 2020. *Potential of Mixed Consortium of Enterobacteriaceae and Serratia marcescens in Synthetic Wastewater Treatment and Power Generation in Microbial Fuel Cell.* University of Nottingham. http://eprints.nottingham.ac.uk/59824.

Liang, J., Nabi, M., Zhang, P., Zhang, G., Cai, Y., Wang, Q., Zhou, Z., and Ding, Y. 2020. Promising biological conversion of lignocellulosic biomass to renewable energy with rumen microorganisms: A comprehensive review. *Renewable and Sustainable Energy Reviews* 134: 110335.

Liu, H., Cheng, S., and Logan, B. E. 2005. Production of electricity from acetate or butyrate using a single-chamber microbial fuel cell. *Environmental Science & Technology* 39(2): 658–662.

Liu, Z., Liu, J., Zhang, S., and Su, Z. 2009. Study of operational performance and electrical response on mediator-less microbial fuel cells fed with carbon-and protein-rich substrates. *Biochemical Engineering Journal* 45(3): 185–191.

Logan, B. E., Hamelers, B., Rozendal, R., Schröder, U., Keller, J., Freguia, S., Aelterman, P., Verstraete, W., and Rabaey, K. 2006. Microbial fuel cells: Methodology and technology. *Environmental Science & Technology* 40(17): 5181–5192.

Logan, B. E., and Rabaey, K. 2012. Conversion of wastes into bioelectricity and chemicals by using microbial electrochemical technologies. *Science* 337(6095): 686–690.

Logan, B. E., and Regan, J. M. 2006. Microbial fuel cells-challenges and applications. *Environmental Science & Technology* 40(17): 5172–5180.

Lovley, D. R. 2006. Bug juice: Harvesting electricity with microorganisms. *Nature Reviews Microbiology* 4(7): 497–508.

Lovley, D. R. 2008. The microbe electric: Conversion of organic matter to electricity. *Current opinion in Biotechnology* 19(6): 564–571.

Lu, N., S.-g. Zhou, Zhuang, L., Zhang, J.-t., and Ni, J.-r. 2009. Electricity generation from starch processing wastewater using microbial fuel cell technology. *Biochemical Engineering Journal* 43(3): 246–251.

Mansoorian, H. J., Mahvi, A. H., Jafari, A. J., and Khanjani, N. 2016. Evaluation of dairy industry wastewater treatment and simultaneous bioelectricity generation in a catalyst-less and mediator-less membrane microbial fuel cell. *Journal of Saudi Chemical Society* 20(1): 88–100.

Marassi, R. J., Queiroz, L. G., Silva, D. C., da Silva, F. T., Silva, G. C., and de Paiva, T. C. B. 2020. Performance and toxicity assessment of an up-flow tubular microbial fuel cell during long-term operation with high-strength dairy wastewater. *Journal of Cleaner Production* 120882.

Mohan, S. V., Mohanakrishna, G., Reddy, B. P., Saravanan, R., and Sarma, P. 2008. Bioelectricity generation from chemical wastewater treatment in mediatorless (anode) microbial fuel cell (MFC) using selectively enriched hydrogen producing mixed culture under acidophilic microenvironment. *Biochemical Engineering Journal* 39(1): 121–130.

Nimje, V. R., Chen, C.-Y., Chen, C.-C., Chen, H.-R., Tseng, M.-J., Jean, J.-S., and Chang, Y.-F. 2011. Glycerol degradation in single-chamber microbial fuel cells. *Bioresource Technology* 102(3): 2629–2634.

Nor, M. H. M., Mubarak, M. F. M., Elmi, H. S. A., Ibrahim, N., Wahab, M. F. A., and Ibrahim, Z. 2015. Bioelectricity generation in microbial fuel cell using natural microflora and isolated pure culture bacteria from anaerobic palm oil mill effluent sludge. *Bioresource Technology* 190: 458–465.

Pandey, P., Shinde, V. N., Deopurkar, R. L., Kale, S. P., Patil, S. A., and Pant, D. 2016. Recent advances in the use of different substrates in microbial fuel cells toward wastewater treatment and simultaneous energy recovery. *Applied Energy* 168: 706–723.

Pant, D., and Adholeya, A. 2007. Biological approaches for treatment of distillery wastewater: A review. *Bioresource Technology* 98(12): 2321–2334.

Pant, D., Singh, A., Satyawali, Y., and Gupta, R. 2007. Effect of carbon and nitrogen source amendment on synthetic dyes decolourizing efficiency of white-rot fungus, *Phanerochaete chrysosporium*. *Journal of Environmental Biology* 29(1): 79.

Pant, D., Singh, A., Van Bogaert, G., Olsen, S. I., Nigam, P. S., Diels, L., and Vanbroekhoven, K. 2012. Bioelectrochemical systems (BES) for sustainable energy production and product recovery from organic wastes and industrial wastewaters. *RSC Advances* 2(4): 1248–1263.

Parmar, N. D., and Shukla, S. R. 2019. Decolourization of dye wastewater by microbial methods-A review. *Indian Journal of Chemical Technology* 25(4): 315–323.

Pham, T. H., Aelterman, P., and Verstraete, W. 2009. Bioanode performance in bioelectrochemical systems: Recent improvements and prospects. *Trends in Biotechnology* 27(3): 168–178.

Rabaey, K., Lissens, G., Siciliano, S. D., and Verstraete, W. 2003. A microbial fuel cell capable of converting glucose to electricity at high rate and efficiency. *Biotechnology Letters* 25(18): 1531–1535.

Rabaey, K., Rodríguez, J., Blackall, L. L., Keller, J., Gross, P., Batstone, D., Verstraete, W., and Nealson, K. H. 2007. Microbial ecology meets electrochemistry: Electricity-driven and driving communities. *The ISME Journal* 1(1): 9–18.

Reguera, G., Nevin, K. P., Nicoll, J. S., Covalla, S. F., Woodard, T. L., and Lovley, D. R. 2006. Biofilm and nanowire production leads to increased current in Geobacter sulfurreducens fuel cells. *Applied and Environmental Microbiology* 72(11): 7345–7348.

Ren, Z., Steinberg, L., and Regan, J. 2008. Electricity production and microbial biofilm characterization in cellulose-fed microbial fuel cells. *Water Science and Technology* 58(3): 617–622.

Ren, Z., Ward, T. E., and Regan, J. M. 2007. Electricity production from cellulose in a microbial fuel cell using a defined binary culture. *Environmental Science & Technology* 41(13): 4781–4786.

Renslow, R., Babauta, J., Kuprat, A., Schenk, J., Ivory, C., Fredrickson, J., and Beyenal, H. 2013. Modeling biofilms with dual extracellular electron transfer mechanisms. *Physical Chemistry Chemical Physics* 15(44): 19262–19283.

Rezaei, F., Richard, T. L., and Logan, B. E. 2009a. Analysis of chitin particle size on maximum power generation, power longevity, and Coulombic efficiency in solid – substrate microbial fuel cells. *Journal of Power Sources* 192(2): 304–309.

Rezaei, F., Xing, D., Wagner, R., Regan, J. M., Richard, T. L., and Logan, B. E. 2009b. Simultaneous cellulose degradation and electricity production by *Enterobacter cloacae* in a microbial fuel cell. *Applied and Environmental Microbiology* 75(11): 3673–3678.

Rismani-Yazdi, H., Carver, S. M., Christy, A. D., and Tuovinen, O. H. 2008. Cathodic limitations in microbial fuel cells: An overview. *Journal of Power Sources* 180(2): 683–694.

Rismani-Yazdi, H., Christy, A. D., Dehority, B. A., Morrison, M., Yu, Z., and Tuovinen, O. H. 2007. Electricity generation from cellulose by rumen microorganisms in microbial fuel cells. *Biotechnology and Bioengineering* 97(6): 1398–1407.

Rodrigo, M. A., Cañizares, P., García, H., Linares, J. J., and Lobato, J. 2009. Study of the acclimation stage and of the effect of the biodegradability on the performance of a microbial fuel cell. *Bioresource Technology* 100(20):4704–4710.

Sciarria, T. P., Merlino, G., Scaglia, B., D'Epifanio, A., Mecheri, B., Borin, S., Licoccia, S., and Adani, F. 2015. Electricity generation using white and red wine lees in air cathode microbial fuel cells. *Journal of Power Sources* 274: 393–399.

Selembo, P. A., Merrill, M. D., and Logan, B. E. 2009. The use of stainless steel and nickel alloys as low-cost cathodes in microbial electrolysis cells. *Journal of Power Sources* 190(2):271–278.

Sevda, S., and Abu-Reesh, I. M. 2019. Improved petroleum refinery wastewater treatment and seawater desalination performance by combining osmotic microbial fuel cell and up-flow microbial desalination cell. *Environmental Technology* 40(7): 888–895.

Shehab, N., Li, D., Amy, G. L., Logan, B. E., and Saikaly, P. E. 2013. Characterization of bacterial and archaeal communities in air-cathode microbial fuel cells, open circuit and sealed-off reactors. *Applied Microbiology and Biotechnology* 97(22): 9885–9895.

Singhania, R. R., Patel, A. K., Christophe, G., Fontanille, P., and Larroche, C. 2013. Biological upgrading of volatile fatty acids, key intermediates for the valorization of biowaste through dark anaerobic fermentation. *Bioresource Technology* 145: 166–174.

Song, T.-s., Wu, X.-y., and Zhou, C. C. 2014. Effect of different acclimation methods on the performance of microbial fuel cells using phenol as substrate. *Bioprocess and Biosystems Engineering* 37(2): 133–138.

Strycharz-Glaven, S. M., Snider, R. M., Guiseppi-Elie, A., and Tender, L. M. 2011. On the electrical conductivity of microbial nanowires and biofilms. *Energy & Environmental Science* 4(11): 4366–4379.

Sulonen, M. L., Kokko, M. E., Lakaniemi, A.-M., and Puhakka, J. A. 2015. Electricity generation from tetrathionate in microbial fuel cells by acidophiles. *Journal of Hazardous Materials* 284: 182–189.

Sun, J., Hu, Y.-y., Bi, Z., and Cao, Y.-q. 2009. Simultaneous decolorization of azo dye and bioelectricity generation using a microfiltration membrane air-cathode single-chamber microbial fuel cell. *Bioresource Technology* 100(13): 3185–3192.

Tong, M., Du, Z., and Gu, T. 2013. Converting low-grade biomass to produce energy using bio-fuel cells. In: *Eco-and Renewable Energy Materials*. Heidelberg: Springer, pp. 73–97.

Umar, M. F., Abbas, S. Z., Mohamad Ibrahim, M. N., Ismail, N., and Rafatullah, M. 2020. Insights into advancements and electrons transfer mechanisms of electrogens in benthic microbial fuel cells. *Membranes* 10(9): 205.

Vijayaraghavan, K., Ahmad, D., and Lesa, R. 2006. Electrolytic treatment of beer brewery wastewater. *Industrial & Engineering Chemistry Research* 45(20): 6854–6859.

Virdis, B., Rabaey, K., Rozendal, R. A., Yuan, Z., and Keller, J. 2010. Simultaneous nitrification, denitrification and carbon removal in microbial fuel cells. *Water Research* 44(9): 2970–2980.

Wang, B., Liu, W., Zhang, Y., and Wang, A. 2020. Bioenergy recovery from wastewater accelerated by solar power: Intermittent electro-driving regulation and capacitive storage in biomass. *Water Research* 115696.

Wang, X., Feng, Y., and Lee, H. 2008. Electricity production from beer brewery wastewater using single chamber microbial fuel cell. *Water Science and Technology* 57(7): 1117–1121.

Wang, X., Feng, Y., Ren, N., Wang, H., Lee, H., Li, N., and Zhao, Q. 2009a. Accelerated start-up of two-chambered microbial fuel cells: Effect of anodic positive poised potential. *Electrochimica Acta* 54(3): 1109–1114.

Wang, X., Feng, Y., Wang, H., Qu, Y., Yu, Y., Ren, N., Li, N., Wang, E., Lee, H., and Logan, B. E. 2009b. Bioaugmentation for electricity generation from corn stover biomass using microbial fuel cells. *Environmental Science & Technology* 43(15): 6088–6093.

Xing, D., Cheng, S., Regan, J. M., and Logan, B. E. 2009. Change in microbial communities in acetate-and glucose-fed microbial fuel cells in the presence of light. *Biosensors and Bioelectronics* 25(1): 105–111.

Yang, Q., Wang, X., Feng, Y., Lee, H., Liu, J., Shi, X., Qu, Y., and Ren, N. 2012. Electricity generation using eight amino acids by air – cathode microbial fuel cells. *Fuel* 102: 478–482.

Ye, Y., Ngo, H. H., Guo, W., Chang, S. W., Nguyen, D. D., Liu, Y., Ni, B.-j., and Zhang, X. 2019. Microbial fuel cell for nutrient recovery and electricity generation from municipal wastewater under different ammonium concentrations. *Bioresource Technology* 292: 121992.

You, S.-J., Ren, N.-Q., Zhao, Q.-L., Kiely, P. D., Wang, J.-Y., Yang, F.-L., Fu, L., and Peng, L. 2009. Improving phosphate buffer-free cathode performance of microbial fuel cell based on biological nitrification. *Biosensors and Bioelectronics* 24(12): 3698–3701.

Yu, Y., Ndayisenga, F., Yu, Z., Zhao, M., Lay, C.-H., and Zhou, D. 2019. Co-substrate strategy for improved power production and chlorophenol degradation in a microbial fuel cell. *International Journal of Hydrogen Energy* 44(36): 20312–20322.

Zhang, F., Cheng, S., Pant, D., Van Bogaert, G., and Logan, B. E. 2009a. Power generation using an activated carbon and metal mesh cathode in a microbial fuel cell. *Electrochemistry Communications* 11(11): 2177–2179.

Zhang, J., Zhao, Q., You, S., Jiang, J., and Ren, N. 2008. Continuous electricity production from leachate in a novel upflow air-cathode membrane-free microbial fuel cell. *Water Science and Technology* 57(7): 1017–1021.

Zhang, L., Liu, C., Zhuang, L., Li, W., Zhou, S., and Zhang, J. 2009b. Manganese dioxide as an alternative cathodic catalyst to platinum in microbial fuel cells. *Biosensors and Bioelectronics* 24(9): 2825–2829.

Zhu, X., and Ni, J. 2009. Simultaneous processes of electricity generation and p-nitrophenol degradation in a microbial fuel cell. *Electrochemistry Communications* 11(2): 274–277.

Zuo, Y., Maness, P.-C., and Logan, B. E. 2006. Electricity production from steam-exploded corn stover biomass. *Energy & Fuels* 20(4): 1716–1721.

10

Oil Price Shocks, Environmental Pollution, Foreign Direct Investment, and Renewable Energy Consumption: An Empirical Analysis in East Asian Countries

Van Chien Nguyen, Thu Thuy Nguyen, Subhadeep Mukherjee

CONTENTS

Abbreviations

Augmented Dickey Fuller: ADF
Autoregressive Distributed Lag: ARDL
Departments of Statistics: DOS
Dynamic Fixed Effects: DFE
Energy Consumption: EC
Error Correction Model: ECM

Foreign Direct Investment: FDI
Gulf Cooperation Council: GCC
IM Pesaran Shin: IPS
Income: INC
Levin Lin Chu: LLC
Mean Group: MG
Nonrenewable Energy: non-RE
Observation: Obs
Oil price: OIL
Panel Unit Root Tests: PURTs
Philips–Perron: PP
Pooled Mean Group: PMG
Population growth: POP
Renewable Energy Consumption: REC
Renewable Energy: RE
Renewable Resources: RRs
Standard Deviation: Std. Dev.
United Nations: UN
Variance Inflation Factor: VIF
World Development Indicators: WDI

10.1   Introduction

Energy has been considered to be one of the main factors that affect economic performance and lifestyle of a nation. Energy has always been the crucial part of many industrial revolutions although it has also been considered in critical thinking about its negative externalities on the atmosphere, climate change, and global warming. In addition, the consumption of energy significantly reflects quality of lives in a country (Tran et al., 2020). In the first stage of economic development, the fossil fuels, for example, coal, gas, and oil have greatly not only affected economic development but also caused environmental degradation. These resources are being criticized as one of the major reasons to affect sustainable development by increasing CO_2 emission and other greenhouse gases such as nitrous oxide and methane (Nguyen et al., 2020c; Tran et al., 2020).

In recent years, a changing climate has significantly impacted natural ecosystems and the environment in the form of the sea-level rise and rise in global temperature, for example. In the literature, the main determinants of renewable energy consumption (REC) are mixed. Additionally, it is reported that environmental degradation could help the human beings, the government, policymakers, and researchers to widely support to transition from a high-carbon to lower-carbon energy system (Zeppini and Bergh, 2020). The literature on the environmental quality and REC has been discussed by numerous empirical studies (Sadorsky, 2009; Doytch and Narayan, 2016; Khan et al., 2017; Bamati and Raoofi, 2020; Gozgor et al., 2020; Zeppini and Bergh, 2020). In addition, fossil fuels and renewable energy (RE) are completely substitute goods. In the context of increasing or decreasing fossil fuels, consumers will prefer using more RE. This suggestion is completely supported that changes in fossil fuels price, such as oil prices, the gas price could be determined to find out this relationship (Sadorsky, 2009; Reboredo, 2015; Shah et al., 2018).

It is noteworthy to conclude that foreign direct investment (FDI) refers to flows of capital between the home countries and host countries. An increase in FDI will promote a positive and significant multiplier effect on the receiving economy, especially technology transfer process (Wang, 1990; Tybout, 2000), access to new technologies (Du at el. 2011; Chaudhry at el. 2013), export promotion (Nguyen et al., 2020a), R&D activities, and on RE in the host countries (Zeppini and Bergh, 2020). However, there is lack of investigating the major factors affecting REC in a specific region like East Asia.

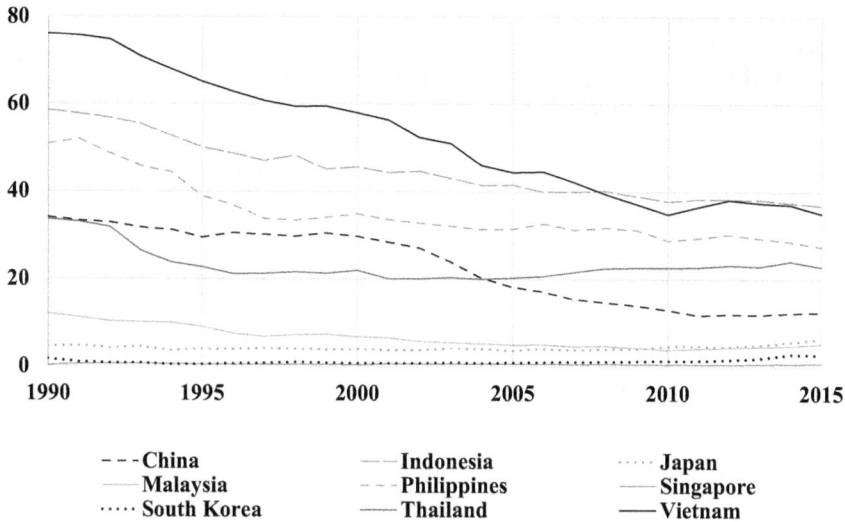

FIGURE 10.1   Trends in REC in East Asia between 1990 and 2015 (source: WDI, 2020).

The prime finding derived from the analysis could help the government, policymakers, and researchers to look up the environmental policies to adopt sustainability. The study aims to fulfill the existing literature gap by investigating the relationship between changes in oil prices, environmental quality, FDI, and REC in an emerging region in Asia. The purpose of this study is to examine how the prime factors affecting REC in the period of 1990–2019. This present study is to discuss panel data, based on the Pooled Mean Group (PMG), Mean Group (MG), and Dynamic Fixed Effects (DFE) approach.

Figure 10.1 describes the trends in REC in East Asia between 1990 and 2015. As suggested in World Development Indicators (WDI), REC is accounted as the proportion of REC of total final energy consumption (EC). Over the years, the proportion of REC of total final EC in most economies as such has continuously declined from approximately 34.1% in 1990 to 12.4% in 2015 for China, 58.6% in 1990 to 36.9% in 2015 for Indonesia, 12% in 1990 to 5.2% in 2015 for Malaysia, 52% in 1990 to 27.5% in 2015 for the Philippines, 33.6% in 1990 to 22.9% in 2015 for Thailand, and 76.1% in 1990 to 35% in 2015 for Vietnam. In the case of South Korea, and Japan, the contribution of REC is limited, and approximately 0.4% and 1.6% for South Korea, 3.6% and 4.6% for Japan in 1990 and 2015, respectively.

10.2   Review of the Literature

A few previous studies have predominantly examined the determinants of REC worldwide, especially in emerging and developed economies (Doytch and Narayan, 2016; Bamati and Raoofi, 2020; Gozgor et al., 2020; Zeppini and Bergh, 2020). However, a few studies have been focused on renewable energy consumption in East Asian economies in the context of environmental pollution and climate change in the region. There is a contradictory result in the previous studies consistent with the relationship between oil price fluctuations, environmental degradation, FDI, and REC (Sadorsky, 2009; Doytch and Narayan, 2016; Khan et al., 2017; Bamati and Raoofi, 2020; Gozgor et al., 2020; Zeppini and Bergh, 2020). The findings indicate that eliminating fossil energy and subsidies for RE is a necessity in the transition trend all over the world.

Doytch and Narayan (2016) compiled a study of 74 countries covering from 1986 to 2012, as a finding of which, foreign investment is not only the financing source that supports a firm to grow, but also the innovative source that enhances the efficiency of energy. Discussing the impact of FDI inflows on renewable and non-RE in 74 economies and followed by Blundell–Bond panel estimator, the study finds that the use of renewable or non-RE resources varies in importance and concernment by FDI inflows.

Therefore, the results indicate that there exists an EC – reducing effect with respect to non-REC and an EC – augmenting effect with respect to REC.

According to Shah et al. (2018), it is believed that oil price, macroeconomic indicators, and government policies can remarkably affect the demand and supply of RE. Using a study on three economies such as Norway, the UK, and particularly the United States, followed on the Vector Autoregression (VAR) model, the results show that Norway and the United States have a strong relation between oil price shocks and RE while no relation for the case of the United Kingdom. This finding can be understood in the context of the United States which is an oil importer and does not much support RE as other similar economies. In addition, the study has some implications, if there is a little bit of support for the RE sector, the investment will be dependent on macroeconomic factors and substitution goods.

Al-Maamary et al. (2017) conducted a study on the effects of oil price shocks on renewable energies in the Gulf Cooperation Council (GCC) in the context of the decline of petroleum resources and the trend of increasing non-RE worldwide, results indicate that increasing oil and gas production can negatively impact on economic growth in GCC countries, particularly the largest oil importer of Saudi Arabia. Furthermore, EC growth is relatively faster than economic growth. Regarding sustainable energy, the CGG countries have considerably implemented policies and projects to promote more investment in sustainable energy. Similarly, using the copula approach, Reboredo (2015) also found that oil price fluctuations significantly provide approximately 30% to the upside and downside of RE risk.

Considering G7 economies, that is, the United States, Japan, Italy, the United Kingdom, Germany, Canada, and France, Sadorsky (2009) indicated that energy security and especially global warming are one of the most important reasons that affect REC by using panel cointegration estimation. The results of empirical research indicate that increases in per capita income and environmental pollution (EP) could be found to be the main operators behind REC in the long run. Oil price shocks have a negative and smaller effect on REC.

By analyzing in developed economies, Ike et al. (2020) conducted on the panel data and especially a causality analysis in G7 economies, the results indicate that a unidirectional causality coming from energy price shocks, income, and trade development to environmental pollution (EP). In contrast, REC has not a direct effect but an indirect effect on EP. Further investigated, Khan et al. (2017) indicate that oil price has steadily declined since 2014 from US $106 per barrel from 2010 to 2014. Therefore, Khan et al. (2017) considerably explained that this downtrend of oil price is caused by the overdemand of domestic oil in the United States and its association with a major exporter as Iraq. The results of this study demonstrate that the recent decline in oil price does not affect the RE sector. However, renewable energies, for example, solar and wind, have been increasingly more competitive in cost than fossil fuel energies. Additionally, the oil price plunge can harm the objectives of clean energy technology in the short run regarding bio-fuel and electric vehicles. Furthermore, a lower oil price could also harm REs and climate policies to encourage investment with low carbon in the long run.

By analyzing metadata of developing and developed countries, Bamati and Raoofi (2020) studied how development level and technology impact on RE production, and using the Generalized Least Squared (GLS) method in the panel data, the study depicts that a high technology export significantly determines RE production in developed economies and insignificantly in enhancing the demand of RE in developing economies. Additionally, changes in oil price have slightly impacted RE production. Per capita income has a positive and significant effect on RE production while environmental degradation totally differs in the developing and developed economies.

In the study that was conducted on the context of economic globalization in OECD economies in the period of 1970–2015, Gozgor et al. (2020) analyzed according to several factors of economic globalization and few techniques. The evidence suggests that a positive and significant impact of income, environmental quality, and oil price fluctuations on RE can be found. More specifically, greater economic globalization could predominantly generate RE. Furthermore, as discussed on the findings, Gozgor et al. (2020) emphasized that enhancing economic aspects of globalization can promote RE in OECD economies. In contrast, there exists an adverse effect of global warming and climate change on the lives of human beings and natural conditions in the long run. It is importantly noteworthy to suggest that the enhancement of knowledge regarding economic globalization and its effect is becoming more important

TABLE 10.1

Measurement of Variables

Dependent Variables	Abbreviation	Source
Renewable Energy Consumption (% of total final energy consumption)	REC	WDI
Independent Variables		
Oil price (WTI in USD per barrel)	OIL	Finance news
Pollution Index (metric tons per capita)	POLLUTION (POLL)	WDI
Foreign Direct Investment ($bn US)	FDI	WDI, DOS
Per capita income ($ US)	INC	WDI, DOS
Population growth (%)	POP	WDI

for all economies worldwide where emphasizing on the importance and use of RE is necessary to protect the environment and ecosystems.

Zeppini and Bergh (2020) studied the relationship between the competitive dynamics of fossil fuels such as oil, gas, coal, and RE sources as the solar and wind. The results suggest that peak oil can cause a transition to use more coal rather than RE and will probably worsen environmental quality. In the case of eliminating subsidies for fossil fuels and supporting RRs, the climate policies for renewable subsidies such as implementation for carbon taxes, market adoption, and R&D activities subsidies which can be focused, the empirical evidence indicates that a potential transition trend from a high-carbon to lower-carbon energy will exist. Also, climate policies can be discussed on two features and will be the channel to solve the climate problem. It concludes that the government should not overpass critical carbon budget and uncertainty of market shares of energy sources.

10.3 Data and Methodology

10.3.1 Data

The data used in this study is retrieved from the database of World Development Indicators (WDI) and Departments of Statistics (DOS) of all East Asian countries that are studied in this paper. The data were collected in nine countries such as Japan, South Korea, China, Vietnam, Thailand, Philippines, Malaysia, Indonesia, and Singapore. The world crude oil price (WTI) in USD per barrel can be daily collected on Finance News. The data were discussed from 1990 to 2019 and analyzed by using Stata 15 software. Table 10.1 indicates the measurement of variables used in this study.

10.3.2 Research Model

The impact of oil price shocks, EP, per capita income, and REC has been investigated and followed by previous studies of Sadorsky, 2009; Doytch and Narayan, 2016; Khan et al., 2017; Bamati and Raoofi, 2020; Gozgor et al., 2020; and Zeppini and Bergh, 2020. However, different from previous studies, we will analyze this study based on theoretical consideration of using Dynamic Fixed Effects Estimates (DFE), Mean Group Estimates (MG), and Pooled Mean Group Estimates (PMG). The study also discusses Panel Unit Root Tests (PURTs) according to the first and the second generation of testing unit roots. Importantly, Cointegration Test will be conducted on the theories of Pedroni, Kao, and Westerlund. Figure 10.2 and Figure 10.3 show the research model and model selection used in this study, respectively.

To find out the aim of this study, we hypothesize the following model:

$$REC_{it} = f\left(OIL_{it}, POLLUTION_{it}, FDI_{it}, INC_{it}, POP_{it}\right)$$

(10.1)

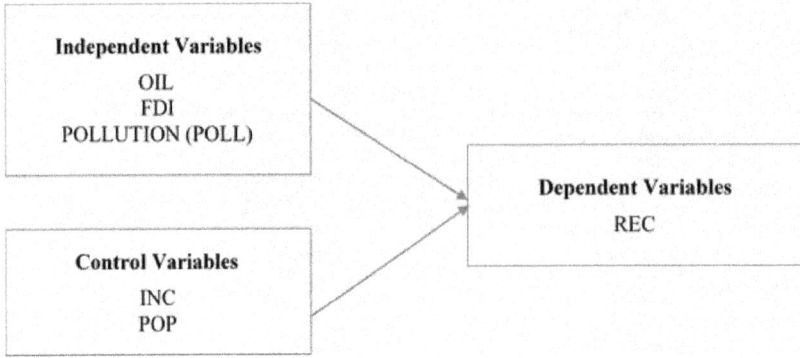

FIGURE 10.2 Research model.

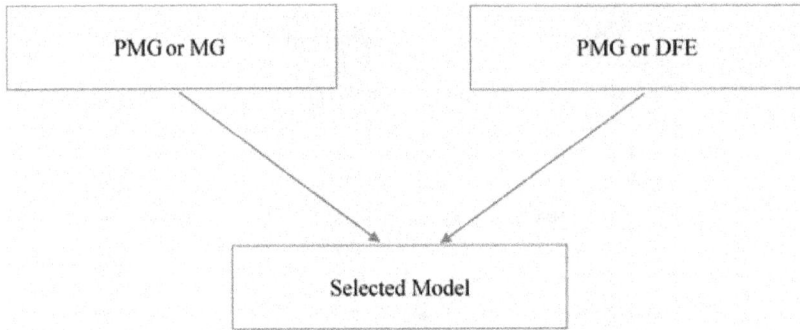

FIGURE 10.3 Model selection.

The specific equation can be written as follows:

$$Y_{it} = \beta_1 X_{it} + \alpha_i + u_{it} \tag{10.2}$$

$$REC_{it} = \alpha + \beta_1 OIL_{it}, \beta_2 POLLUTION_{it}, \beta_3 FDI_{it}, \beta_4 INC_{it}, \beta_4 POP_{it} + u_{it} \tag{10.3}$$

The long-run and short-run relationships corresponding to Eq. 10.3 are given as follows:

$$\Delta REC_{i,t} = \theta_i \left[REC_{i,t-1} - \lambda' X_{i,t} \right] + \sum_{j=1}^{p-1} \xi_{ij} \Delta REC_{i,t-j} + \sum_{j=1}^{q-1} \beta'_{ij} \Delta X_{i,t-j} + \varphi_i + e_{it} \tag{10.4}$$

$$\Delta REC_{i,t} = \theta_i * ECT + \sum_{j=1}^{p-1} \xi_{ij} \Delta REC_{i,t-j} + \sum_{j=1}^{q-1} \beta'_{ij} \Delta X_{i,t-j} + \varphi_i + e_{it} \tag{10.5}$$

Where,
α_i ($i = 1 \ldots n$) is the unknown intercept for each country (n country-specific intercepts),
Y_{it} is the explanatory variable; i and t denote for the country and time, respectively. This is REC (renewable energy consumption),
X_{it} denotes an independent variable and includes OIL, POLLUTION, FDI, INC, and POP;
β_1 is the coefficient for that independent variable;
u_{it} is the error term;
$\theta_i = -(1 - \delta_i)$, group-specific speed of adjustment coefficient (it is expected that $\theta_i < 0$) ;
$\lambda' =$ vector of long-run relationships;

ECT = $REC_{i,t-1} - \lambda' X_{i,t}$, the error correction terms;

ξ_{ij}, β'_{ij} are the short-run dynamic coefficients.

10.4 Results of Economic Modeling

10.4.1 Descriptive Statistics

Tables 10.2 and 10.3 present data description regarding 261 observations of nine East Asian economies covering over 20 years from 1990 to 2018. In fact, for each country, the study collects some main indicators such as mean, minimum, maximum, standard deviation, and the number of observations used in the study. In specific, Table 10.2 shows that REC has remarkably contributed a rate of 19.04% on total EC in East Asian economies from 1990 to 2018, with the minimum value of zero and maximum value of 76.1% of REC on energy consumption, respectively. It indicates that in the early stages, few East Asian countries had not consumed REC because of the very cheap price of energy from hydrocarbons. In fact, in the context of the world's energy, coal, crude oil, and natural gas are the fossil fuels extracted from the ground, therefore this energy comes from primary fuel. In the environmental field, scientists demonstrate that coal, oil, and natural gas are composed mainly out of hydrocarbons (carbon and hydrogen compounds) with other trace elements.

Table 10.3 shows that the price of crude oil has significantly increased from US $14.53 per barrel in the period of 1990–2005 to the highest price of US $98.58 per barrel in the period of 2011–2018, while REC has slowly improved in recent years. In addition, based on the announcement of the United Nations (UN), the global population was estimated at around 2.6 billion people in 1950, reached 5 billion people in 1987 and 6 billion in 1999. Further, the global population was estimated to be 7 billion in October 2011 and to have reached 7.8 billion people as of March 2020. Regarding global POP, it reached a peak in the 1960s with an annual growth rate of 2.2% per year, but since then, it has declined to 1.1% per year in recent years. In the case of East Asian countries, Table 10.2 indicates that POP in this area reached a peak of 5.3% and a minimum value of −1.474% in recent years. It means that POP continuously declined. According to Sherbinin et al. (2007), the population is one of several factors that impact the environment; further, fast population growth (POP) aggravates other externalities such as wars, governance quality, and polluting technologies.

Regarding FDI in the region, China, Japan, and South Korea are the leading FDI receivers with an average value of US $115.67 billion, US $43327.93 billion, and US $30330.45 billion, respectively. In particular, China is known as the second largest FDI recipient in the world after the United States and the largest FDI recipient in Asia in recent years. According to UNCTAD in the 2020 World Investment Report, FDI inflows in the world increased from the US $138 billion in 2018 to the US $141 billion in 2019 with a growth of 2%. This growth is in favor of globalization, free trade agreements in countries, and the fourth industrial and technological revolution. Although FDI to China from the United States and Europe had declined, FDI from other major investors such as Singapore, South Korea, and Japan has been broadly stable. Like China, FDI inflows in South Korea have further grown since the 2000s, and continue to maintain the standard of US $20 billion per year from 2015 to the present.

Global businesses prefer to establish their headquarters in South Korea and broadly expand investment in the high-value-added sectors such as R&D activities, financial and technological field, automotive parts, and advanced healthcare. Japan, one of the major economies in Asia and the world, welcomes foreign investment with an ambitious goal to attract more high-tech sectors in agriculture, forestry, petroleum, financial and technologies, collaborations with world-class corporations, research facilities, and technologies worldwide. FDI firms in Japan have been actively supported by numerous relieving administrative producers to reduce corruption. Based on World Bank's doing business report in "Ease of doing business," Japan has been known on the top in the world in transparency. Additionally, the Japanese government established a new "FDI Promotion Council" that included the ministers and private sector advisors who can suggest improving FDI inward into the country. Furthermore, this country continued to establish an "Investment Advisor Assignment System" in 2016 to select important foreign corporations worldwide

TABLE 10.2

Descriptive Statistics

Country	Variable	REC	OIL	POLE	FDI	INC	POP
China	Obs.	29	29	29	29	29	29
	Mean	20.53	47.68	3.60	115.67	3256.01	0.76
	Std. Dev.	10.70	29.20	2.31	93.67	2205.16	0.30
	Min	11.69	14.53	2.15	3.49	729.16	0.46
	Max	34.08	98.58	7.56	290.93	7752.56	1.47
Indonesia	Obs.	29	29	29	29	29	29
	Mean	40.24	47.68	1.27	7.58	2723.90	1.40
	Std. Dev.	15.34	29.20	0.65	8.90	752.40	0.16
	Min	36.88	14.53	0.82	−4.55	1707.82	1.13
	Max	58.60	98.58	2.56	25.12	4284.65	1.78
Japan	Obs.	29	29	29	29	29	29
	Mean	3.82	47.68	8.11	9.03	43327.93	0.09
	Std. Dev.	1.45	29.20	3.32	10.17	2962.87	0.19
	Min	3.56	14.53	8.62	−2.40	38074.46	−0.20
	Max	6.30	98.58	9.91	40.95	48919.80	0.38
Malaysia	Obs.	29	29	29	29	29	29
	Mean	5.89	47.68	5.22	6.31	7998.95	2.02
	Std. Dev.	3.05	29.20	2.48	3.79	2092.78	0.49
	Min	3.81	14.53	3.14	0.11	4536.97	1.34
	Max	11.98	98.58	8.13	15.12	12120.08	2.82
Philippines	Obs.	29	29	29	29	29	29
	Mean	31.57	47.68	0.74	2.69	1927.46	1.96
	Std. Dev.	12.82	29.20	0.31	2.71	466.06	0.35
	Min	27.45	14.53	0.67	0.23	1445.87	1.40
	Max	51.96	98.58	1.05	10.26	3021.99	2.54
Singapore	Obs.	29	29	29	29	29	29
	Mean	0.44	47.68	9.57	31.92	39330.45	2.26
	Std. Dev.	0.18	29.20	5.03	27.54	10926.45	1.42
	Min	0.19	14.53	4.34	2.20	22571.90	−1.47
	Max	0.71	98.58	18.04	94.81	58247.87	5.32
South Korea	Obs.	29	29	29	29	29	29
	Mean	0.98	47.68	8.04	7.79	17780.56	0.67
	Std. Dev.	0.67	29.20	3.65	4.73	5647.18	0.24
	Min	0.44	14.53	5.76	0.83	8464.94	0.21
	Max	2.84	98.58	11.80	17.91	26761.94	1.04
Thailand	Obs.	29	29	29	29	29	29
	Mean	20.88	47.68	2.81	6.18	4301.63	0.76
	Std. Dev.	8.07	29.20	1.40	4.11	1108.22	0.32
	Min	20.02	14.53	1.61	1.37	2503.80	0.32
	Max	33.64	98.58	4.62	15.94	6361.63	1.40
Vietnam	Obs.	29	29	29	29	29	29
	Mean	47.01	47.68	0.81	4.99	1039.46	1.25
	Std. Dev.	20.93	29.20	0.58	4.66	457.40	0.41
	Min	34.79	14.53	0.30	0.18	433.28	0.92
	Max	76.08	98.58	1.82	15.50	1964.48	2.14

Source: Result from the authors' analysis.

TABLE 10.3

Descriptive Statistics

Country	Variable	Obs.	Mean	Std. Dev.	Min	Max
All East Asian Countries	REC	261	19.040	19.657	0.19	76.081
	OIL	261	47.683	28.748	14.534	98.583
	POLL	261	4.463	4.162	0.30	18.040
	FDI	261	21.350	47.309	–4.550	290.928
	INC	261	13520.7	16244.88	433.283	58247.87
	POP	261	1.241	0.886	–1.474	5.321

Source: Result from the authors' analysis.

regarding investment and technological changes in Japan. In the case of other emerging economies in East Asia such as Malaysia, Philippines, Thailand, and Vietnam, these economies are committed to an ambitious investment plan to welcome FDI inflows to developing overall economies, export promotion, and job creation. This policy is expected to provide significant opportunities for both inbound and outbound investment in the future and maintain the balance between economic development and environmental protection. Additionally, FDI inflows are expected to encourage the consumption of clean energy and encourage the switch to RE in the context of more technological advantages (Doytch and Narayan, 2016).

10.4.2   Panel Unit Root Tests (PURTs)

The first step of PURTs is to examine the stationarity of all series used in this study. In general, there are two methods of testing unit root: first generation and second generation. Theoretically, the first generation of PURTs consists of Levin Lin Chu (LLC) tests and IM Pesaran Shin (IPS) tests. Also, the second generation of PURTs comprises Breitung and Das tests, the Fishers ADF, and the Fishers PP tests. Similar to Breitung tests and Hadri tests, LLC tests are more preferred in the case of strongly balanced panel data. Table 10.4 shows the results of PURTs.

Performing the IPS and LLC tests, the Fisher ADF tests, and the Fishers PP tests, Table 10.4 depicts that the result obtained from the tests indicates that all series are non-stationary at the level. More specifically, at the level, REC, OIL, POLLUTION, FDI, and INC are not stationary at a 5% significance level, but all are stationary at the first difference. Regarding the variable of POP, this variable is stationary at the level based on LLC tests, Fisher's ADF tests, and Fisher's PP tests. Although POP is non-stationary at the level but is stationary at the first difference according to IPS tests.

10.4.3   Multi-Collinearity Tests

According to Gujarati (2004), multi-collinearity can be present when the correlation coefficient among independent variables is over 0.8. Further, a strong multi-collinearity can exist when the absolute value of pairwise collections between variables may be somewhat high. In this study, Table 10.5 depicts that there is less multi-collinearity problem in this study because correlation coefficients do not exceed 0.8. According to the Variance Inflation Factor (VIF) in Table 10.6, we see that this coefficient of all independent variables is less than 2, and it is important to note that no multi-collinearity problem can be found.

10.4.4   Optimal Lag Length and Co-Integration Tests

The study will choose the optimal lag length according to the most common lag for each variable to represent the lags for the model. The lag length, in this case, should be (1, 1, 2, 0, 0, 0). Theoretically, the use of co-integration techniques to find out the presence of long-term relationships among integrated variables

TABLE 10.4

Panel Unit Root Tests (PURTs)

Variable	Order of Integration	IPS Test	LLC Test	Fisher's ADF Test	Fisher's PP Test	Hypothesis
REC	I(0)	2.1393 (0.9838)	2.1559 (0.9845)	−1.2870 (0.9010)	−1.0423 (0.8514)	Not rejected
	I(1)	−1.7244 (0.0423)	−7.4861 (0.0000)	16.5619 (0.0000)	34.1859 (0.0000)	Rejected
OIL	I(0)	0.7358 (0.7691)	−0.9014 (0.1837)	−1.5714 (0.9419)	−1.6269(0.9481)	Not rejected
	I(1)	−3.9719 (0.0000)	−6.1350 (0.0000)	14.1473 (0.0000)	29.5649 (0.0000)	Rejected
POLLUTION	I(0)	0.9506 (0.8291)	2.1775 (0.9853)	−1.8302 (0.9664)	−1.7486 (0.9598)	Not rejected
	I(1)	−4.3815 (0.0000)	−6.8854 (0.0000)	13.8992 (0.0000)	31.2417 (0.0000)	Rejected
FDI	I(0)	3.2208 (0.9994)	1.3313 (0.9085)	−1.0246 (0.8472)	1.0949 (0.1368)	Not rejected
	I(1)	−5.3882 (0.0000)	−10.7542 (0.0000)	36.3869 (0.0000)	51.3225 (0.0000)	Rejected
INC	I(0)	8.0119 (1.0000)	5.4694 (1.0000)	−2.9253 (0.9383)	−2.9017 (0.9981)	Not rejected
	I(1)	−1.9510 (0.0255)	−1.8414 (0.0328)	12.6116 (0.0000)	18.9523 (0.0328)	Rejected
POP	I(0)	−0.8467 (0.1986)	−4.4904 (0.000)	5.5389 (0.000)	3.3375 (0.0004)	Rejected (not for IPS test)
	I(1)	−5.0685 (0.0255)				Rejected (for IPS test)

Source: Results from the authors' analysis.

TABLE 10.5

Correlation Coefficients between Variables

	REC	OIL	POLLUT~N	FDI	INC	POP
REC	1$$					
OIL	−0.1865	1				
POLLUTION	−0.5893	0.0636	1.0000			
FDI	−0.1743	0.2861	0.0237	1.0000		
INC	−0.6118	0.1344	0.5684	0.0254	1.0000	
POP	0.1662	−0.2141	0.0182	−0.2085	−0.1933	1.0000

Source: Result from the authors' analysis.

TABLE 10.6

VIF Coefficients of Independent Variables

Variable	VIF	1/VIF
INC	1.59	0.630101
POLLUTION	1.52	0.658477
POP	1.14	0.878171
OIL	1.13	0.883470
FDI	1.12	0.892199
Mean VIF	1.30	

Source: Result from the authors' analysis.

has enjoyed a growing popularity in the empirical literature. In the case of time-series data, stationarity test should be performed (Chaudhry et al., 2013). If the data is non-stationary at the level, the study needs to check co-integration (Chaudhry et al., 2013). In addition, as reported by Pedroni (1999, 2004), it is evident that co-integration examination will occur when two series exist at a constant co-variance over time. Furthermore, the theoretical framework is roughly the same as in time series; notwithstanding, some tests have been discussed for panel co-integration as the methods of Pedroni (1999, 2004), Kao (1999), and a Fisher-type test using an underlying Johansen methodology (Maddale and Wu, 1999).

TABLE 10.7

Panel Co-Integration Tests

Methods	Co-Integration Tests	Statistics
Pedroni	Panel-v	0.8262
	Panel-rho	−0.1746
	Panel-PP	−2.834***
	Panel-ADF	2.799***
	Group-rho	1.016
	Group-PP	−2.404***
	Group-ADF	3.909***
Kao	t	1.2658
Westerlund	Gt	1.893**
	Ga	2.838***
	Pt	1.841**
	Pa	2.203**

Note: Pedroni, Kao, and Westerlund tests are generated on Stata with the command "xtpedroni, xtcointtest kao, xtwest".
* Significance at 10%.
** Significance at 5%.
*** Significance at 1%.
Source: Results from the authors' analysis.

Accordingly, Pedroni tests are suitable and based on the residuals obtained from a static relationship. Further, simulation studies have confirmed that residual tests have less power compared to those based on a dynamic model. According to the case of a panel that has cross-sectional dependence, Westerlund (2007) panel co-integration tests will be preferred. In addition, the stationary variables can be seen to be a part of short-run dynamics, and therefore co-integration can be analyzed more efficiently.

Table 10.7 indicates the results of cointegration tests. The lag length, in this case, should be (1, 1, 2, 0, 0, 0). However, in the Westerlund test, the number of lags is set to 1 because of the theory of this method. In the analysis of Pedroni, this method discusses four factors as panel-v, panel-rho, panel-PP, and panel ADF statistics. The Pedroni group statistics are also discussed as group-rho, group-PP, and group-ADF statistics. Kao t-statistic using the assumptions of homogeneity in panel data is developed by a least squared dummy variable. Westerlund t-statistic discusses four factors Gt, Ga, Pt, and Pa. Table 10.7 also shows the rejection of Pedroni panel-PP, panel-ADF, group-PP, and group-ADF null hypothesis, which depicts that there is cointegration in this study. Similarly, the rejection of Gt, Ga, Pt, and Pa indicates the existence of cointegration among variables. Regarding Kao t-statistic, variables are not integrated. In conclusion, the results imply the existence of a long-run relationship among variables.

10.5   Estimated Results

10.5.1   Panel Estimations

The DFE, MG, PMG, and PMG estimates, classified by countries, are represented in Table 10.8, Table 10.9, Table 10.10, and Table 10.11, respectively.

10.5.2   Hausman Tests

Testing the null hypothesis of homogeneity through the Hausman test and according to the comparison between MG and PMG estimators, Table 10.12 depicts that we reject the null hypothesis if P-value is less than 0.05. In this case, the probability value is 0.372, and the null hypothesis of homogeneity cannot be rejected. Therefore, the PMG is the best technique in this case.

TABLE 10.8

DFE Estimates

Variable	DFE Estimates			DFE Estimates		
	Coef.	Std. err.	P value	Coef.	Std. err.	P value
Dependent Variable: REC						
Long-run coefficients						
OIL	−2.5459	13.1732	0.847	−1.0637	2.4456	0.664
POLLUTION	−36.2949	202.9203	0.858	−18.0034	48.7174	0.712
FDI	1.156	6.8945	0.867	0.543	1.733	0.754
INC	−0.0035	0.0303	0.907	−0.0003	0.0089	0.971
POP	−101.978	586.226	0.862			
Intercept	−3.819	1.9492	0.05**	−2.8887	1.6624	0.082*
ECM	0.0058	0.0317	0.853	0.0124	0.0308	0.686
Short-run coefficients						
ΔOIL	0.0297	0.0235	0.208	0.0293	0.0232	0.207
ΔPOLLUTION	−0.3989	0.2085	0.056*	−0.3688	0.2052	0.072*
ΔFDI	0.0257	0.0218	0.239	0.0253	0.0216	0.241
ΔINC	−0.0001	0.0005	0.694	−0.0003	0.0004	0.47
ΔPOP	−0.2967	0.7252	0.682			

Source: Result from the authors' analysis. Note: *, **, and *** indicate significance at 10%, 5%, and 1% levels, respectively. Under the long-run slope homogeneity, the Hausman statistic is asymptotically distributed as a chi-square. The lag structure is ARDL (1,1,2,0,0,0), and the order of the variable is REC – renewable energy consumption, oil price, EP – environmental pollution, FDI – foreign direct investment, per capita GDP, POP – population growth.

TABLE 10.9

MG Estimates

Variable	MG Estimates			MG Estimates		
	Coef.	Std. err.	P value	Coef.	Std. err.	P value
Dependent variable: REC						
Long-run coefficients						
OIL	0.0912	0.0523	0.081*	0.1092	0.0474	0.021**
POLLUTION	1.3191	1.2461	0.29	2.9402	1.9337	0.128
FDI	−0.0367	0.216	0.865	0.6949	0.5477	0.205
INC	−0.0089	0.0038	0.021**	−0.0169	0.0075	0.024**
POP	5.8868	4.9746	0.237			
Intercept	24.2302	13.968	0.083*	34.8482	11.7551	0.003***
ECM	−0.8997	0.05	0***	−0.79	0.0576	0***
Short-run coefficients						
ΔOIL	−.067347	0.0301852	0.026**	−0.0506779	0.018615	0.006***
ΔPOLLUTION	−3.48127	1.517935	0.022**	−3.759979	1.622623	0.02**
ΔFDI	0.024304	0.1589755	0.878	−0.3004879	0.1955071	0.124
ΔINC	−.020455	0.0204861	0.318	−0.011998	0.0079207	0.13
ΔPOP	−17.7077	7.732541	0.022**			

Source: Result from the authors' analysis. Note: *, **, and *** indicate significances at 10%, 5%, and 1% levels, respectively. Under the long-run slope homogeneity, the Hausman statistic is asymptotically distributed as a chi-square. The lag structure is ARDL (1,1,2,0,0,0), and the order of the variable is REC – renewable energy consumption, oil price, EP – environmental pollution, FDI – foreign direct investment, per capita GDP, POP – population growth.

TABLE 10.10

PMG Estimates

Variable	PMG Estimates			PMG Estimates		
	Coef.	Std. err.	P value	Coef.	Std. err.	P value
Dependent variable: REC						
Long-run coefficients						
OIL	0.0193	0.0044	0***	0.3401	0.0334	0***
POLLUTION	0.415	0.0453	0***	5.242	1.2384	0***
FDI	0.0094	0.011	0.393	−0.13	0.0901	0.149
INC	−0.0001	0.0001	0.011**	−0.0001	0.0009	0.879
POP				45.0981	3.5844	0***
Intercept	7.8023	4.72499	0.099*	−10.1312	6.8964	0.142
ECM	−0.2602	0.10257	0.011**	−0.178	0.0918	0.053*
Short-run coefficients						
ΔOIL	0.0241	0.0253	0.34	0.0008	0.02958	0.978
ΔPOLLUTION	−2.02765	1.0771	0.06*	−2.7086	1.2749	0.034**
ΔFDI	−0.086	0.2236	0.7	−0.1614	0.2118	0.446
ΔINC	−0.0622	0.0462	0.178	−0.0254	0.0191	0.185
ΔPOP				−1.8742	9.2574	0.84
Number of obs.	9			9		
Number of groups	252			252		
Obs. per group	min = 28			min = 28		
	avg = 28			avg = 28		
	max = 28			max = 28		
Log Likelihood	−420.7775			−409.5517		

Source: Result from the authors' analysis. Note: *, **, and *** indicate significance at 10%, 5%, and 1% level, respectively. Under the long-run slope homogeneity, the Hausman statistic is asymptotically distributed as a chi-square. The lag structure is ARDL (1,1,2,0,0,0), and the order of the variable is REC – renewable energy consumption, oil price, EP – environmental pollution, FDI – foreign direct investment, per capita GDP, POP – population growth.

In this situation, we test for selecting the null hypothesis of homogeneity and the alternative hypothesis of homogeneity according to the Hausman test, and Table 10.12 depicts that we reject the null hypothesis if P value is less than 0.05. In this case, the probability value is 0.999, and so the null hypothesis of homogeneity cannot be rejected. Thus, the PMG is the best technique.

10.5.3   Robustness Check

In this study, we try to apply a robustness check to affirm how certain main regression coefficient estimates meet. More specifically, if regression coefficients are plausible and robust, it means that the findings of the estimated regression are valid and strongly supportive. According to Lu and White (2014), a robustness check can be conducted by changing the regression specification, which is adjusted by adding or removing regressors. Further, if not performed properly, robustness checks could be entirely uninformative and/or completely misleading. Therefore, in this study, we add or remove the variable of POP, then we can compare both scenarios. Tables 10.8, 10.9, and 10.10 show that the study's model is well established. Also, we applied MG and DFE measured for PMG's robustness checks, and the results are similar to those of PMG.

10.6   Critical Discussion

The discussion in this research paper is presented in line with the results obtained in Table 10.10. The estimation of the PMG estimator is finally known as the best model. To test the reliability of the results, this study also analyze the robustness check, and the results are indicated in Tables 10.8 and 10.9.

TABLE 10.11

PMG Estimates and Classified by Countries

Variable	Pooled Mean Group Estimates			Pooled Mean Group Estimates		
	Coef.	Std. err.	Coef.	Std. err.	Coef.	Std. err.
Dependent variable: REC						
Long-run coefficients						
OIL	0.0193	0.0044	0***	0.3401	0.0334	0***
POLLUTION	0.415	0.0453	0***	5.242	1.2384	0***
FDI	0.0094	0.011	0.393	−0.13	0.0901	0.149
INC	−0.0001	0.0001	0.011**	−0.0001	0.0009	0.879
POP				45.0981	3.5844	0***
Short-run coefficients						
China						
ECM	−0.5262	0.1118	0	−0.1212	0.039	0.002
ΔOIL	−0.0675	0.0369	0.067	−0.0035	0.0413	0.931
ΔPOLLUTION	−0.2808	0.2869	0.328	−0.9987	0.4133	0.016
ΔFDI	0.0503	0.0122	0	0.0183	0.0142	0.196
ΔINC	−0.0464	0.0091	0	−0.0012	0.0063	0.844
ΔPOP				−16.9969	11.2483	0.131
Intercept	19.6988	4.292	0	−5.5071	1.714	0.001
Indonesia						
ECM	0.0752	0.0764	0.325	0.0029	0.0865	0.973
ΔOIL	−0.0672	0.096	0.483	−0.0952	0.0945	0.314
ΔPOLLUTION	−1.999	2.9776	0.502	−1.7748	2.9001	0.541
ΔFDI	1.002	0.206	0	0.9285	0.2069	0
ΔINC	−0.0194	0.0105	0.065	−0.0218	0.011	0.049
ΔPOP				65.5654	52.2168	0.209
Intercept	−3.984	3.6398	0.274	1.0289	4.1385	0.804
Japan						
ECM	−0.8964	0.0878	0	−0.031	0.0089	0.001
ΔOIL	−0.0212	0.0083	0.012	0.0142	0.0169	0.4
ΔPOLLUTION	−0.5016	0.0643	0	−0.2272	0.1175	0.053
ΔFDI	−0.0112	0.0095	0.241	−0.0629	0.0134	0
ΔINC	0.0003	0.0001	0.001	0.00008	0.0002	0.692
ΔPOP				0.4915	2.1581	0.82
Intercept	4.8027	2.3694	0.043	−1.8054	1.47	0.219
Malaysia						
ECM	−0.0636	0.058	0.273	−0.0232	0.0077	0.003
ΔOIL	0.0398	0.0212	0.06	0.0499	0.0157	0.002
ΔPOLLUTION	−0.3057	0.1503	0.042	−0.5532	0.1263	0
ΔFDI	−0.1092	0.06	0.069	−0.1617	0.0484	0.001
ΔINC	−0.0006	0.0007	0.409	0.0002	0.0006	0.748
ΔPOP				−5.2754	2.6568	0.047
Intercept	−0.0625	0.3523	0.859	−3.7908	1.0595	0
Philippines						
ECM	−0.2188	0.1017	0.031	−0.8272	0.085	0
ΔOIL	0.0809	0.0885	0.361	−0.112	0.0504	0.026
ΔPOLLUTION	−5.26	5.6891	0.355	−9.6946	2.9636	0.001
ΔFDI	−1.556	0.9232	0.092	−1.1396	0.4626	0.014
ΔINC	−0.05799	0.0222	0.009	−0.0371	0.0118	0.002
ΔPOP				−35.5224	16.0847	0.027

(Continued)

TABLE 10.11 (Continued)

Variable	Pooled Mean Group Estimates			Pooled Mean Group Estimates		
	Coef.	**Std. err.**	**Coef.**	**Std. err.**	**Coef.**	**Std. err.**
Intercept	8.4887	4.2116	0.044	−62.785	9.6733	0
Singapore						
ECM	−0.0264	0.0104	0.012	−0.0007	0.0004	0.089
ΔOIL	0.0016	0.0027	0.552	0.0015	0.0029	0.595
ΔPOLLUTION	−0.0162	0.011	0.141	−0.0146	0.0119	0.221
ΔFDI	0.0014	0.0032	0.652	0.0006	0.0036	0.855
ΔINC	−0.00002	0.0001	0.407	−0.00001	0.00003	0.734
ΔPOP				−0.0261	0.0279	0.349
Intercept	0.0387	0.0695	0.578	−0.1243	0.1	0.214
South Korea						
ECM	−0.2138	0.0471	0	−0.0111	0.004	0.006
ΔOIL	0.007	0.007	0.316	0.0185	0.0094	0.049
ΔPOLLUTION	−0.1115	0.0467	0.017	−0.0621	0.0508	0.222
ΔFDI	−0.052	0.0248	0.036	−0.0878	0.0313	0.005
ΔINC	−0.0001	0.0001	0.372	−0.0002	0.0002	0.309
ΔPOP				−1.0239	0.6913	0.139
Intercept	−0.1224	0.2749	0.656	−0.8738	0.4342	0.044
Thailand						
ECM	−0.0474	0.1213	0.696	−0.3897	0.0779	0
ΔOIL	0.0771	0.0764	0.312	−0.0542	0.0511	0.289
ΔPOLLUTION	−0.3769	1.0763	0.726	−2.099	0.6978	0.003
ΔFDI	0.279	0.1827	0.127	0.1648	0.1156	0.154
ΔINC	−0.0084	0.0062	0.179	0.0055	0.0048	0.247
ΔPOP				−16.1315	7.367	0.029
Intercept	0.6695	2.63	0.8	−18.819	3.644	0
Vietnam						
ECM	−0.4245	0.1494	0.005	−0.2011	0.0859	0.019
ΔOIL	0.167	0.1148	0.146	0.1881	0.1182	0.112
ΔPOLLUTION	−9.396	4.781	0.049	−8.9535	4.9118	0.068
ΔFDI	−0.3795	1.1479	0.741	−1.1135	1.1793	0.345
ΔINC	−0.4273	0.1354	0.002	−0.1745	0.07	0.013
ΔPOP				−7.948	23.5474	0.736
Intercept	40.691	14.1754	0.004	1.496	4.0653	0.713

Source: Result from the authors' analysis. Note: *, **, and *** indicate significance at 10%, 5%, and 1% levels, respectively. Under the long-run slope homogeneity, the Hausman statistic is asymptotically distributed as a chi-square. The lag structure is ARDL (1,1,2,0,0,0), and the order of the variable is REC – renewable energy consumption, oil price, EP – environmental pollution, FDI – foreign direct investment, per capita GDP, POP – population growth.

Regarding the estimated results, our analysis shows that the relationship between oil price shocks, EP, FDI, and REC in the case of East Asian economies could be carefully analyzed.

10.6.1 Oil Price Shocks

The regression coefficient of this variable is positive and statistically significant in the short run but statistically insignificant in the long run. To be precise, the impact of oil price shocks on REC is statistically significant and positive. These findings indicate that higher oil price shocks will significantly enhance REC in East Asian economies in the short run.

TABLE 10.12

Hausman Test for MG and PMG

Variable	Coefficients			1/VIF
	(b) mg	**(B)** pmg	**(b-B)** Difference	**sqrt(diag(V_b - V_B))** S.E.
OIL	0.0912	0.3401	0.2489	0.2684
POLLUTION	1.3191	5.242	−3.9229	3.3165
FDI	−0.0367	−0.13	0.0933	1.1124
INC	−0.0089	−0.0001	−0.0087	0.0199
POP	5.8868	45.0981	−39.2113	7.5444

    b = consistent under Ho and Ha; obtained from xtpmg
    B = inconsistent under Ha, efficient under Ho; obtained from xtpmg
    Test: Ho: difference in coefficients not systematic.
    $chi^2(5) = (b − B)'[(V\_b − V\_B)^{\wedge}(−1)](b − B) = 1.5129$
    $Prob > chi^2 = 0.3720$

Source: Result from the authors' analysis.

TABLE 10.13

Hausman Test for DFE and PMG

Variable	Coefficients			1/VIF
	(b)dfe	**(B)pmg**	**(b-B)** Difference	**sqrt(diag(V_b – V_B))** S.E.
OIL	−2.5459	0.3401	−2.8861	4.0896
POLLUTION	−36.2949	5.242	−41.537	3.7107
FDI	1.156	−0.13	1.2861	8.6075
INC	−0.0035	−0.0001	−0.0033	0.0555
POP	−101.9783	45.0981	−147.0764	9.7201

    b = consistent under Ho and Ha; obtained from xtpmg
    B = inconsistent under Ha, efficient under Ho; obtained from xtpmg
    Test: Ho: difference in coefficients not systematic.
    $chi^2(5) = (b − B)'[(V\_b − V\_B)^{\wedge}(−1)](b − B) = 0.03$
    $Prob > chi^2 = 0.9999$

Source: Result from the authors' analysis.

As mentioned in the microeconomics, oil, gas, and RE are substitute goods. As suggested in Tran et al. (2020), substitute goods are at least two products that can be used for the same purpose by the same consumers. If the price of one of the products falls or rises, the demand for substitute goods is likely to decline or increase. In this case, if the oil price rises, the demand for REC will be likely increase. The finding is strongly supported by this theory. Additionally, Shah et al. (2018) also found a similar result in the case of Norway but no relationship in the case of the UK. Reboredo (2015) found that oil price fluctuations significantly provide approximately 30% changes in RE. In contrast, Sadorsky (2009) conducted a study in G7 economies and indicated that oil price shocks have a negative and smaller effect on REC.

On further examining in specific countries, changes in oil prices do not impact REC in the cases of China, Indonesia, Japan, Singapore, Thailand, and Vietnam but strongly effect in the case of Malaysia, Philippines, and South Korea in the short run. This evidence is consistent with Sadorsky (2009) who conducted a study in G7 economies. However, Gozgor et al. (2020) also conducted a study in OECD economies, and the findings support for the countries Malaysia, Philippines, and South Korea.

10.6.2 Environmental Pollution

In theory, the relationship between EP and the demand for REC can be found. The empirical investigations in this analysis are carried out by identifying the long and short-run relationship from PMB estimates, and results show that regression coefficient is positive and statistically significant in the short run but negative and statistically significant in the long run. As discussed on the specific countries, a negative and statistically significant impact can be found in the cases of China, Japan, Malaysia, Philippines, Thailand, and Vietnam; and no impact was found for Indonesia, Singapore, South Korea in the short run.

Khan et al. (2017) further explained that renewable energies, e.g., solar, and wind, have been increasingly more competitive in cost than fossil fuel energies, especially in the long run. In the long run, countries should enhance its environmental quality by investing on more renewable sources and reducing fossil fuel energies in order to avoid climate change, sea-level rise, global temperature rise in the world. Zeppini and Bergh (2020) analyzed that the contribution of R&D in the energy industry can reduce the high-carbon to low-carbon energy consumption.

10.6.3 Inward Foreign Direct Investment

In the short run, the results of the study indicate that FDI has a significant negative effect on REC. This finding can be explained on the ground that changes in FDI cannot impact REC in both the short and long run. This can be emphasized that although FDI inflows have significantly played a crucial role in the industry sector in addition to the technology transfer process (Wang, 1990; Tybout, 2000), access to new technologies (Du at el. 2011; Chaudhry at el. 2013), export promotion (Nguyen et al., 2020b), or even local firms in host countries get the opportunity to enhance productivity and production efficiency and join the global value chain, but FDI in the host countries cannot promote REC. We cannot find the evidence in the sample data of nine East Asian countries in the short and long runs.

Further discussed on specific countries in East Asian countries, Table 10.10 indicates that FDI inflows can have a positive and significant effect on REC in the short run for the case of Indonesia; negative and significant effect for Philippines, Malaysia, Japan, and South Korea; and insignificant effect for China, Singapore, Thailand, and Vietnam. In general, this result depicts that the finding can be divided into two scenarios. FDI in countries (especially in high-income economies) can reduce REC. In addition, energy has always been a vital part of many industrial revolutions, but it remains critical discussing its adverse effects on environmental quality. More specifically, negative externalities in the environment can hurt sustainable development by causing EP and affecting human lives, and therefore a reduction in the use of fossil fuels for increasing RE should be advocated. This process is dependent on government policy and the specific context of each country.

10.6.4 Economic Growth and Population Growth

In agreement with economic growth, it is evident that the effect of economic growth on REC is also negative but insignificant in both the short and long runs. More specifically, a greater or worse economic performance will not affect REC in the case of East Asian countries. This finding is not related to the theoretical expectations and other empirical studies. As shown in the descriptive statistics, developed countries have a higher per capita income than developing countries, but Figure 10.1 indicates that the proportion of REC in developed and emerging economies such as China and Japan have sharply declined while it slightly declined in developed economies as South Korea and Singapore. This finding is not supported by Sadorsky (2009) on a study in G7 economies with the suggestion that increases in per capita income could be found as the main driver behind REC in the long run.

Furthermore, the empirical investigations in the analysis are carried out by identifying short-run association among economies in East Asia, and it is established that the effect of economic performance on REC is insignificant in the short run for the case of high and upper-middle-income economies, for example, China, Japan, Malaysia, Singapore, South Korea, and Thailand; but it is negative and statistically significant in the short run for the low and lower-middle-income economies, for example, Vietnam, Philippines, and Indonesia. In the low and lower-middle-income economies, there is a J curve relationship between income and EP. In economics, the environmental Kuznets curve is a hypothesis of discussing the relationship between environmental quality and income. In this regard, environmental quality tends to become worse in parallel to greater per capita income until average income reaches a certain point in the development process. Under the pressure of economic development, the low and lower-middle-income economies are not likely to shift from using fossil fuels to REC, which has a higher cost. However, regulatory changes, knowledge, and technological innovations can lead to reducing EP by shifting from hydrocarbons with other trace elements to consume more RE.

Regarding POP, it is evident to confirm that the regression coefficient of POP is positive and statistically significant in the long run. It concludes that higher growth of population in East Asian countries can drive the use of REC. Further, as for the short-run relationship, the study failed to find the short-run relationship between POP and REC. Focusing on eight countries, the study identifies that the negative robustness could be found for the case of middle- income economies as Malaysia, Philippines, Thailand but the insignificant impact can be found for the case of high and upper-middle-income economies as China, South Korea, Vietnam, Japan, and Singapore in the short run. The POP in East Asian countries significantly declined in recent years, especially in high-income and upper-middle-income economies. For example, POP of China, South Korea, Japan, and Singapore steadily declined from 1.47%, 1.04%, 0.38%, and 5.32% in the 1990s to 0.46%, 0.21%, -0.20%, -1.47%, respectively, while REC in these countries gradually decreased in 1990 to the present. It meant that a large volume of energy has been contributed from fossil fuel energy. In contrast, the demand for REC in middle-income economies has seen an increase in this period because, in the 1990s, the supply of REC in these countries was somewhat low. In the case of Malaysia, this country has contributed a large share of ASEAN's fossil fuel resources in the past because of relying on domestic gas production. Therefore, Malaysia is more likely to switch from fossil energy to renewables, the proportion of gas in the power mix declined from 67% in 2005 to 47% in 2015. Furthermore, Malaysia also has set a target of 20% REC by 2025. This finding in East Asian countries could be supported by studies in the past. As shown by Sherbinin et al. (2007) in a study, the population is one of several factors that impact the environment.

10.7   Implications of Renewable Energy Consumption and the Use of Biofuels in East Asian Countries

Producing and consuming RE are an inevitable trend of modern times. The use of non-RE resources in the context of a rapidly growing world population and ever-depleting natural resources, leading to rising world oil prices, will lead to habitat destruction and adversely affect world sustainable growth. At present, the world is facing the crucial problem of climate change, and it indicates that making a big effort to create more biofuels has been necessary. In addition, Xiong et al. (2017) concluded that using more biodiesel production from biomass, pollutant into the environment can be reduced.

Simultaneously, an approximation of 18% of the world's energy is presently created from renewable resources, for example, hydroelectric power (United Nations, 2019). However, fossil fuels' demand is still at a high level and it is positively consistent with economic growth. Because of the high demand for nonrenewable resources (RRs), CO_2 emissions from 2005 to 2017 in developing Asian countries rose by 6.26 billion tons. According to International Energy Outlook (2016), in the East Asian region, China has risen to be the country with highest CO_2 emissions, increasing to 7.54 billion tons in 2014 from 4.52 billion in 2005. Approximately 60% of the world's population inhabits in Asia and confronts many challenges brought about by environmental degradation from exorbitant consumption of energy produced from fossil resources.

International Energy Outlook (2016) also indicates that in the period between 1990 and 2015, total CO_2 emissions originated from fossil fuels' usage in Asia increasing to 5.158 billion from 2.002 billion tons. Additionally, to reach great growth targets, many countries in Asia have been controlling their credit on RE produced from hydro, wind, geothermal, biogas, solid biofuels waste, marine resources, and liquid biofuels. At the regional level, the proportion of RE in aggregate energy use is around 6.28% in East Asia, 20.75% in Central Asia, 6.3% in West Asia, 41.44% in South Asia, and 37.73% in South East Asia (Figure 10.4).

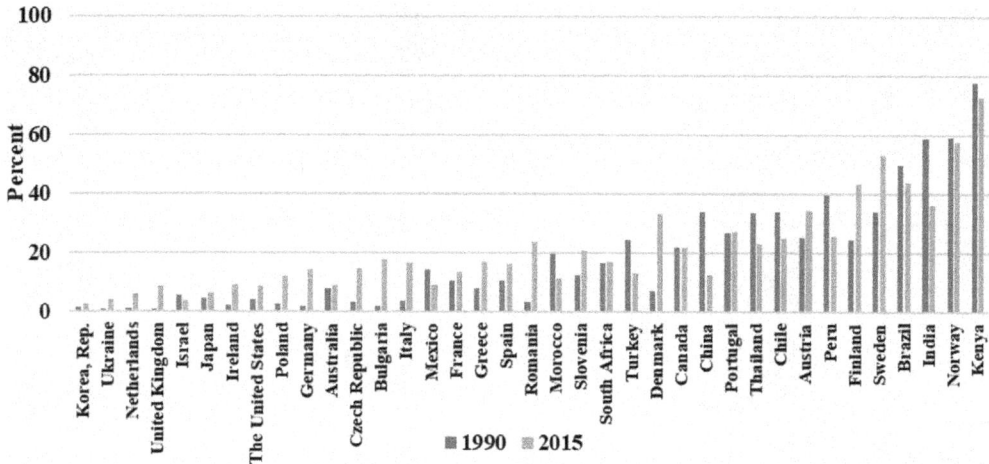

FIGURE 10.4 The proportion of RE resources to energy consumption in 1990 and 2015 (data source: World Bank, 2020).

In the demand of global energy consumption, there exists a rising distribution of renewables in decreasing climate change and supporting affordable costs for resides who are trapped in poverty. Based on United Nations Environment Program (2017), approximately 19.1% of international energy consumption was resourced from RE in 2013. The present growth of electricity production is mainly led by solar, photovoltaics, hydropower and wind energies. The heating capacity grows gradually, and the biofuel production for transportation has nowadays increased after a plump during the 2 years of 2011 and 2012. As per the study promoted by the International Energy Agency (2019), the renewable distribution of electricity production is expected to rise from 18.3% in 2002 to about 39% by 2050. Further, it harbors a target to make global CO_2 emissions to reduce by 50% by 2050 and that RE will have an essential role in restricting the long-run mean world temperature to between 2°C and 2.4°C.

Figure 10.4 describes the contribution of RE in the period of 1990–2015, showing a positive change has been made. This is explicit that RE can be predominantly produced from biomass in East Asian countries compared to other resources such as geothermal, solar, hydropower, wind and waste in other areas. Developing biomass demand lowers the amount of import of other non-renewable resources, it can increase employment in rural areas, and lead to greener energy consumption.

Together with the United States and Europe, present RR creativities have originated in some other countries in Asia, Latin America, and Africa. The distribution of RE has risen dramatically in the cooling and heating, electricity, and transportation sectors. According to the report of the RE Policy Network for the twenty-first century, China ranked the first in international investment in RRs, followed by the United States, Japan, the United Kingdom, and Germany. China was in the leading position in utilizing hydropower, water heating, and solar photovoltaics. Rankings for different resources based on "capacity per capita" are shown in the following Table 10.14.

In most developing countries, goods and services demand is overloaded, due to the elevated population growth rate and ineffectual industries. Therefore, the high demand for goods and services puts

TABLE 10.14

Leading Countries Investing in Renewable Fuels and Power

Renewable Energy	Countries
Hydropower	China, Canada, Brazil, Russia, and the United States
Solar photovoltaics	Germany, Belgium, Italy, Czech Republic, and Greece
Wind power	Denmark, Germany, Sweden, Ireland, and Spain
Geothermal	Iceland, Hungary, New Zealand, Japan, and Turkey

Source: International Energy Outlook (2016).

TABLE 10.15

Average Value of Key Energy, Environmental and Economic Indicators of Some East Asian Countries during 1990 and 2015

Countries	Pollution Index%	Percentage of Renewable Energy	Carbon Dioxide Emissions (Kilotons)	Fossil Fuels Energy (Kilogram of Oil Equivalent)	Per Capita GDP	Forest Area (km²)
China	88.96	18	4751043	1124.43	2490.7	1840389.21
Indonesia	76.41	37	296428	1124.43	1642.1	1012323.08
Japan	60.24	4	1189878	3223.4	37125.6	249297.69
Malaysia	67.08	7	135415	315.04	5951.5	217944.42
Philippines	70.25	29	63986.2	387.95	1508.1	70171.15
Singapore	65.18	1	46561	5034.93	31980.5	163.5
South Korea	68.15	1	450256	3383.32	16602.2	62743.08
Thailand	73.23	23	19058	388.17	3320.3	159827.69
Vietnam	88.1	36	71203.8	407.14	769.8	122997.31

Source: World Bank (2020).

those countries ahead of the need to boost production by increasing the exploitation of natural resources themselves. Increasing the production of goods and services will lead to an increase in energy demand, putting pressure on energy generated by nonrenewable or renewable sources. Average economic growth per capita, RE, non-RE, CO_2 emissions, and several other predominant factors including environmental pollution index, and forest area in some East Asian countries recently can be summarized in the following Table 10.15.

As suggested by the International Energy Agency (2013), in recent years, developing emerging countries have been investing at a faster rate in renewable energy. International investment in RE increased nearly four times (from the US $36 billion to the US $139 billion) in developed countries, whereas for developing countries, the growth rate was more than 14 times (from the US $9 billion to the US $131 billion) and increased from the US $45 billion to the US $270 billion globally from 2004 to 2014 as reported in International Energy Outlook 2016. In addition, China was the country with the largest investment in RRs in 2014 with the US $83.3 billion, the United States ranked second with the US $38.3 billion, and Japan was at the third position with the US $35.7 billion. Among developing countries, China with the investment of US $138.9 billion, Brazil with the US $7.6 billion, India with the US $7.4 billion, and South Africa with the U.S. $5.5 billion ranked among the top countries of investment, while investment in Indonesia, Chile, Mexico, Kenya, and Turkey was more than the US $1 billion. Solar, wind, biomass, geothermal, marine power, and small hydroelectricity contributed about 9.1% of the world's electricity output in 2014, an increase of 6% compared to the level of 8.5% of 2013. Coming to 2014, the global market achieved a record financing in the wind and solar power, accounting for 92% of total investment in renewables. As shown by the United Nations Environment Program (2020), investment for solar energy reached 29% at the US $149.6 billion while investment for wind rose to 11%, a peak of the US $99.5 billion. In the case of Malaysia, RE is estimated to account for 20% of its total energy source by 2025, and it is estimated that Malaysia will need an investment of the US $8 billion in renewable energy. Investments are expected not only from the government but also from public and private partnerships. The implementation of Malaysia's Energy Supply Industry may push the country to achieve its goals. Green energy transactions that can be done through the grid will create more competition.

Among the effort of almost countries all over the world, including developing emerging countries, especially East Asian ones, they are spending their social and economic strength in developing RE. For example, the Vietnamese Government recently revised the Power Development Plan to improve the proportion of RRs like solar, wind, and biomass. The recent figure reveals that a total of 332 solar projects has already been successfully registered with the capability of 27 megawatts in total, including 121 solar projects (7,234 megawatts) that will start to create electricity by 2020 and 211 solar projects (13,069 megawatts) in progress to be approved (VIR, 2019). According to the national Ministry of Industry and

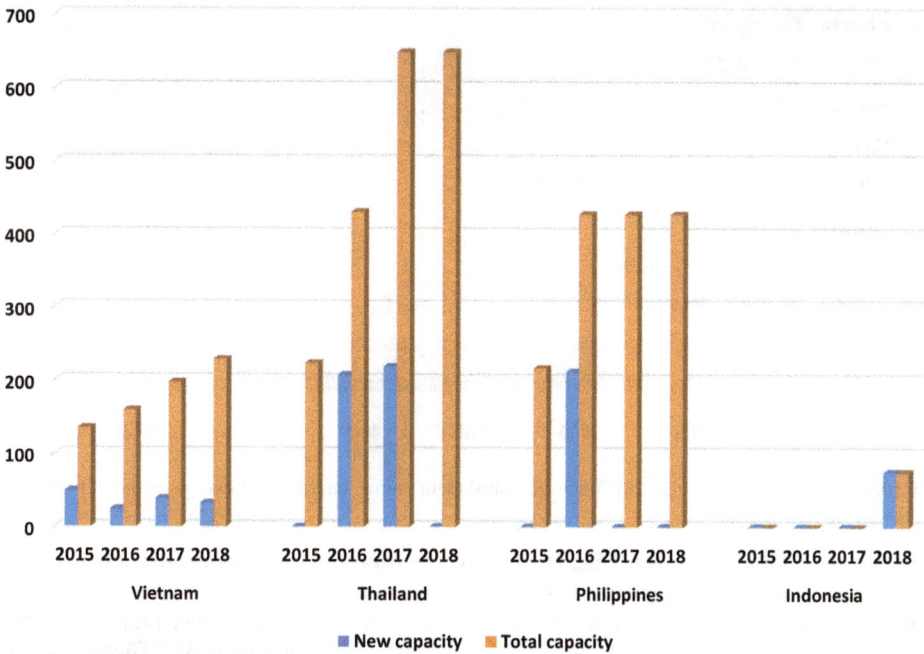

FIGURE 10.5 East Asian wind installations and targets (onshore and offshore) (data source: UN Environment, 2017).

Trade, hydropower occupied 40% of the country's total electricity output during 2016–2017. The total hydropower production will increase by 21,600 megawatts in 2020, 24,600 megawatts in 2025, and 27,800 megawatts in 2030. It can be illustrated partially in Figure 10.5.

Another country with a fast-growing RE industry in East Asia is South Korea. Recently, liquefied natural gas accounted for the largest portion of Korea's energy industry, followed by coal, accounting for 27.1% of the total energy source. Nuclear energy accounts for 19.2% of the energy output. By 2030, plans related to nuclear power in the total energy industry are expected to be withdrawn by 11.7% and then dropped deeply to 9.9% by 2034. After that, by 2030, the percentage of energy is expected to increase to 33.1% before reaching the target of 40% 4 years later (Slav, 2020), and Japan is a special case in which RE plays a vital role, due to natural disasters here. It is necessary to consider the following factors. Japan relies on energy imported from overseas sources and has recently abandoned nuclear power. The World Nuclear Association's 2019 report reveals that Japan needs to import 90% of its energy. After the Fukushima disaster in 2011, the share of Japan's energy from nuclear resources has decreased from 30% in 2011 to only an expected 20% in 2030.

With this strategy, Japan is focusing on RE resources such as wind, tidal and solar power to lower the dependency on foreign energy imports and promote innovation in the domestic energy sectors (Casey, 2020). Also, Singapore relies heavily on imported fuel – 95% of its electricity is supplied with natural gas. This requires the needs of Singapore to optimize energy use. The Singaporean government has committed to using 100% greener electric energy or hybrid public buses by 2040. Infrastructure for the national electric vehicles charging network is being installed, and 2,000 charging points are being located throughout the country under the electric car-sharing program named BlueSG, and one-fifth of that will be used for public transport (Wah and Lei, 2019).

Consider a typical example of the situation of China, a leading Asia country in producing RE. The era of Chinese dam construction began in the 1950s and has peaked in the past two decades. When Baihetan reaches its maximum capacity by the end of 2022, China will have completed five of the world's largest hydroelectric plants in just 10 years. In 2017, Chinese hydroelectric dams were thought to generate more electricity than the combined supply of all other countries, leaving out the United States and India. The turning point that marks the stage of the construction of huge hydroelectric projects is the Three

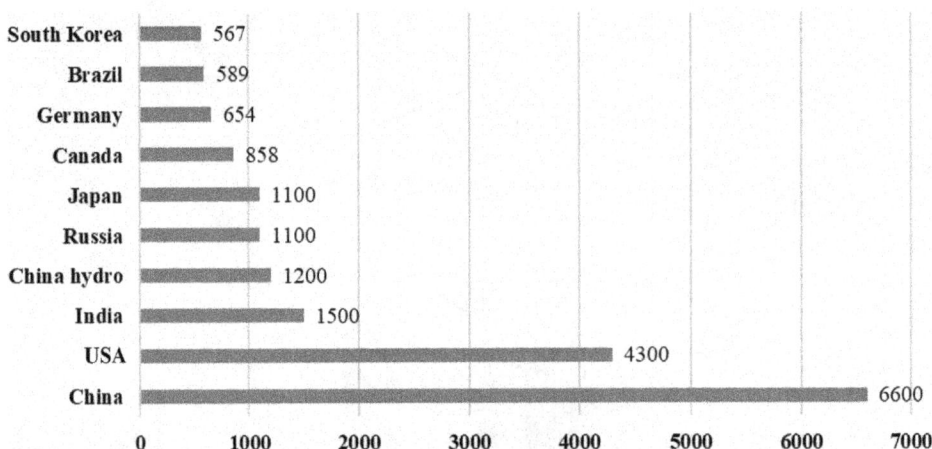

FIGURE 10.6   Electricity produced in 2017 (terawatt-hours) (data source: International Energy Agency, 2019).

Gorges Dam project that blocks the Yangtze, the river that runs through the mountains with the largest length in China. Construction began in 1994 and was not completed until 2012. With a capacity of 22.5 gigawatt, this is the largest hydroelectric project in the world. Two larger super projects, Xiangjiaba with a dominant capacity of 6.4 Gigawatt and Xiluodu Dam with a capacity of 13.9 Gigawatt (Khe Lac Do), were also completed later in 2014 on the Jinsha River. China's hydroelectric dams produce more energy than the entire national grid, as shown in Figure 10.6.

The development of RE in recent years has been aided by many positive programs of the government such as tax credits, grants, and some incentives. It has surged the cost competitiveness of renewables. In several different countries, RE can compete extensively with conventional energy sources. In general, making an investment environment, developing professional expertise, and unfastening political and financial barriers are good key steps in the deployment of RE, and almost all the East Asian countries have participated in this process. Financially, establishing quotas, import duties, credit incentives or tax, investment subsidies, and sales tax releases for clean products such as green certificate transactions and solar cells are the main tools toward that goal. As far as policy is concerned, governments, international cooperation agencies, energy planners, and other relevant agencies must participate in the implementation of strategies to develop RE all over the world. Specifically, the tax rates on petroleum products and coal consumption proposed by the Government of Vietnam, respectively, increased by 50% and 33% in recent years (Das, 2018).

The main purpose of the tax policies was to promote greener energy production and consumption and to reduce CO_2 emissions since Vietnam had committed to lower CO_2 levels at the Paris Climate Change Conference. These are the two main sources of energy for most areas to raise taxes that will significantly affect the country. To promote sustainable and environmental economic growth, one can easily see that it is essential for developing economies to shift away from fossil fuels to focus on renewables. To achieve the goals of the Kyoto Protocol, renewable sources can be used to lower CO_2 emission in developing countries. The infrastructure needed for RRs can be quite costly at an early stage, requiring each country to invest heavily in the private sector. Therefore, it is necessary to offer more traditional investment-based access in regions – for instance, the innovation of the green bond market. In many developing countries, persuading financial institutions to lend investors can be very complicated and cost a lot of time; therefore, the green bond markets may be an ideal solution to bring financial support to many countries that have a high demand for green infrastructure (Chapman et al., 2018). Additionally, developing countries should participate in global conventions to exchange effective modern technologies and reinforce low CO_2 emission policies to encourage RRs. They should also implement unilateral policies under tax incentives and subsidies to install equipment needed to generate RE. Overall, steps need to be taken to control CO_2 emission at both national and regional levels, toward green technology, cleaner energy production, carbon taxation, and subsidized green projects. In developing countries, creativity to offer renewables can reduce fossil fuel consumption and promote an environmentally sustainable energy supply. Thanks to technological innovation, production methods can also be promoted to control CO_2 emission. Also, energy conservation can further improve energy efficiency, promote economic growth, and warrant sustainable development. In the end, in East Asian economies, CO_2 is not restricted to national borders; to limit its emission, it is necessary to communicate through regional dialogues and promote a policy foundation for cooperation between different partners. In agreement with policies regarding CO_2 reduction, few East Asian economies have targeted biofuel consumption. According to Elder and Hayashi (2018), China, Indonesia, and Malaysia have greatly produced a huge amount of biofuels meanwhile a small contribution of biofuels in the energy demand in Vietnam, Thailand, and the Philippines. However, the trends in using more biofuels should be increased more as it relates to per capita income and the level of economic development.

10.8 Conclusions

In the context of increasing EP and trends in the potential transition from polluted and less polluted energy and using more RE, the studies on oil price shocks, environmental quality, FDI, and REC have been conducted in the case of East Asian covering the period of 1990–2019. Additionally, the past investigations have lacked advanced examination methods of panel data – e.g., the Pooled Mean Group, Mean Group, and Dynamic Fixed Effects approach, and cointegration test. The empirical studies demonstrate that the proportion of REC of total final EC in most East Asian economies has increasingly declined, except for a few non-advantage economies as South Korea, Japan, and Singapore. Further, changes in oil prices and level of EP are positive and statistically significant in the short run but statistically insignificant in the long run for oil prices and negative and significant for EP factor. In terms of FDI, it has a significant and negative effect on REC. Also, POP could positively promote REC in the long run. In contrast, the impact of per capita income on REC is negative but insignificant in both the short and long run. The study also indicates that it is a global challenge in sustainable development posed by increasing CO_2 emission and other greenhouse gases such as nitrous oxide and methane. Additionally, as climate change and global warming have predominantly hurt ecology, health, and lifestyle, the trends of transition from fossil fuels to renewables will occur worldwide.

REFERENCES

Al-Maamary, H. M. S., Kazem, H. A., and Chaichan, M. T. 2017. The impact of oil price fluctuations on common renewable energies in GCC countries. *Renewable and Sustainable Energy Reviews* 75: 989–1007.

Bamati, N., and Raoofi, A. 2020. Development level and the impact of technological factor on renewable energy production. *Renewable Energy* 151: 946–955.

Casey, J. P. 2020. New laws and new targets: Renewable power in Japan. *Power Technology*. www.power-technology.com/features/new-laws-and-new-targets-renewable-power-in-japan (accessed 21 June 2020)

Chapman, A., Fujii, H., and Managi, S. 2018. Key drivers for cooperation toward sustainable development and the management of CO_2 emissions: Comparative analysis of six Northeast Asian countries. *Sustainability* 10: 244.

Chaudhry, N. I., Mehmood, M. S., and Mehmood, A. 2013. Empirical relationship between foreign direct investment and economic growth: An ARDL co-integration approach for China. *China Finance Review International* 3(1): 26–41.

Das, K. 2018. Vietnam proposes higher environmental protection taxes. *Vietnam Briefing*. www.vietnam-briefing.com/news/vietnam-proposes-higher-environmentalprotection-taxes.html/ (accessed 13 June 2020)

Doytch, N., and Narayan, S. 2016. Does FDI influence renewable energy consumption? An analysis of sectoral FDI impact on renewable and non-renewable industrial energy consumption. *Energy Economics* 54: 291–301.

Du, L., Harrison, A., and Jefferson, G. H. 2011. Testing for horizontal and vertical foreign investment spillovers in China. *Journal of Asian Economics* 23(3): 234–243.

Elder, M., and Hayashi, S. 2018. A regional perspective on biofuels in Asia. In: *Biofuels and Sustainability,* (Eds.) K. Takeuchi, H. Shiroyama, O. Saito, and M. Matsuura. Science for Sustainable Societies. Tokyo: Springer, pp. 223–246.

Gozgor, G., Mahalik, M. K., Demir, E., and Padhan, H. 2020. The impact of economic globalization on renewable energy in the OECD countries. *Energy Policy* 139: 111365.

Gujarati, D. 2004. *Basic Econometrics.* 4th Edition. New York: McGraw-Hill Companies.

Ike, G. N., Usman, O., Alola, A. A., and Sarkodie, S. A. 2020. Environmental quality effects of income, energy prices and trade: The role of renewable energy consumption in G-7 countries. *Science of The Total Environment* 721: 137813.

International Energy Agency. 2013. *World Energy Outlook 2013.* Capella Festa, Paris. www.wec-france.org/DocumentsPDF/Evenements/06-12-13_AIE.pdf

International Energy Agency. 2019. World energy outlook 2019 reports. www.iea.org/topics/world-energy-outlook.

International Energy Outlook. 2016. *Key World Energy Statistics.* US Energy Information Administration, Washington, DC 20585. www.eia.gov/outlooks/ieo/pdf/0484(2016).pdf

Kao, C. 1999. Spurious regression and residual-based tests for cointegration in panel data. *Journal of Econometrics* 90(1): 1–44.

Khan, M. I., Yasmeen, T., Shakoor, A., Khan, N. B., and Muhamad, R. 2017. 2014 oil plunge: Causes and impacts on renewable energy. *Renewable and Sustainable Energy Reviews* 68(1): 609–622.

Lu, X., and White, H. 2014. Robustness checks and robustness tests in applied economics. *Journal of Econometrics* 178(1): 194–206.

Maddale, G. S., and Wu, S. 1999. A comparative study of unit root tests with panel data and a new simple test. *Oxford Bulletin of Economics and Statistics* 61(S1): 631–652.

Nguyen, T. T., Nguyen, V. C., and Tran, T. N. 2020a. Oil price shocks against stock return of oil and gas-related firms in the economic depression: A new evidence from a copula approach. *Cogent Economics & Finance* 8(1): 1799908.

Nguyen, V. C., Nguyen, T. T., and Nguyen, H. T. 2020b. Government ability, bank-specific factors and profitability: An insight from banking sector of Vietnam. *Journal of Advanced Research in Dynamical and Control Systems* 12(4): 415–424.

Nguyen, V. C., Thanh, H. P., and Nguyen, T. T. 2020c. Do Electricity consumption and economic growth lead to environmental pollution? Empirical evidence from association of Southeast Asian Nations countries. *International Journal of Energy Economics and Policy* 10: 297–304.

Pedroni, P. 1999. Critical values for cointegration tests in heterogeneous panels with multiple regressors. *Oxford Bulletin of Economics and Statistics* 61: 653–670.

Pedroni, P. 2004. Panel cointegration: Asymptotic and finite sample properties of pooled time series tests with an application to the PPP hypothesis. *Econometric Theory* 20: 597–325.

Reboredo, J. C. 2015. Is there dependence and systemic risk between oil and renewable energy stock prices? *Energy Economics* 48: 32–45.

Sadorsky, P. 2009. Renewable energy consumption, CO_2 emissions and oil prices in the G7 countries. *Energy Economics* 31(3): 456–462.

Shah, I. H., Hiles, C., and Morley, B. 2018. How do oil prices, macroeconomic factors and policies affect the market for renewable energy? *Applied Energy* 215: 87–97.

Sherbinin, A., Carr, D., Cassels, S., and Jiang, L. 2007. Population and environment. *Annual Review of Environment and Resources* 32: 345–373.

Slav, I. 2020. South Korea embarks on an ambitious RE plan. https://oilprice.com/Latest-Energy-News/World-News/South-Korea-Embarks-On-An-Ambitious-Renewable-Energy-Plan.html (accessed 21 June 2020)

Tran, T. N., Nguyen, T. T., Nguyen, V. C., and Vu, T. T. H. 2020. Energy consumption, economic growth and trade balance in East Asian – A panel data approach. *International Journal of Energy Economics and Policy* 10(4): 443–449.

Tybout, J. R. 2000. Manufacturing firms in developing countries: How well do they do, and why? *Journal of Economic Literature* XXXVIII: 11–44.

UN Environment. 2017. REN21 – RE policy network for the 21st century. http://climateinitiativesplatform.org/index.php/REN21_(Renewable_Energy_Policy_Network_for_the_21st_Century) (accessed 12 June 2020)

UN Environment Programme. 2020. *Global Trends RE Investment.* Frankfurt School FS-UNEP Collaborating Center for Climate and Sustainable Energy Finance. www.fs-unep-centre.org/wp-content/uploads/2020/06/GTR_2020.pdf

United Nations. 2019. Global issues – population, world population prospects 2019. www.un.org/en/sections/issues-depth/population/index.html

Vietnam Investment Review (VIR). 2019. Rooftop solar to ensure REfuture. www.vir.com.vn/rooftop-solar-to-ensure-renewable-energy-future-65746.html

Wah, T. S., and Lei, Z. 2019. Singapore well-positioned to build a sustainable, smart-energy future. *The Business Times.* www.businesstimes.com.sg/opinion/singapore-well-positioned-to-build-a-sustainable-smart-energy-future

Wang, J. 1990. Growth technology transfer, the long-run theory of international capital movements. *Journal of International Economics* 29(3–4): 255–271.

Westerlund, J. 2007. Testing for error correction in panel data. *Oxford Bulletin of Economics and Statistics* 69(6): 709–748.

World Bank. 2020. *World Development Indicators.* Washington, DC: World Bank. https://databank.worldbank.org/source/world-development-indicators

Xiong, X., Yu, I. K. M., Cao, L., Tsang, T. C. W., Zhang, S., and Ok, Y. S. 2017. A review of biochar-based catalysts for chemical synthesis, biofuel production, and pollution control. *Bioresource Technology* 246: 254–270.

Zeppini, P., and Bergh, J. C. J. M. 2020. Global competition dynamics of fossil fuels and renewable energy under climate policies and peak oil: A behavioral model. *Energy Policy* 136: 110907.

Index

For Product Safety Concerns and Information please contact our EU
representative GPSR@taylorandfrancis.com
Taylor & Francis Verlag GmbH, Kaufingerstraße 24, 80331 München, Germany

9 780367 566104